ALLE ZEIT WACH 1842

Surendra K. Saxena Nilanjan Chatterjee
Yingwei Fei Guoyin Shen

Thermodynamic Data on Oxides and Silicates

An Assessed Data Set Based on Thermochemistry and High Pressure Phase Equilibrium

With 34 Figures

Springer-Verlag
Berlin Heidelberg New York
London Paris Tokyo
Hong Kong Barcelona
Budapest

Dr. Surendra K. Saxena
Department of Theoretical Geochemistry
Institute of Earth Sciences
Uppsala University
Box 555
S-75236 Uppsala
Sweden

Dr. Nilanjan Chatterjee
Department of Geological Sciences
University of Lexington
Lexington, KY 40506-0059
USA

Dr. Yingwei Fei
Geophysical Laboratory
Carnegie Institution of Washington
Washington, D.C. 20015
USA

Dr. Guoyin Shen
Department of Theoretical Geochemistry
Institute of Earth Sciences
Uppsala University
Box 555
S-75236 Uppsala
Sweden

ISBN-13:978-3-642-78334-0 e-ISBN-13:978-3-642-78332-6
DOI: 10.1007/978-3-642-78332-6

Production Editor: Herta Böning, Heidelberg
Cover Design: Erich Kirchner, Heidelberg
Typesetting: Camera-ready by author
32/3145-5 4 3 2 1 0 – Printed on acid-free paper

Preface

This book aims to provide an internally consistent thermodynamic data base which can be used globally at any earthly pressure and temperature. The data base would be useful not only to scientists who need phase equilibrium computation at room temperature and pressure but also to scientists (chemists, physicists, metallurgists, material scientists, geologists) who need high temperature or both high temperature and high pressure properties of solids.

In the last thirty years or so and especially during the last decade, we have seen some of the most profound developments in the experimental techniques to study high pressure phase equilibrium (e.g., the high pressure piston-cylinder, the multi-anvil presses and the ultra high pressure diamond-anvil cells). These developments have been matched by similar developments in our ability to measure themodynamic properties of solids and fluids. In addition to the advanced calorimetric techniques (scanning and flash calorimetry), we have the synchrotron light source and infra-red, Raman and Brillouin spectroscopic techniques to help us measure useful solid properties contributing to an improved knowledge of the solid state behavior. The technical development has been further matched by our increased sophistication in applying theoretical methods (e.g., computer assessment of data by optimization of several variables, simulation of energetics of substances by several semi-empirical models) to obtain new data or improve the quality of the existing ones. With such prolific scientific activity in the solid state field, we naturally expect that a data base should continue to improve rapidly with time. While little change is expected in the standard state thermochemical data on most solids, we should expect improvements in the pressure-volume-temperature properties and in the high temperature heat capacities of many solids in the near future. In the mean time, the data base presented here should serve as an important tool in designing new phase equilibrium experiments.

In addition to the co-authors, many other students and colleagues have contributed to a successful completion of this book. The authors thank in particular Jianzhong Zhang, Pingfang Shi and Anatoly Belonoshko. An earlier version of the book was reviewed by Prof. Mats Hillert (Royal Technical University, Stockholm) and M. Roots. In working on problems of thermodynamic data, I feel personally priviledged to be associated with G. Eriksson and B. Sundman. Financial support for the research has come from the Swedish Natural Science Research Council (NFR) and Nutek (CAMPADA project).

Uppsala, July, 1993 S. K. Saxena

Contents

Chapter 1

Introduction: Thermodynamic Data Systematics

There are different sources of data available for establishing an internally consistent data set: calorimetric measurements, phase equilibrium data, measured thermophysical properties of a phase, and heat capacities and entropies estimated from lattice vibrational models, see, for example, Chopelas (1990a, b), Kieffer (1979a, b, c), Ghose et al. (1992, pers. commun.) and Hofmeister and Chopelas (1991). The data base presented in this study is evaluated by considering all the available information in a system of oxides and silicates. The emphasis of this work is on critically evaluating the available data and on obtaining an internally consistent set of thermochemical and thermophysical data. An internally consistent data set is one which permits the computation of phase equilibrium relations as established through experimental studies and is at the same time compatible with calorimetric and other measurements of thermophysical properties of the phases. Generating such a data base is complicated by the large uncertainties associated with the high pressure phase equilibrium studies, both because of the errors in pressure and temperature calibration and because many experiments may be for unreversed reactions. In spite of these problems, it is important to review the present status of the existing thermochemical data set and its relation to the experimental phase studies. If a thermochemical data set produces calculated results which are reasonably consistent with experimental work, it can be used for computing possible phase equilibrium relations over a wide range of pressure, temperature and composition conditions. Such predictions are useful in planning future phase equilibrium experiments, which may then be used to better constrain the preliminary data base. Conversely, if the attempt to create a data base exposes major inconsistencies, experiments to resolve these discrepancies would be important.

In metallurgical studies, the work of data assessment has been in progress for many years. The research of the present authors has been particularly influenced by the programs and programming techniques of Eriksson (1975 and several later program versions), and of the group at the Royal Technical University (B. Sundman, B. Jansson, M. Hillert). Needless to say that this is not the only study of this type from the geochemists. It is difficult to state when the first attempt to assess thermochemical data was made by a geochemist; the work of E-an Zen (e.g. Zen, 1969) is a good example of the beginning of such studies. A major effort was made by Helgeson and coworkers (Helgeson et al., 1978) to systematize a large data base. Since then important progress has been made by Berman (e.g. Berman, 1988) and Holland and Powell (1990). Several other authors (e.g. Gasparik, N.D.Chatterjee, Day, Chernosky, Newton, Navrotsky, see references throughout the text) have individually contributed to the data assessment efforts.

The recent studies of Berman (1988) and Holland and Powell (1990) have resulted in thermochemical data bases which work very well in computing phase equilibrium relations at low to medium pressure and temperature systems. Ideally, one would like to start from these data sets and extend the base to include high pressure phases. We did not succeed in using these data for many of the important phases, e.g. enstatite ($MgSiO_3$) and forsterite (Mg_2SiO_4) in systematizing the data so that the constraints from the high pressure phase equilibrium experiments could be satisfied. On the other hand it was relatively easy to choose enthalpy and entropy data for such phases within the calorimetric uncertainties and satisfy the phase equilibrium constraints for all available data both at low and high pressures and temperatures. This is mostly because there are tightly reversed experimental data on many high pressure-temperature reactions in contrast to the data at low pressures and temperatures. It was also necessary for us to use valid C_P extrapolations to high temperatures that are internally consistent with the available data on thermal expansion α and compressibility β. The pressure and temperature dependence of the latter two parameters and the use of Birch-Murnaghan from ambient to very high temperatures are the exclusive features of the present data base.

The mineral names and abbreviations for the phases are as follows:

SiO_2:

α-Quartz (α-Q), β-quartz (β-Q), coesite (Coe), stishovite (St).

MgO:

Periclase (Per)

TiO_2:

Rutile (Rut)

Fe-O:

Iron (Fe), wüstite-FeO (Wu-FeO), wüstite (Wu-$FeO_{1.5}$), magnetite (Mt), high-magnetite (H-mt), hematite (Hm), high-hematite (Hhm).

Al_2O_3:

Corundum (Cor)

CaO:

Lime (CaO)

MgO-SiO_2:

Forsterite (Fo), orthoenstatite (Oen), low-clinoenstatite (Cen-l), high pressure clinoenstatite (H-Cen), β-forsterite (β), γ-forsterite (γ), proetoenstatite (Pen), Mg-garnet (Mg-Gar), perovskite (Pv).

MgO-SiO_2-H_2O:

Antigorite (Ant), anthophyllite (Anth), talc (Tc), chrysotile (Chr), cummingtonite (Cum).

FeO-SiO_2:

Fayalite (Fa), orthoferrosilite (Fer), Fe-perovskite (Fe-Pv), β-fayalite (β-Fa), γ-fayalite (γ-Fa)

FeO-SiO_2-H_2O:

Grunerite (Grun)

CaO-SiO_2:

Wollastonite (Wol), pseudo-wollastonite (Psw).

FeO-Al_2O_3:

Hercynite (Hc)

CaO-MgO-SiO_2:

Monticellite (Mo), orthodiopside (Odi), clinodiopside (Cdi)

CaO-MgO-SiO_2-H_2O:

Tremolite (Trem).
$CaO-Al_2O_3-SiO_2$:
Anorthite (An), gehlenite (Ge), grossular (Gr), merwinite (Mer), Ca-tschermakite (CaTs).
$CaO-Al_2O_3-SiO_2-H_2O$:
Lawsonite (Law).
$CaO-Al_2O_3-SiO_2-H_2O$:
Zoisite (Zo), clinozoisite (Clz), margarite (Marg).
$MgO-Al_2O_3-SiO_2$:
Cordierite (Cord), pyrope (Py).
$MgO-Al_2O_3-SiO_2-H_2O$:
Hydrous cordierite (H-Cord), clinochlor (Cln)
$FeO-Al_2O_3-SiO_2$:
Ferrocordierite (Fe-Cord), almandine (Alm).
$Na_2O-Al_2O_3-SiO_2$:
Albite (Ab), jadeite (Jd).
$Na_2O-Al_2O_3-SiO_2-H_2O$:
Paragonite (Pg).
K_2O- $Al_2O_3-SiO_2$:
Sanidine (San), low-T K-feldspsr (Kspar).
$K_2O-Al_2O_3-SiO_2-H_2O$:
Muscovite (Ms).
$K_2O-Al_2O_3-SiO_2-H_2O-MgO$:
Phlogopite (Phl).
$K_2O-Al_2O_3-SiO_2-H_2O-FeO$:
Annite (Ann).
$FeO-TiO_2$:
Ilmenite (Ilm), ulvospinel (Fe_2TiO_4).
$CaO-FeO-SiO_2$:
Andradite (Andr), clinohedenbergite (Chd).
$MnO-SiO_2$:
Tephroite (Teph).
Carbon:
Graphite (C), diamond (Di).
Carbonates:
Calcite (Calc), magnesite (Mag), siderite (Sid), aragonite (Arag).
Others:
Steam (W), glaucophane (Gl).

Chapter 2

Thermodynamic Relations

Temperature and Pressure Dependence of the Gibbs Free Energy

The change in Gibbs free energy (ΔG) as a function of temperature (T) and pressure (P) can be obtained from the relations:

$$\left(\frac{\partial^2 \Delta G}{\partial T^2}\right)_P = -\frac{C_P(T)}{T} \tag{2.1}$$

and

$$\left(\frac{\partial \Delta G}{\partial P}\right)_T = V(P,T) \tag{2.2}$$

where $C_P(T)$ is heat capacity at constant P and at a temperature T and $V(P,T)$ is the molar volume at a temperature T and a pressure P. The free energy of a reaction is given by

$$\Delta G(P,T) = \Delta H^0_T - T\Delta S^0_T + \int_1^P \Delta V(P,T)dP \tag{2.3}$$

where $\Delta H^o{}_T$ and $\Delta S^o{}_T$ are the standard enthalpy and entropy of a reaction, respectively, at temperature T and 1 atm. They are given by

$$\Delta H^0_T = \Delta H^0_{298} + \int_{298}^T \Delta C_P dT \tag{2.4}$$

and

$$\Delta S^0_T = \Delta S^0_{298} + \int_{298}^T \frac{\Delta C_P}{T} dT \tag{2.5}$$

where $\Delta H^o{}_{298}$ and $\Delta S^o{}_{298}$ are the standard enthalpy and entropy of reaction at 298.15 K, ΔC_P is the heat capacity difference between products and reactants, and

$\Delta V(P,T)$ is the volume change for the reaction.

The extrapolation of temperature and pressure dependence of the Gibbs free energy of a solid or a liquid phase to high temperature and pressure is complicated by the nature of the heat capacity, C_P, and the behavior of molar volume, $V(P,T)$. In order to obtain a suitable form of the Gibbs free energy for solids at high pressure and high temperature, it is necessary to discuss the heat capacity at high temperature and the equation of state in detail.

Heat Capacity at High Temperature

The general relations

The form of the heat capacity (C_P) for high temperature extrapolation has been discussed by Haas and Fisher (1976), Lane and Ganguly (1980), Holland (1981), Berman and Brown (1984) and Fei and Saxena (1987).

In general, C_P for a solid can be expressed as

$$C_P = C_V + \alpha^2 V K_T T + C_P' \tag{2.6}$$

where C_V is the heat capacity of a crystal at constant volume. α is the coefficient of thermal expansion, V is the molar volume, and K_T is the isothermal bulk modulus, all at temperature T. C_P' is the contribution from cation disordering and anharmonicity (other than those incorporated in the $\alpha^2 V K_T T$ term).

Einstein (1907) and Debye (1912) derived equations for the temperature dependence of C_V, based on the assumption that atoms behave as harmonic oscillators in a crystal lattice. A reasonable expression for C_V can be derived by assuming that a crystal is composed of a system of atoms which vibrate as harmonic oscillators all with the same frequency, ν. In such a solid C_V is expressed by

$$C_V = 3Rn\left[\frac{x^2 e^x}{(e^x - 1)^2}\right] \tag{2.7}$$

where $x = q_E/T$. q_E is the Einstein temperature given by $q_E = h\nu/k$. R, h, k and ν are the gas constant, Planck constant, Boltzmann constant and the frequency of atoms, respectively. A precise expression for C_V may be given by assuming that the crystal has a whole spectrum of frequencies from ν_1 to ν_m, where ν_m represents some maximum frequency for a particular crystal,

$$C_V = \frac{9Rn}{x^3}\int_0^{x?}\left[\frac{x^4 e^x}{(e^x - 1)^2}\right]dx \tag{2.8}$$

where $x = h\nu_m/(kT)$. Equation (2.7) is difficult to use in fitting experimental data at intermediate temperatures because of the assumption of a single oscillator frequency. Equation (2.8) is even more difficult to apply to thermodynamic calculations not only because of the complexity of the expression but also because of the difficulty in determining the frequency distribution. Kieffer's (1979a) review on the applicability of the Debye theory of lattice vibrations shows that heat capacities of silicates show large deviations from the behavior expected from the theory.

The isobaric thermal expansion α_p and isothermal compressibility β_T, respectively, are given by:

$$\alpha_p = \frac{1}{V_T}\left(\frac{\partial V}{\partial T}\right)_p \tag{2.9}$$

$$\beta_T = -\frac{1}{V_T}\left(\frac{\partial V}{\partial P}\right)_T \tag{2.10}$$

The inverse of β_T is the bulk modulus K_T. It is experimentally convenient to measure the heat capacity of solids at constant pressure (C_p) rather than at constant volume (C_V). Many different polynomial equations have been used for fitting the measured C_p data (Saxena, 1988 and Saxena and Zhang, 1989). In this book, the following equations are used:

$$C_V = c_0 + c_1 T + c_2 T^{-1} + c_3 T^{-2} + \ldots\ldots. \tag{2.11}$$

$$\alpha_p = \alpha_0 + \alpha_1 T + \alpha_2 T^{-1} + \alpha_3 T^{-2} + \ldots\ldots. \tag{2.12}$$

$$K_T = 1/(\beta_0 + \beta_1 T + \beta_2 T^2 + \beta_3 T^3 + \ldots) \tag{2.13}$$

where c_i are the coefficients of the heat capacity equation, α_i the coefficients in the expansion for isobaric thermal expansion, and β_i the coefficients of compressibility. The choice of a suitable polynomial expression for C_V is the most difficult. For a Debye solid, the low-temperature limit is given by the 'Debye T^3' law ($C_V = A[T/\Theta^3]$ where A is a constant and Θ is the Debye temperature. At high temperature C_V for a Debye solid is to approach the 3nR limit. In a non-Debye solid the departure from this limit must be considered. The empirical expression (2.11) is obviously not suitable for fitting the C_V data at low temperature. Therefore we have used the Kieffer model to calculate C_V to temperatures of about 1000 K. At high temperatures, the applicability of the equation has been tested by Saxena (1989) for silicates for which data on thermal expansion and compressibility are available.

Estimation of C_V

The measured heat capacities of minerals show strong deviations from Debye-like calorimetric behavior at all temperatures. The deviations increase with the complexity of the structure. The Debye model does not account for the measured calorimetric properties of minerals because it does not allow for the details of the vibrational spectra. Kieffer (1979a, b, c, 1980) has developed a model based on lattice dynamics that allows calculation of heat capacity at constant volume from vibration spectra and acoustic velocities. Kieffer (1979c, 1980) calculated heat capacities of many minerals with the use of the model. The model generally yields good estimates of the heat capacity. Many researchers (e.g., Akaogi et al 1984; Hofmeister 1987; McMillan and Ross 1987; Fei et al., 1990) used this model to estimate the heat capacities and entropies of minerals. With more experimental data, a set of related models has been developed with small differences in band assignments, partitioning of modes into optic continua and Einstein oscillators, and dispersion relations for each phase such that these models are consistent with the observed spectra, acoustic velocities and constraints of crystallographic symmetry. Their results show that the calculated heat capacities are insensitive to the details of modelling. Kieffer (1979c) pointed out that the model has different accuracies in different temperature ranges: for C_V, ±30-50% below 50 K, ±4% at 298 K and ±1% at 700 K. Kieffer's model demonstrates that the heat capacities of minerals at the temperature range from 300 to 1000 K are relatively sensitive to the distribution of optic modes; the insufficient vibrational parameters clearly constitute a weak point of the modelling. Here, we take the lower and upper cutoff frequencies w_l, w_u of the optic continuum as calculation parameters based on the spectroscopic data to put some bounds on the thermodynamic C_p data over the temperature range of 200-400 K in our high temperature (>300 K) C_p-C_V calculation. With the C_V data calculated from Kieffer's model to a temperature of 1000 K, it becomes possible to assess the internal consistency of C_p, α_p, β_T with the application of non-linear programming techniques, taking care of the C_p, α_p, β_T data treated in equilibrium calculations. Since Kieffer's model provides a framework to interrelate thermodynamic, spectroscopic, acoustic, structural, and compositional data, we may calculate C_V for some compounds by using the available experimental data on polymorphs and isostructural solids.

In this study the upper limit of the applicability of Kieffer's model was taken somewhat generously as 1000 K. Kieffer's model does not include the effect of thermal expansion (Kieffer, 1979c, p.43). Kieffer quoted Barron and Munn (1968) in emphasizing that for nonisotropic solids, the heat capacity is under the double restraint of constant volume and isotropic stress. Approximate correction for the effect of thermal expansion could be made using the thermal Grüneisen parameter and the calculated Debye temperatures as done by Barron et al. (1959) who found the effect for MgO as 0.5% at 270 K. The effect could be higher at 1000 K which is our chosen limit for the Kieffer model. For a number of substances for which there are measured data at high temperatures on α_p, K_T and C_p, the C_V may be calculated directly. Table 2.1 shows that for several phases, our assumption that Kieffer's model may be used to as high temperature as 1000 K does not introduce any significant error. The maximum error found is about 2% which is better than that found in many experimental measurements and theoretical estimations.

Method of Optimization

Procedure: Saxena (1988) discussed the procedure for systematizing the thermochemical data on C_p, α_p and K_T and estimation of C_V (if not calculated theoretically, see below) by minimizing the function

$$F = S(C_{ex} - C_{cal})^2 \tag{2.14}$$

where C_{ex} denotes the experimental data of C_p; $C_{cal} = C_V + \alpha_p^2 V_T T/\beta_T$. There are very few solids for which the data on C_p, α_p and K_T are available over a range of temperatures. For such solids, the major source of uncertainty in our systematization lies in the measured data on thermal expansion. Anderson and Zou (1989) noted a spread of 20% at 1400 K in the data on α_p for forsterite in the available experimental data. In contrast the errors in the measured value of (dK_T/dT) are less than 1% (Anderson et al., 1982). The errors in the fit of the C_p data are also quoted as less than 1% (e.g., Robie et al. 1982a). The actual error in

Table 2.1. Comparison of experimental and calculated C_V for some minerals

Minerals	T (K)	Cv(Exp)	Cv(Kieffer)
	300	117.77	116.57
Forsterite	700	159.14	159.03
	1000	167.42	166.65
	300	78.92	79.08
Corundum	700	113.86	113.59
	1000	120.43	119.03
	300	37.18	37.06
Periclase	700	46.68	47.02
	1000	47.42	48.45
	300	323.40	324.68
Pyrope	700	449.94	451.04
	1000	471.47	473.63

Note: All C_p data are from Richet and Fiquet (1991); the data on α_p and K_T are from Anderson et al. (1991).

the experiments may be larger. For most solids, the data generally available are C_p over a limited range of temperature and α_p and K_T at 298.15 K. While the statistical errors of fitting the model given by equation (2.6) to the C_p data will be of the same magnitude as of the experimental C_p data, the errors in the model values of α_p and K_T will be unknown. It is, however, necessary to obtain such data for the calculation of the Gibbs free energy $\Delta G(1,T)$ over a wide range of temperatures in phase equilibrium studies, while the present methods of computing phase equilibrium at high pressure and temperature either ignore the temperature and pressure dependence of α_p and K_T or use extrapolated values without regard to equation (2.6).

In this book, the procedure in the internal systematization of the data consists of the constraints from

a) the available C_p data which in many cases have been extrapolated to high temperatures by data systematics involving phase equilibrium experimental data (e.g., Holland and Powell 1990 or Berman 1988);
b) ıne available data on thermal expansion and bulk modulus;
c) the calculated C_V data based on Kieffer's model to 1000 K; and
d) the enthalpy/bulk modulus correlation.

Choice of the polynomials, error of extrapolation and internal systematization: We shall consider the data on corundum as an example to show how the choice of the form of an equation affects our extrapolated data. Equation (2.12) may be used with two or more of the four constants α_0, α_1, α_2 and α_3 for thermal expansion. As is obvious from Fig. 2.1 (see Saxena and Shen, 1992), the two constants (straight line, not shown) will not fit the experimental data as well as the equation with three or four constants. Anderson et al. (1991) do not quote an error on the experimental data. Assuming that it could be as much as that for forsterite (20%) (see Anderson et al. 1982), the equation with two constants would still be unacceptable, but the equation with three or four constants will work well.

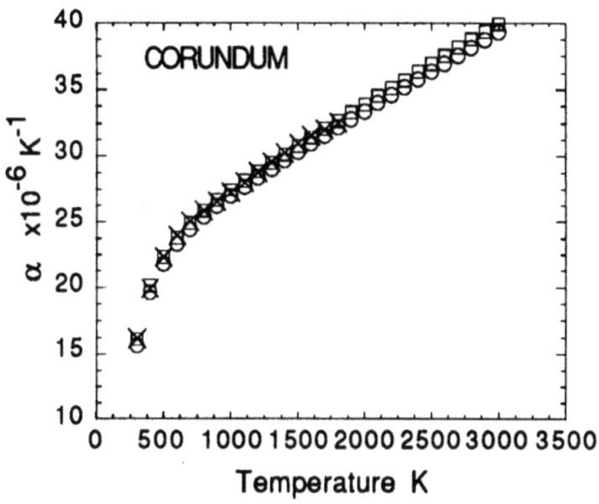

Fig.2.1. Illustration of the effect of choosing different polynomials from the extrapolated data on thermal expansion of corundum. Crosses: experimental data; squares: four-constant fit; circles: three-constant fit.

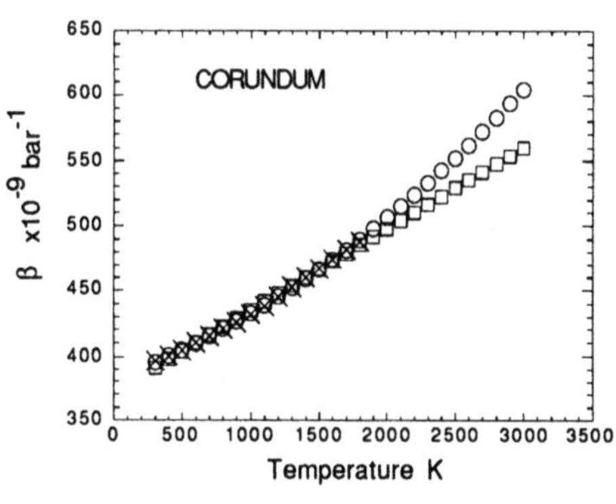

Fig. 2.2. The fitted and extrapolated data on the compressibility of corundum. Crosses: experimental data; squares: four-constant fit; circles: three-constant fit.

Figure 2.2 (see Saxena and Shen, 1992) shows the fitted and extrapolated data on the compressibility of corundum. The three constant equation provides us with a superior fit to the experimental data. Unlike the error on the thermal expansion data, the errors on bulk modulus data are quoted as less than 1% (Anderson et al. 1982). Therefore the curvature in the plotted data in Fig. 2.2 could be significant.

Figure 2.3 (see Saxena and Shen, 1992) shows various properties of corundum as a function of temperature. The C_p data are fitted with the general 5 constant expression used in this study:

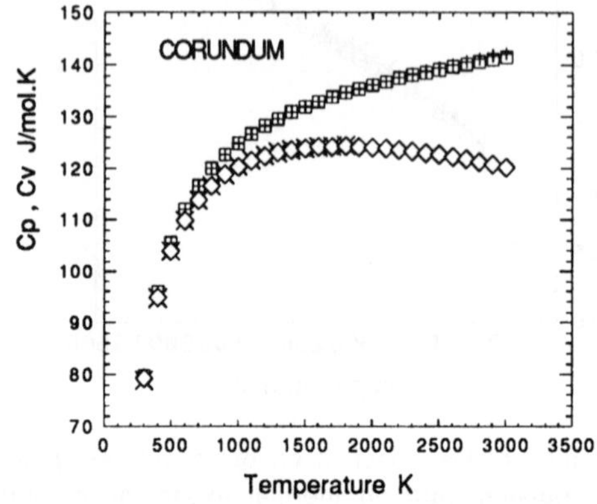

Fig. 2.3. The experimental data and extrapolations on C_p and C_V of corundum. Squares: C_p data of Richet and Fiquet (1991) and extrapolation by Eq (2.16); plusses: C_p data of Richet and Fiquet (1991) and extrapolation by Eq. (2.15); crosses: experimental C_V from available C_p, α_p and K_T (Anderson et al.,1991); diamonds: calculated C_V data from the extrapolated data on C_p, α_p and K_T.

$$C_p = a + bT + cT^{-2} + eT^{-3} + gT^{-1} , \tag{2.15}$$
$$C_p = a + b\ln T + cT^{-1} + dT^{-2} + eT^{-3} \tag{2.16}$$

Equation (2.16) is from Richet and Fiquet (1991). Both expressions produce a perfect fit of the experimental data (to 1800 K; the data points not shown in Fig. 2.3 because they fall on the fitted data points). There is also no significant difference between the extrapolated data to 3000 K.

While the polynomials with certain chosen constants appear to yield reasonable extrapolations, there is still a need for further constraints on the extrapolated data as demonstrated below. We may, for example, assume that our extrapolated data ($T >$ 1800 K) for C_p, α_p and K_T are correct and, using equation (1), a) calculate C_V and compare the data with a modeled C_V or b) from modeled C_V, calculate C_p and compare it with the extrapolated C_p. The crosses in Fig.2.3 represent 'experimental' C_V based on the experimental C_p from Richet and Fiquet (1991) and the data on α_p and K_T from Anderson et al. (1991). The diamonds in Fig.2.3 represent the calculated C_V data from the extrapolated ($T>$1800 K) data on C_p, α_p and K_T. We note in Fig. 2.3 that such a C_V decreases with increasing temperature. If for the sake of argument we adopt the Kieffer model even at temperatures higher than 1000 K, the calculated C_p from the extrapolated α_p and K_T appears as shown in Fig. 2.4.(see Saxena and Shen, 1992). Such a C_p exceeds the melt C_p and also does not fit well over a large temperature range.

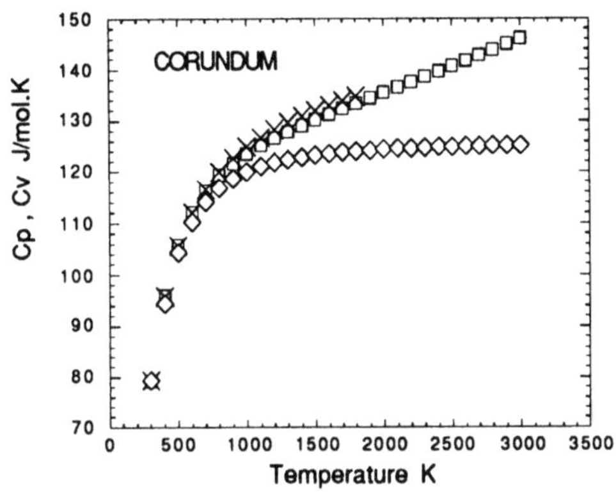

Fig. 2.4. Calculated C_p of corundum from C_v (Kieffer) and experimental data (to 1800 K) and extrapolated α_p and K_T. Crosses: experimental data from Richet and Fiquet (1991); diamonds: C_v data from Kieffer's model; squares: calculated C_p from extrapolated α_p, K_T and Kieffer's C_v, plusses: experimental C_v as in Fig. 2.3.

To internally systematize the C_p, α_p and K_T data, we use the optimization procedure as discussed above. The final results on corundum using such a procedure are shown in Fig. 2.5 (see Saxena and Shen, 1992). In this case, we use the experimental data on C_p, α_p and K_T to 1800 K, take the C_v from Kieffer's model to 1000 K and then optimize the coefficients for α_p. K_T and C_v. The resulting C_p data are regressed again to yield the C_p coefficients. Figure 2.5 shows that the new C_p data fit the experimental C_p data with the same degree of high accuracy as is necessary in fitting the calorimetric data (see also Table 2.7 later where the fits of α_p and K_T are also shown).

Thus the C_p is formulated by taking into account the important role of the $\alpha^2 V K_T T$ term and, therefore, links the measured heat capacity with the measured physical properties of solids at high temperature. This provides an additional constraint for evaluating an internally consistent thermochemical and thermophysical data set based on data from the calorimetric measurements, from the experimental phase equilibria and from the measurements of solid physical properties.

We have used equation (2.9) in assessing and refitting the C_p data. Such assessed data are presented with a flexible C_p formulation adopted at many centers (e.g. A.D.Pelton with the F*A*C*T system at Ecole Polytechnique, Montreal or B.Sundman with THERMOCALC at the Royal Technical University at Stockholm) working with large data bases. The formulation is:

$$C_p = a + bT + cT^{-2} + dT^2 + eT^{-3} + fT^{-0.5} + gT^{-1} \quad (2.17)$$

Anderson's Enthalpy/Bulk Modulus Relationship

Anderson (1989) discussed an important correlation between enthalpy and adiabatic bulk modulus (K_S). For the three solids MgO, Al_2O_3 and Mg_2SiO_4 (forsterite), he found that the relation is almost linear. Over the temperature range for which we have the C_p data, the following relation may be introduced as an additional constraint in the optimization:

$$K_T = \frac{C_v}{C_p}(a + b\Delta H) \tag{2.18}$$

here a, b are the coefficients of the linear relationship with $a = K_{S,300}$, b close to $-\delta_S\gamma/V_0$ (δ_S: adiabatic Anderson-Grüneisen parameter); $\Delta H = H(1,T) - H(1,300)$. Note that this constraint does not require that K_T follow such temperature dependence beyond the range of the 'experimental' C_p data.

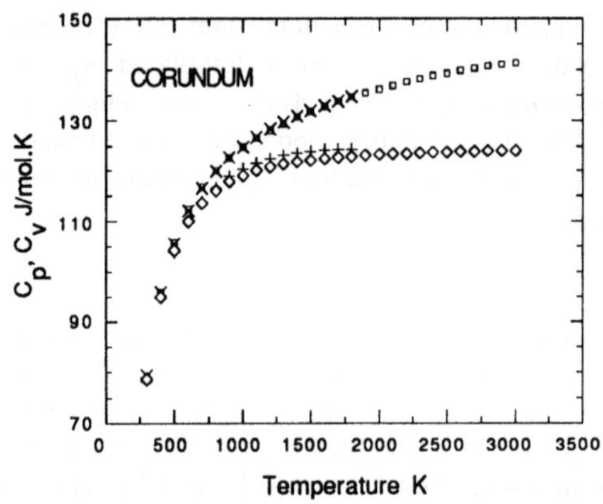

Fig. 2.5. Optimized data on C_p and C_V of corundum. Crosses: experimental data from Richet and Fiquet (1991); squares: optimized C_p data from the relation ((2.6); diamonds: optimized data on C_V which are closely silmilar to the calculated data from Kieffer's model for corundum, plusses: experimental C_V as in Fig. 2.4.

Some Comparisons

The C_p/C_v systematization has been performed for many solids for which there are sufficient spectroscopic data to calculate C_v using Kieffer's model. Several minerals have been added to Kieffer's C_v data base by considering the w_l and w_u frequencies to be similar to those of the structurally analogous compounds. The sources for the original data on C_p are listed in the Appendix. For some minerals high temperature C_p data are available (e.g., Richet and Fiquet, 1991). For others the data were fitted by Chatterjee (1989) and Fei et al. (1990) with the expression as discussed by Fei and Saxena (1987) and for all minerals are consistent with the phase equilibrium experiments. The data on α_p and β_T are listed in the Appendix. The coefficients for Cv are listed in Table 2.2 (from Saxena and Shen, 1992).

Table 2.2. Coefficients for the heat capacity at constant volume*

	Mineral	c_0 (10^{-3})	c_1 (10^{+3})	c_2 (10^{-4})	c_3 (10^{-8})	c_4 (10^{-10})
1	Albite	0.3310	-1.2944	-1.0508	-0.1836	0.3053
2.	Andalusite	0.2006	-0.2393	-0.1041	-0.1274	0.1803
3.	Annite	0.5573	-0.4317	-3.7937	-0.1102	0.1779
4.	Anorthite	0.3304	-1.1820	-0.9419	-0.1779	0.2968
5.	Anthophyllite	1.0346	-1.2053	-4.1003	-0.4047	0.5856
6.	Brucite	0.1269	0.7402	-2.7492	0.0812	-0.1309
7.	Ca-Tschermak	0.2511	-0.3639	-0.2192	-0.1344	0.1900
8.	Coesite	0.0782	-0.1946	-0.4849	-0.0404	0.0073
9.	Corundum	0.1243	0.0616	0.1168	-0.0731	0.0857
10.	Diopside	0.2512	-0.3677	-0.2245	-0.1318	0.1868
11.	Enstatite	0.1315	-0.4865	-0.3364	-0.0826	0.11415
12.	Fayalite	0.1755	2.6177	-0.1075	-0.0694	0.0956
13.	B-Fayalite	0.1755	3.8103	-0.1195	-0.0628	0.0884
14.	Ferrosilite	0.1256	-0.2349	-0.0865	-0.0474	0.0443
15.	Forsterite	0.1915	0.1639	-2.4711	0.0239	-0.0479
16.	B-Forsterite	0.1836	-0.4886	-1.8436	-0.0011	-0.0279
17.	G-Forsterite	0.1651	3.3164	0.6328	-0.1268	0.1620
18.	Garnet(Mg)	0.1256	0.3273	-0.1025	-0.0689	0.0970
19.	Glaucophane	1.0347	-1.2411	-4.0577	-0.4430	0.6364
20.	Grossular	0.5015	-0.5667	-0.2895	-0.2677	0.3680
21.	Grunerite	1.0351	-1.2896	-4.2252	-0.3503	0.5188
22.	Hedenbergite	0.2600	-2.2236	-1.5014	-0.0624	0.0769
23.	Hercynite	0.1740	0.0886	0.1324	-0.0712	0.0727
24.	Ilmenite	0.1263	-0.3474	-0.2056	-0.0204	0.0054
25.	Ilmenite(Mg)	0.1246	0.2946	0.0696	-0.0828	0.1072
26.	Jadeite	0.2512	-0.3759	-0.2176	-0.1449	0.2076
27.	Kyanite	0.2005	-0.2155	-0.0855	-0.1273	0.1760
28.	Lime	0.0518	-0.1302	-0.1090	-0.23E-4	-0.0177
29.	Monticellite	0.1753	-0.1564	-0.0722	-0.0846	0.1141
30.	Muscovite	0.5314	-0.3377	-3.5038	-0.1608	0.2400
31.	Paragonite	0.5314	-0.3321	-3.4897	-0.1700	0.2526
32.	Periclase	0.0463	0.4148	0.2192	-0.0149	0
33.	Perovskite	0.1246	0.0044	0.0690	-0.0740	0.0938
34.	Phlogopite	0.5716	-3.5181	-5.8147	-0.0346	0.0445
35.	Pyrope	0.5022	-0.6987	-0.4037	-0.2782	0.3892
36.	Pyrophyllite	0.5059	-0.2853	-3.2826	-0.1804	0.2589
37.	Sanidine	0.3310	-1.2933	-1.0491	-0.1844	0.3062
38.	Sillimanite	0.2011	-0.3297	-0.1949	-0.1268	0.1831
39.	Spinel	0.1740	0.0777	0.1474	-0.0941	0.1089
40.	Stishovite	0.0753	-0.1109	-0.0494	-0.0556	0.0814
41.	Talc	0.5308	-0.1967	-3.4362	-0.1675	0.2462
42.	Tephroite	0.1754	-0.1759	-0.0995	-0.0726	0.0992
43.	Tremolite	1.0344	-1.1849	-4.0689	-0.4198	0.6037
44.	Wollastonite	0.1252	-0.1046	-0.0663	-0.0648	0.0937

* $C_v = c_0 + c_1 T + c_2 T^{-1} + c_3 T^{-2} + c_4 T^{-3}$, C_v in J/mol·K, T in K.

Table 2.3. Comparison of the optimized and available data for corundum

T (K)	C_p J/mol.K		C_v J/mol.K		α 10^{-6}		K_T GPa	
	Exp	Model	Exp	Model	Exp	Model	Exp	Model
300	79.420	79.187	78.92	78.723	16.200	15.499	252.11	251.925
400	95.790	95.940	94.78	94.945	19.920	19.760	249.94	248.759
500	105.59	105.948	104.02	104.281	22.310	22.993	247.17	245.629
600	112.02	112.364	109.87	109.945	23.970	25.416	243.83	242.522
700	116.57	116.807	113.86	113.591	25.050	27.269	240.87	239.427
800	119.99	120.096	116.73	116.061	25.860	28.717	237.76	236.335
900	122.68	122.666	118.84	117.804	26.580	29.873	234.81	233.235
1000	124.87	124.763	120.43	119.076	27.270	30.810	231.47	230.121
1100	126.71	126.536	121.61	120.032	28.020	31.583	228.23	226.986
1200	128.28	128.074	122.50	120.767	28.770	32.226	224.60	223.823
1300	129.65	129.437	123.13	121.343	29.550	32.768	220.90	220.629
1400	130.87	130.664	123.60	121.803	30.270	33.228	217.22	217.399
1500	131.96	131.782	123.92	122.176	30.930	33.621	214.20	214.131
1600	132.95	132.809	124.15	122.482	31.530	33.958	210.95	210.822
1700	133.86	133.758	124.35	122.737	32.010	34.250	207.71	207.471
1800	134.69	134.640	124.41	122.950	32.520	34.503	204.87	204.079
1900		135.460		123.132		34.722		200.646
2000		136.225		123.287		34.913		197.172
2100		136.939		123.420		35.080		193.661
2200		137.604		123.537		35.225		190.113
2300		138.223		123.638		35.351		186.534
2400		138.798		123.727		35.461		182.925
2500		139.331		123.806		35.556		179.291
2600		139.823		123.877		35.639		175.637
2700		140.276		123.940		35.709		171.968
2800		140.691		123.996		35.770		168.287
2900		141.068		124.047		35.821		164.602
3000		141.410		124.094		35.863		160.915

Note: The data on C_p is from Richet and Fiquet (1991). The data on thermal expansion and bulk modulus are those adopted by Anderson et al. (1991). The experimental C_v is calculated from Eq. (1). See also C_p calculation by McMillan and Ross (1987).

Complete data on α_p and K_T to high temperatures are available on six minerals: periclase, pyrope, corundum, CaO, grossular and forsterite (Anderson et al. 1991). Since the grossular data are for a solid solution, the data will not be used here. We shall consider these minerals below in some detail. The calculated results will be compared to the measured data and it will be demonstrated with some figures and tables that the extrapolated data to 3000 K show reasonably smooth trends of variation.

Corundum: Figure 2.5 shows the optimized results on C_p and C_V. Table 2.3 (from Saxena and Shen, 1992) shows a comparison of the data to 1800 K and the extrapolation to 3000 K. The major misfit here is represented by the data on thermal expansion, with the maximum difference of 12% at 1000 K which is within the experimental error of such measurements. We note that such an error in thermal expansion also corresponds to only about 1% error in C_p. This may be seen by optimization involving a different set of C_p data for corundum. The C_p values of Robie et al (1978) differ by about 1% from those of Richet and Fiquet (1991). By using the data of Robie et al (1978), we get optimized α_p values of 26.3x10^{-6} at 1000 K and of 31.5x10^{-6} at 1800 K. Compared to the data in Table 2.3, there is a difference of about 14% at 1000 K and about 10% at 1800 K. For bulk modulus, because the Anderson's enthalpy/ bulk modulus relationship has been used as an additional constraint, the uncertainties are much less than that on thermal expansion.

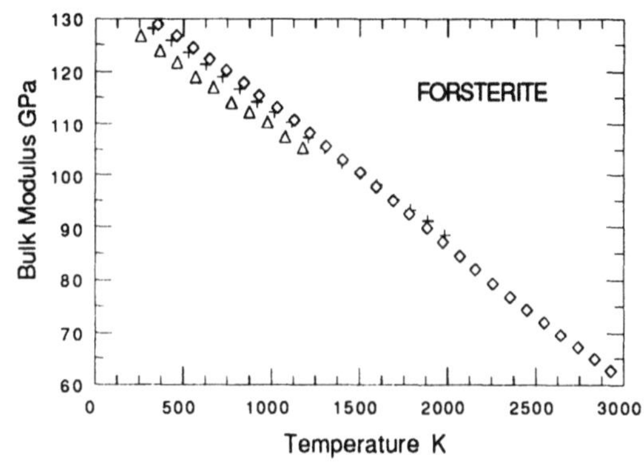

Fig.2.6. Bulk modulus data of forsterite plotted against temperature. Diamonds: systematized data of this study; plusses: Isaak et al. (1989); triangles: Manghnani and Matsui (1981).

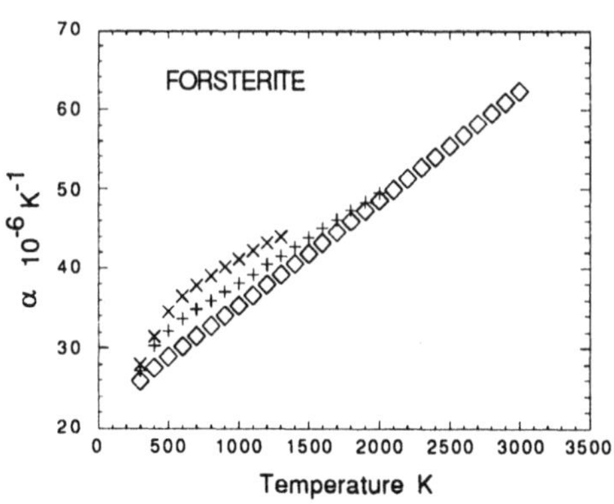

Fig.2.7. Thermal expansion data of forsterite plotted against temperature. Crosses: Matsui and Manghnani (1985); plusses: Isaak et al. (1989) based on Kajiyoshi (1986), diamonds: this study.

Forsterite: Table 2.4 (from Saxena and Shen, 1992) shows the optimized heat capacity and other data. Systematized heat capacity at constant pressure is consistent with the data of Robie et al. (1978, 1982b) within 1% and with the data of Richet and Fiquet (1991) within 0.5%. Kieffer's model gives a good estimation of heat capacity at constant volume up to about 1000 K and that above 1000 K anharmonic effects become apparent on C_V. The anharmonic behavior as revealed in the study by Gillet et al. (1991) is at least partly reproduced by adopting the present procedure of systematization of the data. If the trend of the anharmonic C_V of Gillet et al (1991) were to continue, we will run into the problem of a very small bulk modulus at high temperature. The fits of the systematized data on bulk modulus and thermal expansion are shown, respectively, in Figs. 2.6 and 2.7. Figure 2.6 compares three sets of bulk modulus data; the data of Manghnani and Matsui (1981) and of Isaak et al. (1989) are shown along with the systematized data. Isaak et al (1989) calculated the isothermal bulk modulus at constant pressure from $K_T=K_S/(1+\alpha\gamma T)$. Their recommended values of thermal ex-

Table 2.4. Comparison of the optimized and available data for forsterite

T (K)	C_p J/mol.K		C_V J/mol.K		α 10^{-6}		K_T GPa	
	Exp	Model	Exp	Model	Exp	Model	Exp	Model
300	119.00	119.130	117.77	118.002	27.2	25.938	129.0	128.040
400	139.01	138.931	137.00	137.250	30.3	27.615	127.0	125.862
500	150.75	150.173	147.96	147.897	32.2	28.952	123.0	123.673
600	158.22	157.774	154.63	154.844	33.6	30.213	121.0	121.463
700	163.57	163.454	159.14	159.804	34.8	31.464	119.0	119.223
800	167.80	167.992	162.50	163.551	35.9	32.726	116.0	116.948
900	171.41	171.798	165.18	166.494	37.0	34.000	114.0	114.633
1000	174.64	175.113	167.42	168.874	38.1	35.289	112.0	112.274
1100	177.62	178.087	169.35	170.841	39.2	36.590	110.0	109.870
1200	180.43	180.820	170.98	172.497	40.5	37.901	107.0	107.421
1300	183.11	183.380	172.50	173.911	41.6	39.222	105.0	104.929
1400	185.69	185.812	173.88	175.134	42.7	40.550	102.0	102.396
1500	188.17	188.152	175.03	176.202	43.9	41.884	100.0	99.825
1600	190.58	190.423	176.14	177.145	45.0	43.225	98.0	97.223
1700	192.92	192.641	177.08	177.983	46.2	44.570	95.0	94.594
1800		194.818		178.733		45.920		91.944
1900		196.963		179.408		47.273		89.282
2000		199.081		180.020		48.629		86.613
2100		201.176		180.578		49.988		83.945
2200		203.249		181.087		51.350		81.285
2300		205.301		181.556		52.713		78.641
2400		207.331		181.988		54.079		76.019
2500		209.340		182.387		55.447		73.427
2600		211.326		182.759		56.816		70.871
2700		213.287		183.104		58.186		68.355
2800		215.223		183.427		59.558		65.887
2900		217.133		183.729		60.931		63.469
3000		219.014		184.013		62.305		61.106

Note: The data on C_p is from Richet and Fiquet (1991). The data on thermal expansion and bulk modulus are those adopted by Anderson et al. (1991). The experimental C_v is calculated from Eq. (1). Some other data on bulk modulus and thermal expansion (Matsui and Manghnani, 1985) are shown, respectively, in Figs. 2.6 and 2.7. Additional data on calculated C_v are available in Chopelas (1990a) and Gillet et al. (1991).

pansion (Kajiyoshi, 1986) are close to the systematized values here as seen in Fig. 2.7. The figure also shows the data on thermal expansion of Matsui and Manghnani (1985). We note that these data are within the experimental range of uncertainties of the thermal expansion data as plotted in the figure and are also not inconsistent with the data of Suzuki (1975) and White et al. (1985) (not plotted). Finally, we have checked the correlation between ΔH $(H_T\text{-}H_{300})$ and the optimized bulk modulus over the temperature range of 300 to 3000 K. While Anderson's (1989) model involved the adiabatic bulk modulus, we find that such a relationship also exists closely between isothermal bulk modulus and ΔH for forsterite. However, this relationship between K_T and ΔH may not always hold for other solids.

Table 2.5 Comparison of the optimized and available data for periclase

T (K)	C_p J/mol.K		C_v J/mol.K		α 10^{-6}		K_T GPa	
	Exp	Model	Exp	Model	Exp	Model	Exp	Model
300	37.35	37.703	36.82	37.176	31.20	31.183	162.0	160.780
400	42.80	43.553	41.89	42.633	35.70	35.927	159.0	157.698
500	45.58	46.228	44.29	44.931	38.40	38.475	156.0	154.639
600	47.21	47.734	45.53	46.063	40.20	40.188	153.0	151.600
700	48.32	48.729	46.28	46.681	41.40	41.521	150.0	148.579
800	49.17	49.474	46.74	47.044	42.60	42.658	147.0	145.575
900	49.89	50.091	47.04	47.269	43.80	43.687	144.0	142.587
1000	50.55	50.637	47.31	47.417	44.70	44.650	141.0	139.616
1100	51.16	51.146	47.51	47.518	45.60	45.573	138.0	136.660
1200	51.74	51.636	47.68	47.590	46.50	46.469	135.0	133.721
1300	52.30	52.115	47.87	47.644	47.10	47.347	132.0	130.800
1400	52.85	52.592	47.99	47.686	48.00	48.212	129.0	127.897
1500	53.38	53.069	48.08	47.721	48.90	49.069	126.0	125.014
1600	53.90	53.549	48.15	47.752	49.80	49.920	123.0	122.152
1700	54.41	54.033	48.28	47.779	50.40	50.766	119.0	119.314
1800	54.91	54.521	48.31	47.805	51.30	51.609	117.0	116.501
1900		55.014		47.829		52.449		113.714
2000		55.511		47.853		53.288		110.957
2100		56.012		47.877		54.124		108.230
2200		56.516		47.901		54.960		105.536
2300		57.024		47.925		55.795		102.877
2400		57.535		47.950		56.629		100.253
2500		58.048		47.975		57.463		97.669
2600		58.563		48.001		58.296		95.123
2700		59.079		48.027		59.129		92.620
2800		59.595		48.054		59.962		90.159
2900		60.112		48.082		60.795		87.742
3000		60.629		48.110		61.628		85.370

Note: The data on C_p is from Richet and Fiquet (1991). The data on thermal expansion and bulk modulus are those adopted by Anderson et al. (1991). The experimental C_v is calculated from Eq. (1).

Periclase: Table 2.5 (from Saxena and Shen, 1992) shows the internally systematized data on periclase. The heat capacity (C_p) data of Richet and Fiquet (1991) are reproduced within 0.5% and of Robie et al. (1978) with somewhat larger errors. As the table shows all other data, e.g. the data on bulk modulus as reviewed by Anderson et al. (1991) and the thermal expansion data of Suzuki (1975), are closely reproduced. The calculated data on C_v by Chopelas (1990b) and the data of Isaak et al. (1990) are also quite similar to the data in Table 2.5.

Molar Volume at High Pressure and High Temperature

The molar volume at temperature T and 1 atm can be calculated from

$$V(1,T) = V^0_{298}\exp\left(\int_{298}^{T} \alpha dT\right) \tag{2.19}$$

where $V^o{}_{298}$ is the molar volume at 298 K and 1 atm..

The contribution of VdP to Gibbs free energy becomes significant at high pressure. Therefore, the use of a suitable equation of state for the solids at high pressure and temperature is necessary for the phase equilibrium calculation. Experimental data on molar volumes are available either at high pressure and room temperature or at high temperature and 1 atm. The temperature dependence of molar volume can be represented by equation (2.12). The Murnaghan equation of state, which is based on the assumption that the bulk modulus changes linearly with pressure, and the Birch-Murnaghan equation of state, which is based on the finite-strain theory, are often used to represent experimental data on molar volume at high pressure. The Murnaghan equation of state is given by

$$P_M = \frac{K_{T,0}}{K'_{T,0}}\left[\left(\frac{V^0}{V}\right)^{K'_{T,0}} - 1\right] \tag{2.20}$$

and the Birch-Murnaghan equation of state is

$$P_{B\text{-}M} = \frac{3}{2}K_{T,0}\left[\left(\frac{V^0}{V}\right)^{7/3} - \left(\frac{V^0}{V}\right)^{5/3}\right]\left\{1 - \frac{3}{4}(4 - K'_{T,0})\left[\left(\frac{V^0}{V}\right)^{2/3} - 1\right] + \ldots\right\} \tag{2.21}$$

where $K_{T,o}$ and $K_{T,o}'(= [\partial K_{T,o}/\partial P]_T)$ are the isothermal bulk modulus and its pressure derivative at 298 K, respectively. (We drop the 'o' from the subscript in the labels and stipulate that K_T and K'_T are values at 0 pressure). As discussed by Jeanloz and Knittle (1986), the Birch-Murnaghan equation of state may be considered as a better expression than the Murnaghan equation of state for the pressure-volume relation at very high pressures.

The method of extrapolation of the existing data to high P-T space is still controversial. Anderson and Zou (1989) and Heinz and Jeanloz (1984) have expounded on the relationships among the various thermochemical and thermophysical variables. In Anderson and Zou's model, the simultaneous effect of pressure and temperature on volume is taken care of by the thermal pressure (P_{th}), *i.e.*

$$P = P_{B\text{-}M} + P_{th} \tag{2.22}$$

where P_{th} is calculated by

$$P_{th} = \int_{298}^{T} \alpha K_T dT \tag{2.23}$$

Equation (2.23) has been simplified by assuming the product of α and K_T is constant.

The Use of the Birch-Murnaghan (BM) Equation at High Temperature

With all the information on the temperature dependence of α and β available, one may proceed to apply the isothermal form of the BM equation (see Saxena and Zhang, 1990) at different temperatures as follows.

The $\int VdP$ is calculated by adopting the third order Birch-Murnaghan equation of state [equation (2.21)] where the temperature dependence of the isothermal bulk modulus is included and V^0/V is replaced by $V(1,T)/V(P,T)$. The temperature dependence of all variables, except $(dK_T/dP)_T$, is known from the data systematization at conditions of 1 bar and T as discussed in the previous section. By using the experimental data on in-situ P-V-T determinations or with the help of the phase equilibrium experimental data (less preferably the shock-wave data, see Saxena and Zhang, 1990 for the method), we may determine the temperature coefficients by expressing $(dK_T/dP)_T$ with an appropriate polynomial, e.g.

$$(\delta K_T/\delta P)_T = K'_{300} + K''(T-300)\ln(T/300) \tag{2.24}$$

where K'_{300} is the pressure derivative in the Birch - Murnaghan equation and K"T the temperature coefficient (not to be confused with K'p the second derivative of the bulk modulus). We refer to this model as *the high-temperature Birch-Murnaghan (HTBM)* model. For computational convenience $\int PdV$ may be calculated from equation (2.21), instead of from $\int VdP$. The relation between $\int PdV$ and $\int VdP$ is given by

$$\int_{1}^{P} VdP = \int_{V(P,T)}^{V(1,T)} PdV + V(P-1) \tag{2.25}$$

where

$$\int_{V(P,T)}^{V(1,T)} PdV =$$
$$\frac{3}{2}K_T V(1,T)\left[\frac{3}{4}(1+2x)(Y^{4/3}-1) - \frac{3}{2}(1+x)(Y^{2/3}-1) - \frac{1}{2}x(Y^2-1)\right] \tag{2.26}$$

where

$$x = \frac{3}{4}\left[4 - \left(\frac{\partial K_T}{\partial P}\right)_T\right] \quad \text{and} \quad Y = \frac{V(1,T)}{V(P,T)}$$

The temperature dependence of the bulk modulus accounts for part of the simultaneous effect of pressure and temperature on volume. The calculated results for MgO are close to that calculated by Anderson and Zou's thermal pressure model. However, the traditional method, in which the temperature dependence of the isothermal bulk modulus is not included, results in a significant overestimation of volume at high pressure and high temperature.. This modification does not significantly affect the phase equilibrium calculation where the phases in the reaction have similar thermal expansion and compression because only the volume change is of concern in the calculation, but it may play an important role in the reaction where the phases involved have very different thermophysical behavior.

Comparison with Anderson's Model

Equation (2.24) is sufficient to completely determine the pressure part of the Gibbs energy of any phase (within the range of pressure of the applicability of the equation of state); however, in view of the lack of in-situ type P-V-T data and the assumptions involved in the *Cp/Cv* optimization (e.g. empirical polynomial forms and/or Kieffer's model of Cv), it is useful to calculate certain characteristic parameters as used in other P-V-T models (e.g. Anderson and Goto, 1989) and compare the data. One such parameter is the Grüneisen parameter γ given by:

$$\gamma = \alpha_P K_T V / C_V \tag{2.27}$$

and its functional dependence on temperature. The latter information is useful in deciding the applicability of the Mie-Grüneisen equation. As discussed by Anderson and Goto (1989), we may consider

$$(\Delta\gamma/\Delta T)_V = (\Delta\gamma/\Delta T)_P + (\Delta\gamma/\Delta P)_T \alpha_P K_T \tag{2.28}$$

with volume dependence of γ expressed as

$$\gamma = \gamma_O (V/V_O)^{\gamma} \tag{2.29}$$

where γ_O and V_O are γ and V respectively at 0 K and 1 bar.

To obtain the pressure variation of γ one may use

$$(\Delta\gamma/\Delta P)_T = -(\gamma\gamma_O/K_T)(V/V_O)^{\gamma} \tag{2.30}$$

In the tables of thermodynamic data (see Appendix), we have calculated the Grüneisen parameter (γ) and the Anderson-Grüneisen constant (δ) given by:

$$\delta_T = -(1/\alpha K_T)(dK_T/dT)_P \tag{2.31}$$

to facilitate comparison of the data with those calculated using Anderson's model.

Chapter 3

Techniques for Data Assessment

Computational Techniques

Various techniques are adopted in the systematization, depending upon the complexity of a problem, to construct the internally consistent data base. The phases considered in this study can be classified into four distinct groups:

i) Phases with well determined calorimetric data, e.g., MgO.
ii) Phases whose thermochemical properties have been measured calorimetrically by different workers with different results, e.g., enstatite, forsterite.
iii) Phases whose thermochemical properties are not available, but experimental reversals, including data on distribution of major elements between coexisting phases, are available, e.g., ferrosilite.
iv) Phases for which neither calorimetric nor experimental data are available.

In the first case, the calorimetric data are accepted as they are and the phases are used as a foundation for the data base. In the second and third cases, a method of optimization is adopted to find the best set of data which reproduces the experimental phase equilinbrium data within their uncertainty limits. In the fourth case, an estimate based on theoretical considerations is made.

The Manual Adjustment Approach Using Free Energy Minimization

This method is mainly followed when experimental data are on a single reaction and are too few to run an optimization. Only enthalpy and entropy are adjusted within their stated uncertainties. The method is also useful to obtain start values for an optimization when a lot of experimental data are available on a single reaction.

For computation of phase equilibrium, the method of minimization of the total Gibbs free energy of an assemblage as discussed by Eriksson (1975), Smith and Missen (1982) and used by Saxena and Eriksson (1983a,b, 1985), Saxena and Chatterjee (1986) and Fei and Saxena (1986) is used. This computational procedure has the advantage of considering the entire data set in each calculation and therefore the results reflect the internal consistency of the whole data set. Thus each datum is tested in a multireaction environment. For example, in the system MgO-SiO_2, we have the following possible reactions:

$$\underset{\text{periclase}}{MgO} + \underset{\text{quartz}}{SiO_2} = \underset{\text{enstatite}}{MgSiO_3} \tag{3.a}$$

$$\underset{\text{periclase}}{2\,MgO} + \underset{\text{quartz}}{SiO_2} = \underset{\text{forsterite}}{Mg_2SiO_4} \tag{3.b}$$

$$\underset{\text{forsterite}}{Mg_2SiO_4} + \underset{\text{quartz}}{SiO_2} = \underset{\text{enstatite}}{2MgSiO_3} \tag{3.c}$$

$$\underset{\alpha\text{-quartz}}{SiO_2} = \underset{\beta\text{-quartz}}{SiO_2} \tag{3.d}$$

$$\underset{\text{enstatite}}{MgSiO_3} = \underset{\text{clinoenstatite}}{MgSiO_3} \tag{3.e}$$

Let us assume that there exist experimental phase equilibrium relations for the above system and a set of thermochemical data which reproduces some of the experimental results but not all. If the experiments are all reversed and the thermochemical data do not reproduce the equilibrium pressure and temperature of the reactions (3.a), (3.c) and (3.e), the offending species is enstatite and its data should be adjusted. If, in addition to the three reactions, reaction (3.b) is also not reproduced, then the data on MgO would also need adjustment. While the above procedure can easily be performed by hand calculation, the total Gibbs free energy minimization method expedites the computation on multicomponent systems with additional information on divariant fields. Furthermore, each computation involves all the data on the phases of the system (not just the ones used in each individual reaction). When we compute the equilibrium of a single reaction, only the total free energy of the phases participating in the reaction is minimized. Such a procedure may result in a spurious value of enthalpy of a phase. This cannot happen with the method used here, since all phases that can form compositionally in the system are included in the possible assemblages for which the total free energy is computed and compared by the minimization technique.

Note that this procedure in certain cases is not apparently more useful than simple hand calculation. Such may be the case, for example, if we wish to determine the standard free energy of formation of Al_2O_3 orthopyroxene according to the reaction:

$$\underset{\text{Orthopyroxene}}{MgSiO_3 + Al_2O_3} = \underset{\text{Pyrope}}{MgAl_{2/3}SiO_4} \tag{3.f}$$

With the standard free energy of formation of pyrope and enstatite known and the standard free energy change of the reaction being $RT\ln(X_{Al}X_{En})$, where X_{Al} and X_{En} are mole fractions of Al_2O_3 and $MgSiO_3$ in orthopyroxene, the standard free energy of formation of Al_2O_3-orthopyroxene can be determined if experimental data on X_{Al} and X_{En} are available. However, with hand calculation, the condition that the standard free energy of formation of Al_2O_3-orthopyroxene should always be greater than the standard free energy of formation of corundum is not automatically enforced as it is with the total free energy minimization method. The situation becomes more complex in multicomponent-multiphase systems. Since it is equally easy to compute both simple and complex systems with the installed program, all computations are done with the free energy minimization program.

The Regression Technique

Thermochemical data for phases are mostly determined from calorimetric experiments. In theory, one should be able to reproduce experimentally observed equilibria using these thermochemical data. However, due to errors in both calorimetric and equilibrium experiments, this is often not the case. Many techniques of curve fitting have been suggested by different workers such as Haas and Fisher's (1976) regression analysis method and the linear and mathematical programming methods of Gordon (1973), Halbach and Chatterjee (1982), Berman et al. (1986), Day and Kumin (1980), etc. In this study available thermochemical data are systematized by adjusting them within their uncertainties. The method followed is described in the following paragraph. The final data set is compatible with most of the experimental equilibrium data.

In some cases, thermochemical data for phases are not available. In such cases, they are calculated from equilibrium data involving the phase. When data for only one phase in a reaction are unknown, it is calculated as follows. The change in free energy of a reaction

$$A + B = C + D \tag{3.g}$$

at equilibrium is zero and can be written as

$$\Delta G^o = G^o{}_{f,C} + G^o{}_{f,D} - G^o{}_{f,A} - G^o{}_{f,B} = 0 \tag{3.1}$$

where, $G^o{}_{f,A}$ is the free energy of formation of phase A etc. If $G^o{}_{f,C}$ is unknown, it is easily calculated from equation (3.1). If several equilibrium pressures and temperatures are known for reaction (3.g), $G^o{}_{f,C}$ is regressed according to equation (3.1) and enthalpy, entropy and heat capacity functions for phase 'C' are obtained. In many cases, one or two quantities among H, S, and C_p are known from calorimetry. This makes the regression procedure much simpler.

In some cases there may be more than one uncertain quantity. For example, suppose in reaction (3.g) 'D' is a component of a solid solution and the excess property of the solution is not well determined. The free energy of 'D' will then be expressed as $G^o{}_{f,D} + RTln\ a_D$, where a_D is the activity of 'D' in the solid solution. Suppose, the free energy of 'C' is available, but is also not well determined. Then it is necessary to adjust both quantities (activity of 'D' and free energy of 'C') by trial and error until a reasonable result is achieved. Let us call the two uncertain quantities 'x' and 'y' with errors of δx and δy associated with them. First, the value of y is calculated keeping x constant. If y_c (the calculated value of y) lies within δy, the result is accepted. If it does not, x is adjusted (to let us say, x_c) within δx and y_c is redetermined. The procedure is repeated several times. In the final calculation, both xc and yc should lie within δx and δy, respectively.

The data set obtained is finally tested in a multiphase-multicomponent environment for global consistency. This is done by the method of minimization of total Gibbs free energy of an assemblage as discussed before.

The Optimization Technique

An internally consistent data base is constructed by considering all the experimentally determined data and theoretically or empirically extrapolated values for the system under consideration. These include phase equilibrium data, calorimetric and volume data and values extrapolated by theoretical methods such as that of heat capacity. Thermodynamic relationships described above link these data together and constrain as a set. Evaluation of thermodynamic properties of a phase by refinement involves a large number of data. A standard optimization technique may be used to solve this problem.

The problem of minimizing the difference (e.g. chi-square) between predictions from a theoretical form of an expression and the experimental data has been commonly solved in geochemical literature through regression analyses which use the middle points of the error bars. Recently, however, the linear and non-linear programming techniques have been popularly used to perform optimization procedures to obtain thermodynamic consistency of the optimized parameters with all valid experimental data (e.g. Halbach and Chatterjee, 1982; Berman et al., 1986).

Today, many optimization or minimization routines are available (e.g. in numerical recipes). The method adopted here is based on the minimizing subroutines (MINUIT) as discussed by James and Roots (1975). The first of these is the Monte Carlo search routine, in which all of the variable parameter values are chosen randomly according to uniform distributions centered at the best previous values with widths equal to the starting parameter errors. The routine is extremely useful for a system with little information on parameters. The second is based on the SIMPLEX method which is one of the most successful stepping methods to approach the minimum. The third, MIGRAD, is based on Fletcher's (1970) method which has the advantage of producing a full covariance matrix, whereas the SIMPLEX only gives estimates of the diagonal elements. Therefore, the MIGRAD minimizer can perform the error analysis and give both the individual and global correlation coefficients. The three methods successively lead to a closer approach to the minimum. Combined use of these methods always gives the best performance in function minimization.The program MINUIT has the advantage of flexibility. The functions, the data base and the parameters to be optimized can be entered in a convenient way. Errors (including half brackets), if available, can be considered for all the data points and the parameters to be optimized. These parameters include thermochemical and thermophysical data and phase equilibrium data. The errors in pressure and temperature in the phase equilibrium data can be directly entered. In multireaction optimizations, the inconsistencies in equilibrium P-T data between different reactions can be easily detected through chi-square values. Discrepancies between phase equilibrium and thermodynamic data can be identified by either high chi-square values or optimized parameter values beyond the given limits.

The minimization may take the form of minimizing a chi-square function built from several different types of data. For example, we may define a function for ΔG^o of a heterogeneous reaction in which there are subfunctions corresponding to thermochemical and thermophysical properties of individual phases of the reaction, as discussed in Chapter 2. A typical problem may consist of minimizing the ΔG^o of a reaction with experimentally determined brackets for equilibrium pressures and temperatures and with some known and unknown thermodynamic properties of the reacting phases. Mainly two types of optimization procedure have been performed in this study, (1) optimizing thermodynamic properties of individual phases in the multi-reaction environment and (2) optimizing solution properties in the binary system.

To demonstrate how the method works, the following examples have been chosen:

Example 1. To evaluate thermodynamic properties of Mg_2SiO_4(β-phase) and SiO_2 (stishovite) from reactions:

Mg_2SiO_4 (olivine) = Mg_2SiO_4 (β-phase)
$2MgSiO_3$ (pyroxene) = Mg_2SiO_4 (β-phase) + SiO_2 (stishovite)
SiO_2 (coesite) = SiO_2 (stishovite)

Suppose that the thermochemical and thermophysical data for olivine, pyroxene and coesite have been well established from other evaluations; the thermophysical properties of β-phase and stishovite have been determined; the calorimetric measurements for β-phase and stishovite have been reported with some uncertainties; and the equilibrium pressures and temperatures of the three reactions above have been experimentally determined within certain brackets. If enthalpies and entropies of the β-phase and stishovite are chosen to be optimizing parameters, the reported uncertainties of calorimetric data may be used as the lower and upper limits of the parameters. The uncertainties in equilibrium pressure may be treated in a similar way. The best values of enthalpies and entropies for both phases can be obtained through an optimization procedure if there is no major inconsistency among the experimental data. On the other hand, in the case of multireaction optimization, the inconsistency may be easily detected by a high chi-square value. When many parameters need to be optimized through a limited number of reactions, it is important to give the lower and upper limits for the parameters, because several solutions may result with equally good fit to the experimental data.

Example 2: To obtain solution parameters from equilibrium distribution data between coexisting phases, such as magnisiowustite (Mw) and olivine, Mw and β-phase, Mw and spinel, and Mw and perovskite.

Thermodynamic data of end members of all the solid solutions are assumed to be well evaluated. The distribution of Mg or Fe between two coexisting solid solutions is considered an ion-exchange reaction, as discussed in Chapter 2. The equilibrium distribution data determined at various pressure and temperature conditions between various pairs of coexisting solutions may be simultaneously fitted with Margules formulation. If many ion-exchange reactions with one or more common solid solutions are taken into consideration, the optimized solution parameters should be well constrained. In addition, the solution calorimetric data should also be used in the optimization.

It is also possible to optimize solution parameters and certain properties of the end member at the same time, but the optimized results should be carefully evaluated. In such a case the end member data and the interaction parameters become mutually dependent.

Phase Equilibrium Calculation by SOLGASMIX

The multicomponent and multireaction equilibrium calculations are best done by adopting the method of minimizing the total Gibbs free energy of a chosen system. The linear algebraic techniques were discussed by Eriksson (1975) and Smith and Missen (1982). They form the basis of the computer program SOLGASMIX (Eriksson, 1975) which is used for all chemical equilibrium calculations in this study.

The total Gibbs free energy of a chemical system can be expressed as:

$$G=\sum \mu_i n_i \quad (3.2)$$

where μ_i is the chemical potential. Chemical potential and activity are interrelated by the expression

$$\mu=\mu^0+RT\ln a \quad (3.3)$$

where μ^0 denotes the standard state potential. For an ideal gas phase species, a species in a non-ideal solution phase and a stoichiometric phase, respectively, we find

$$\mu_i=\mu_i^0+RT\ln P+RT\ln x_i \quad (3.4)$$

$$\mu_i=\mu_i^0+RT\ln \gamma_i+RT\ln x_i \quad (3.5)$$

$$\mu_i=\mu_i^0 \quad (3.6)$$

In these equations x is the mole fraction and γ is the activity coefficient.

The minimization of *G* in equation (3.2) at constant pressure and temperature is achieved with the constraints imposed by the mass balance relationships represented as

$$\sum a_{ij}n_i=b_j \quad (j=1,2,......,l) \quad (3.7)$$

where a_{ij} is the number of atoms of the jth element in a molecule of the ith species, *l* is the total number of elements and b_j is the total amount of the jth element. This is a simple form of the constrained optimization problem which may be solved by the Lagrange method of undetermined multipliers. For this, we define a function,

$$F=G+\sum_{j=1}^{l} \lambda_j(b_j-\sum a_{ij}n_i) \quad (3.8)$$

where λ_j denotes the Lagrangian multipliers, for which the necessary conditions, at an extremum of *F*, are

$$\left(\frac{\partial F}{\partial n_i}\right)_{n_{k\neq i},\lambda}=\mu_i-\sum a_{ij}\lambda_j=0 \quad (3.9)$$

$$\left(\frac{\partial F}{\partial \lambda_j}\right)_{\lambda_{k\neq j},n}=b_j-\sum a_{ij}n_i=0 \quad (3.10)$$

$$n_i\geq 0 \quad (3.11)$$

Combination of equations (3.4) to (3.6) with equation (3.9) gives

$$\mu_i^0 + RT\ln P + RT\ln x_i - \sum a_{ij}\lambda_j = 0 \tag{3.12}$$

$$\mu_i^0 + RT\ln\gamma_i + RT\ln x_i - \sum a_{ij}\lambda_j = 0 \tag{3.13}$$

$$\mu_i^0 - \sum a_{ij}\lambda_j = 0 \tag{3.14}$$

Equation (3.12) is valid for gas phase species while equations (3.13) and (3.14) hold for components of solution phases and stoichiometric phases, respectively.

The system consisting of equations (3.7) and (3.12) to (3.14) with the unknowns n_i and λ_j is non-linear because of the logarithmic term in equations (3.12) and (3.13). The next step is therefore a linearization of these equations by expansion in a Taylor series around an estimated equilibrium composition up to and including the term of the first order. This is equivalent to making a quadratic approximation to the free-energy surface, and we can obtain an expression which relates n_i linearly to λ_j and the estimated equilibrium amounts. Incorporation of this expression into equation (3.7) gives then the final linear system of equations. The number of unknowns is reduced to the sum of elements and phases assumed to be present at equilibrium.

Parametrized activity-coefficient expressions to be inserted into equation (3.13) are supplied by the user. In order to avoid the need of also specifying derivatives, $\ln\gamma_i$ is treated as constant when calculating the partial derivative with respect to n_i in the Taylor expansion. The partial derivative is therefore approximate as long as the estimated equilibrium composition does not correspond to a free-energy minimum.

The approximation to the free-energy surface implies an iterative algorithm and, if positive, the calculated n_i values are used as improved estimates in the subsequent iteration cycle. If some n_i values are negative, these are set to zero before being used as the starting point for a new Taylor expansion. The iterative procedure ends when the calculated values coincide with the starting estimates.

The condensed phases included in the initial estimate are constrained by the Gibbs phase rule but need not necessarily be the correct ones of the final equilibrium state. Another phase combination might yield a lower free energy, and condensed phases need to be withdrawn from or added to the previous combination until the set of equilibrium phases is found. This set has the characteristic feature that the activity for an omitted stoichiometric phase must be less than one as must the sum of mole fractions for an omitted solution phase.

For a more detailed description of the equations used in SOLGASMIX, see Eriksson and Rosén (1973) and Eriksson (1975). The original SOLGASMIX program has been revised many times. The latest version is ChemSage.

Chapter 4

Assessment of Data on Some Oxides and Silicates: Calorimetry and Low-Pressure Phase Equilibrium Experiments

Recently, through experimental study, much data have accumulated on the stability of oxides and silicates. In this chapter, an internally consistent thermodynamic data base on some oxides and silicates is presented which can be used to calculate phase equilibrium relations which are consistent with most of the well reversed phase equilibrium experiments. The techniques used in data assessment are discussed in Chapter 3. Details of the assessed data for each phase are given below. For some oxides, the thermodynamic data are directly adopted from calorimetric sources. These are taken as standards for assessing properties of other phases for which there is an inconsistency between thermochemical data and phase equilibrium experiments. The data on gases, steam, CO_2, CH_4, CO, H_2 and O_2 and on the oxides CaO, FeO (stoichiometric), MgO and Al_2O_3 are from Robie et al. (1978).

The System SiO_2

Recently, Grønvold et al. (1989) determined heat capacities of α-quartz between 298.15 and 847.3 K and β-quartz between 847.3 and 1000 K. Their values do not differ appreciably from the values in Robie et al. (1978). All values presented here were determined through calorimetry except the enthalpy and entropy of β-quartz, which are determined through optimization of the phase equilibrium data on the transformation

$$\alpha\text{-quartz} = \beta\text{-quartz} \tag{4.a}$$

The experimental data on this transformation are from Fenner (1913), Cohen and Klement (1967) and Mirwald and Massonne (1980). We have found it convenient to treat the two forms as totally independent phases. The α-quartz is adopted as a reference state with two C_p expressions one below and one above 848 K. This procedure is not totally satisfactory since the lambda transition at pressures higher than 1 atm is ignored. However, we must consider that the α-quartz above 848 K is fictive and it is designed to effectively permit the calculation of the high pressure phase equilibrium. Since the behavior of the lambda transition at high pressures is unknown, we adopt for a simpler solution than to attempt to model the pressure variation of the lambda Cp.

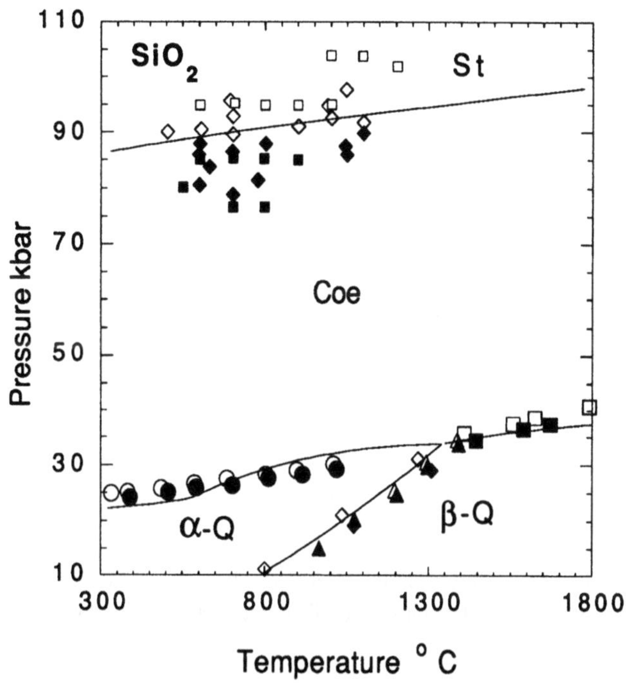

Fig. 4.1. Comparison of phase equilibrium computations with experimental data in the system SiO_2: circles, Bohlen and Boettcher (1982); squares and triangles, Mirwald and Massonne (1980); diamonds, Cohen and Klement (1967); small squares, Suito (1977); small diamonds, Yagi and Akimoto (1976).

For coesite and stishovite, we have used the constraints from the phase equilibrium experiments (Bohlen and Boettcher, 1982) and the calorimetry data of Akaogi and Navrotsky (1984). The calculated curves for reaction (4.a) and for the quartz-coesite and the coesite-stishovite transformations are shown in Fig. 4.1.

Andalusite, Kyanite and Sillimanite

Experimental results on the following equilibria:

andalusite = kyanite (4.b1)
kyanite = sillimanite (4.b2)
sillimanite = andalusite (4.b3)

as obtained by different workers (Evans,1965, Newton,1966, 1969, Althaus, 1967, Richardson et al.,1968, 1969; Holdaway,1971) are significantly different. The reason for such a discrepancy is the small difference in the free energies of formation of the three polymorphs.

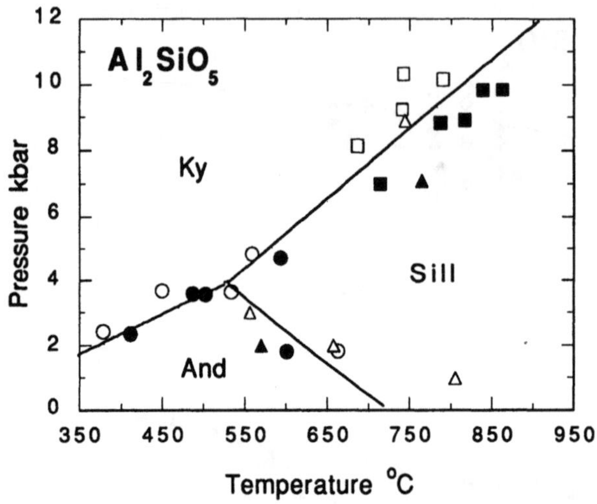

Fig. 4.2. Comparison of phase equilibrium computations with experimental data in the system Al_2SiO_5: circles, Holdaway (1971); triangles, Kerrick and Henninger (1984); squares, Richardson et al. (1968).

Robie and Hemingway (1984b) determined the heat capacities of the three aluminosilicates in the range 10-380 K. The calculated slopes of phase boundaries among these phases match the experimental data of Evans (1965), Newton (1966, 1969) and Holdaway (1971). Robie and Hemingway concluded that since the results of Evans, Newton and Holdaway are consistent with their independently derived calorimetric data, the equilibrium results of these three workers are to be preferred over those of others. The thermodynamic properties given by Robie and Hemingway are adopted in this work. Hemingway also concluded that Al and Si are ordered in sillimanite to at least 1100 K. Calculated positions of these polymorphic transformations and the experimental data are shown in Fig. 4.2.

Wollastonite, Gehlenite, Akermanite and Monticellite

Wollastonite and gehlenite participate in reactions involving the breakdown of anorthite. These reactions, studied by various workers (see Figs. 4.3 and 4.4), are as follows:

An + Wol = Gr + Q (4.c)

An + Geh = Gr + Cor (4.d)

An + Geh + Wol = Gr (4.e)

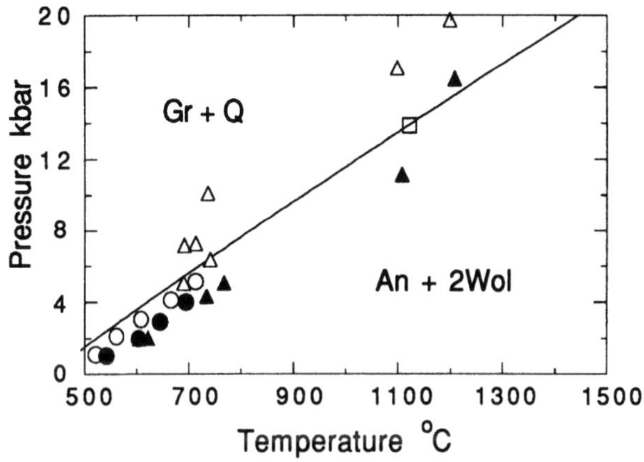

Fig. 4.3. Comparison of phase equilibrium computations with experimental data of the reaction An + Wol = Gr + Q (4.c): circles, Huckenholz et al. (1975); triangles, Hays (1966); squares, Windom and Boettcher (1976).

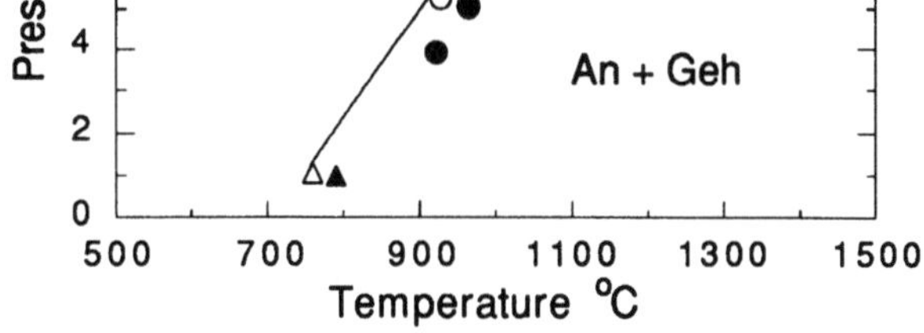

Fig.4.4. Comparison of phase equilibrium computations with experimental data of the reaction An + Geh = Gr + Cor (4.d): circles, Huckenholz et al. (1975); triangles, Boettcher (1970).

Reaction (4.c) was also studied by Huckenholz et al. (1975). Stability of these minerals and of akermanite and monticellite was also studied by Yoder (1968). The properties of wollastonite are as listed in Haas et al. (1981) and are closely similar to those of Charlu et al. (1981).

For the reaction (4.d) involving gehlenite, Gasparik (1984b) evaluated the experimental data of Boettcher (1970) and Huckenholz et al. (1975) and found the equilibrium pressure as

$$P = 23.8T\,(^{o}C) - 17440 \tag{4.1}$$

Using these data, $\Delta C'_p$, the difference between the experimental and the calculated heat capacities, for gehlenite is found to be

$$\Delta C'_p = 7.5491(\pm 0.134) + 0.0027126(\pm 0.00019)T \tag{4.2}$$

and a corresponding ΔH^o_f as -3980763 J/mol. This value is comparable to -3979046 J/mol derived from the $\Delta H^o_{f,970}$ value of Charlu et al. (1981) using the C_p expression from Robie et al. (1978). The entropy of gehlenite is selected as 204.35 J/mol.K following Woodhead (1977), who considers a progressively increasing disorder in gehlenite with temperature. In Figs. 4.3, 4.4 and 4.5, respectively, the calculated curves for reactions (4.c), (4.d) and (4.e) along with the experimental data are shown.

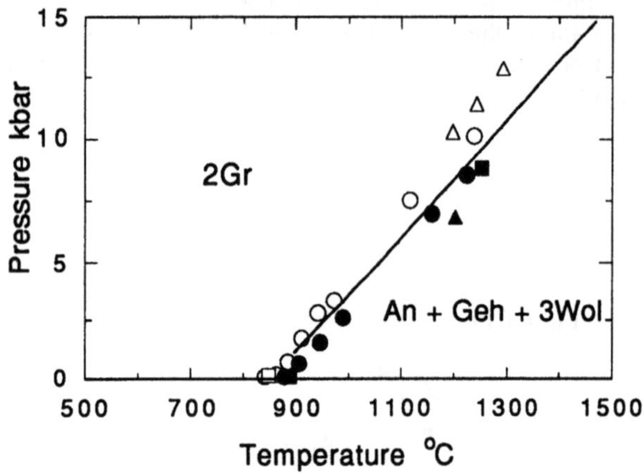

Fig.4.5. Comparison of phase equilibrium computations with experimental data of the reaction An + Geh + Woll = Gr (4.e): circles, Huckenholz et al. (1975); triangles, Hays (1966); squares, Shmulovich (1974).

The experimental data of Yoder (1968) involving wollastonite, monticellite and akermanite are used to calculate the enthalpy of akermanite. The entropy and heat capacity data are from Robie et al. (1978). The values are closely similar to the calorimetric data of Haas et al. (1981) and Charlu et al. (1981).

The enthalpy of monticellite is adjusted from Brousse et al. (1984) so that an improved fit to Yoder's (1968) experimental data involving wollastonite, monticellite and akermanite is obtained. The entropy is from Robie et al. (1978) and the heat capacity equation is as recommended by Helgeson et al. (1978).

Enstatite and Pyrope

For pyrope, the enthalpy is from Charlu et al. (1975), the entropy is from Haselton and Westrum (1980) and the heat capacity is from Haselton (1979). The values are calorimetric. The enthalpy of enstatite is from Brousse et al. (1984). Although there are other determinations of the enthalpy, the value of Brousse et al. is close to the average of all the values. So, their value is adopted. Entropy and heat capacity of enstatite are from Haselton (1979). All these values are calorimetric. The enstatite is the orthoenstatite and has been used for data systematics at high pressure (see Chapter 5).

Forsterite

Brousse et al. (1984) discussed the calorimetric data on enthalpy of forsterite which range from a high of -57.823 to a low of -63.262 (kJ/mol, $\Delta H^o_{f,298}$ from oxides). The value recommended here is -60.68 kJ/mol. Most of the calorimetric data are close to this value. The entropy and C_p of forsterite are from Robie et al. (1978). Robie et al. (1982b) determined the heat capacity of forsterite in the range 5-380 K. The C_p recommended here is closely similar to the C_p determined by them. The forsterite data is used for the systematics at high pressure (Chapter 5).

Fayalite and Ferrosilite

Ferrosilite breaks down to fayalite and quartz according to the following reaction:

$$\text{Fer} = \text{Fay} + \text{Q} \tag{4.f}$$

This reaction was studied experimentally by Bohlen et al. (1980). However, when orthopyroxene solutions break down there is a three-phase stability field where the orthopyroxene solution is present along with the olivine solution and quartz. This phenomenon was investigated by Bohlen and Boettcher (1981). Free energy of fayalite is adopted from Robie et al. (1982a) which is determined by calorimetry.

Free energy of ferrosilite is closely constrained by reaction (4.f). Enthalpy (adjusted by 300 J) and entropy are the recommended values of Bohlen and Boettcher (1981) and the heat capacity is calculated from reaction (4.f). In Fig. 4.6, a calculated curve for the reaction (4.f) along with the experimental data is shown.

Orthodiopside

The property of this fictive component is adopted from Saxena et al. (1986). Their evaluations are based on the phase equilibrium data on the enstatite-diopside join.

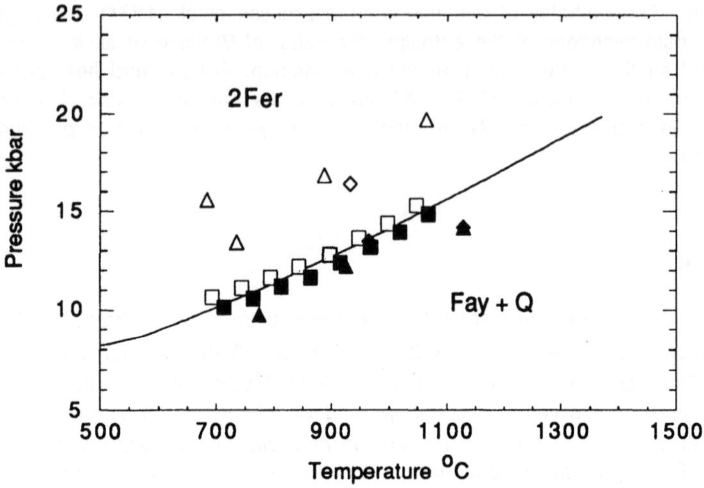

Fig. 4.6. Comparison of phase equilibrium computations with experimental data of the reaction Fer = Fay + Q (4.f): squares, Bohlen et al. (1980); triangles, Smith (1971); diamonds, Akimoto et al. (1965).

Clinoenstatite

As discussed in the next chapter on polymorphs of $MgSiO_3$, this is the low-clinoenstatite. The enthalpy and entropy are evaluated from the experimental data of Lindsley et al. (1981) on enstatite-diopside equilibria. These values are close to the values of Kiseleva et al. (1979). The heat capacity is assumed to be the same as that of enstatite.

Diopside

The properties listed in Robie et al. (1978) are adopted. These data are from calorimetric sources.

Hedenbergite

The values are as recommended by Saxena et al. (1986). All values in their study were adjusted from the recommended values of Helgeson et al. (1978) to fit the equilibrium data on quadrilateral pyroxenes from Lindsley (1983). Recently, Haselton et al. (1987) determined the heat capacity of hedenbergite by low-temperature adiabatic and differential scanning calorimetry in the range 9-647 K.

The heat capacity recommended by Saxena et al. (1986) is consistent with the experiments of Haselton et al. The difference between the heat capacity of Saxena et al. and Haselton et al. at 647 K is 0.8 J/cation.

Ca-Tschermak, Anorthite and Grossular

Ca-tschermak is related to grossular and anorthite by the following equilibrium reactions:

Gr + Cor = CaTs (4.g)

An + Geh + Cor = CaTs (4.h)

Reaction (4.g) was studied by Gasparik (1984b), Hays (1966) and Wood (1978), and reaction (4.h) was studied by Hays (1966). The equilibria involving anorthite and grossular are given by the reaction:

An = Gr + Ky + Q (4.i)

which was studied by many workers (see Figure 4.7). Koziol and Newton's (1988a) results constrain the boundary within tight reversal brackets. Gasparik's (1984c) results place the reaction boundary a few kbars above that of Goldsmith's (1980).

Another reaction involving anorthite and grossular is for the silica unsaturated system and is:

Gr + Ky = An + Cor (4.j)

This reaction was studied experimentally by Gasparik (1984b)(see Fig. 4.8). Equilibria of anorthite with gehlenite and wollastonite are described later.

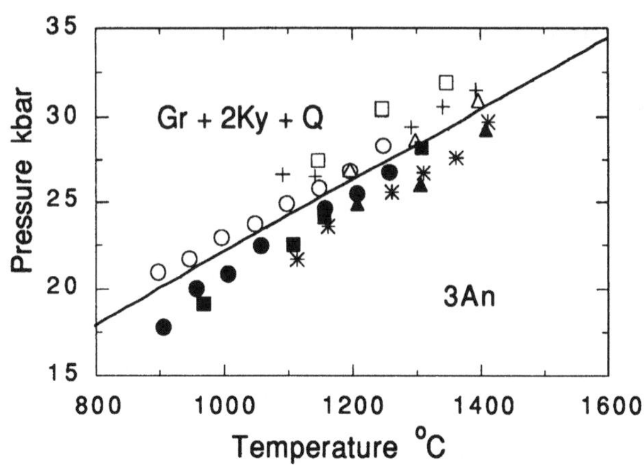

Fig.4.7. Comparison of phase equilibrium computations with experimental data of the reaction An = Gr + Ky + Q (4.g): circles, Koziol and Newton (1986); triangles, Gasparik (1984a); squares, Hariya and Kennedy (1968); stars and plusses, Goldsmith (1980).

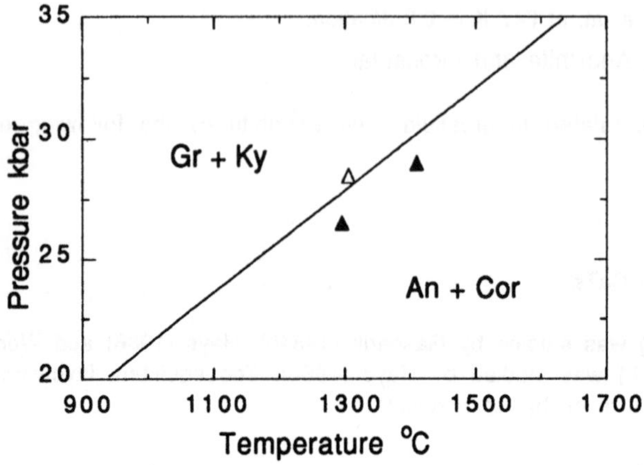

Fig. 4.8. Comparison of phase equilibrium computations with experimental data of the reaction Gr + Ky = An + Cor (4.j): experimental data from Gasparik (1984c).

For grossular, the enthalpy is from Charlu et al. (1975), the entropy is from Haselton and Westrum (1980) and the heat capacity is from Haselton (1979). The values are calorimetric.

Free energy of Ca-tschermak was calculated by Saxena and Chatterjee (1986) using Gasparik's (1984b) data on reactions (4.g) and (4.h) which yielded

$$\Delta C'_p = 2220824(\pm 130)T^{-2} \tag{4.3}$$

with a corresponding $\Delta H^o_{f,298}$ of -3300902 J/mol, where, $\Delta C'_p$ is the difference between the experimental and the calculated heat capacities. This $\Delta H^o_{f,298}$ is comparable to the $\Delta H^o_{f,298}$ (calculated from $\Delta H^o_{f,970}$) of -3297994 J/mol determined by Charlu et al. (1978). The enthalpy and entropy needed adjustment in this study while the heat capacity of Saxena and Chatterjee (1986) is retained. The data on *Cp* represents the addition of $\Delta C'_p$ to the ΔC_p expression of Haselton et al. (1984).

The heat capacity of anorthite, as listed in Robie et al. (1978) up to 1800 K, shows an anomalous extrapolation above 1200 K. In this study, the data up to 1200 K are fitted to the Fei-Saxena equation for high-temperature extrapolation (subsequently refitted by Saxena and Shen as discussed in Chapter 2). The resulting equation gives very similar results when compared to the equation of Holland (1981). Entropy of anorthite is from Robie et al. (1978) and enthalpy is calculated from reaction (4.j). Another equilibrium involving anorthite in conjunction with wollastonite and gehlenite is shown in Fig. 4.5.

Almandine and Hercynite

The following reaction involving almandine and hercynite was recently investigated by Bohlen et al. (1986):

Alm + Sill = Her + Q (4.k)

Bohlen et al. (1983a) also studied the following reaction involving the garnet composition grossular$_1$almandine$_2$:

garnet = Fay + An (4.l)

Heat capacity of almandine is estimated from the Fe-Mg exchange reaction

En + Alm = Fer + Pyr (4.m)

by assuming that ΔC_p of the reaction is zero. Although this is normally not true, it is preferred over the C_p determined by Metz et al. (as in Bohlen et al. 1986) because when their determined Cp values are extrapolated to high temperatures such as the range of experiments of Lee and Ganguly (1987) (see Ganguly and Saxena, 1988) they become too high to fit the observed equilibria. Note that the upper limit of calorimetric experiment of Metz et al. is 1000 K.

Distribution data for the exchange reaction (4.m) were determined from the synthesis experiments of Kawasaki and Matsui (1983) and the reversal experiments of Lee and Ganguly (1987) (see Ganguly and Saxena, 1988). They both suggest a relatively ideal behavior of the orthopyroxene and garnet solutions. Therefore the free energy of almandine is calculated from reaction (4.m) according to the following equilibrium relation

$\Delta G^o + RT\ln K_D$ (4.m) = 0 (4.4)

Enthalpy and entropy of almandine are available from calorimetric experiments of Chatillon-Colinet et al. (1983a) (-1749.9±3.6 kJ/mol of cation, an average of two different methods) and Metz et al. (1983) (113.39 J/mol.K of cation). Enthalpy and entropy of almandine are adjusted in order to reproduce the available distribution data by satisfying relation (4.5).

Free energy of hercynite is determined from the reaction (4.k). The heat capacity expression for hercynite used by Saxena and Eriksson (1983b) is adjusted here. The entropy is from Robie et al. (1978). Enthalpy is adjusted within the error limits of Robie et al. (1978) (-1966.48±8.5 KJ/mol). Calculated position of reaction (4.k) and the experimental results of the reaction are shown in Fig. 4.9.

Spessartine

Free energy is from Bokreta (1989). The values were calculated from lattice energy consideration.

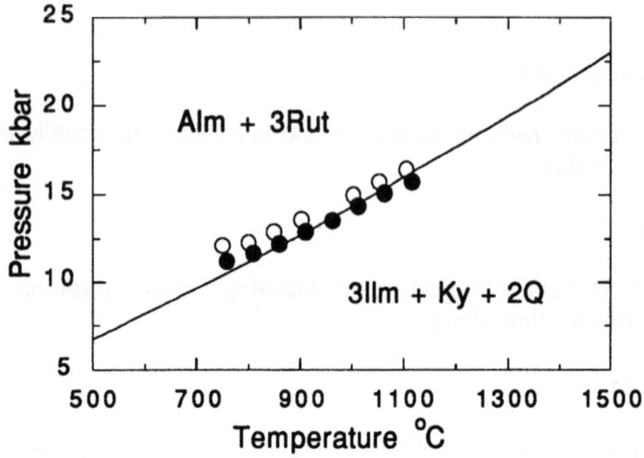

Fig. 4.9. Comparison of phase equilibrium computations with experimental data of the reaction Alm + Ru = Ilm + Ky + Q (4.n): experimental data from Bohlen et al. (1983b).

Rutile and Ilmenite

The following reactions involving rutile and ilmenite were experimentally reversed by Bohlen et al. (1983a) and Bohlen and Liotta (1986) respectively:

Alm + Ru = Ilm + Ky + Q (4.n)

garnet + Ru = Ilm + An + Q (4.o)

Reaction (4.o) involves the garnet composition grossular$_1$almandine$_2$. Enthalpy and entropy of rutile are adjusted within the error limits of Robie et al. (1978) to fit the experimental data on reaction (4.n). The heat capacity of rutile is from Robie et al. (1978). Entropy and heat capacity of ilmenite are from Anovitz et al. (1985). The enthalpy of both ilmenite and rutile and the entropy of rutile are adjusted within their stated errors (Robie et al.,1978). The data on ilmenite and rutile thus adjusted, along with the solution data on the grossular-almandine solution discussed later, reproduce the experimental data on reaction (4.o) well. The calculated curve for reaction (4.n) and with the experimental data are shown in Fig. 4.10.

Sanidine and Muscovite

The equilibrium relations between these two minerals have been studied by many workers. Important reactions are as follows:

Ms = Cor + San + H_2O (4.p)

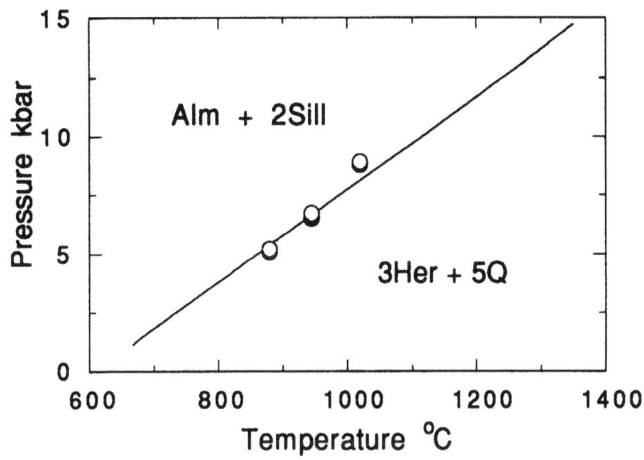

Fig.4.10. Comparison of phase equilibrium computations with experimental data of the reaction Alm + Sill = Her + Q (4.k): experimental data from Bohlen et al. (1986).

$$Ms + Q = San + Al_2SiO_5 + H_2O \qquad (4.q)$$

The muscovite breakdown reactions mark the boundary between medium and high grade metamorphism. Experimental work has been done on reaction (4.p) by Velde (1966) and Chatterjee and Johannes (1974) and on reaction (4.q) by Evans (1965), Althaus et al. (1970), Day (1973) and Chatterjee and Johannes (1974). Experimental information available on the stability of sanidine and muscovite are from the reactions (4.p) and (4.q). Krupka et al. (1979) attempted to calculate the enthalpy and free energy of muscovite from the experimental data of Chatterjee and Johannes (1974) on the above reactions. Their calculations were based on the fugacity values of steam from Chatterjee and Johannes (1974) and Fisher and Zen (1971) and the heat capacity data on muscovite determined by them and the unpublished data on sanidine listed in Robie et al. (1978). In this work the more general equation for water fugacity of Saxena and Fei (1987a,b) and the published heat capacity data on sanidine by Hemingway et al. (1981) are used. The enthalpy of sanidine is from Robie et al. (1978). The thermal expansion of muscovite is also optimized from a combination of the phase equilibrium and heat capacity data. The optimized entropy of sanidine is found to be 7.09 J/mol.K higher than that determined by Openshaw et al. (1976). This amount can be considered as the disorder entropy of sanidine. Sanidine shows an Al/Si disorder and since, at very low temperatures, the heat capacity is not well characterized, this treatment is reasonable. However, it should be kept in mind that the zero-point Al/Si disorder entropy of muscovite is also unknown. Calculated curves for reactions (4.p) and (4.q) (for sillimanite and andalusite) along with the experimental data are shown respectively in Figs. 4.11,4.12a, 4.12b.

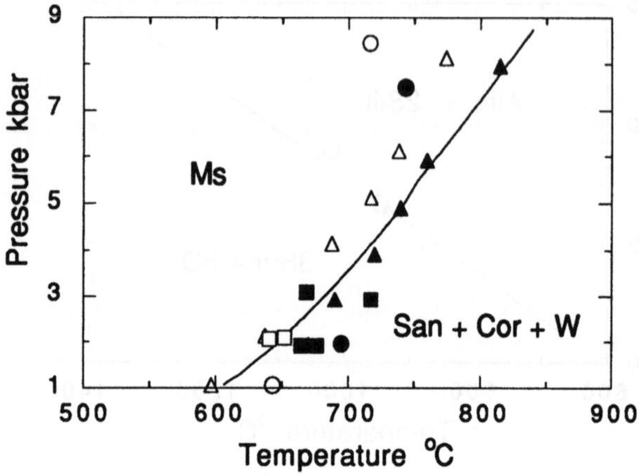

Fig. 4.11. Comparison of phase equilibrium computations with experimental data of the reaction Ms = Cor + San + H_2O (4.p): circles, Velde (1966); triangles, Chatterjee and Johannes (1974); squares, Evans (1965).

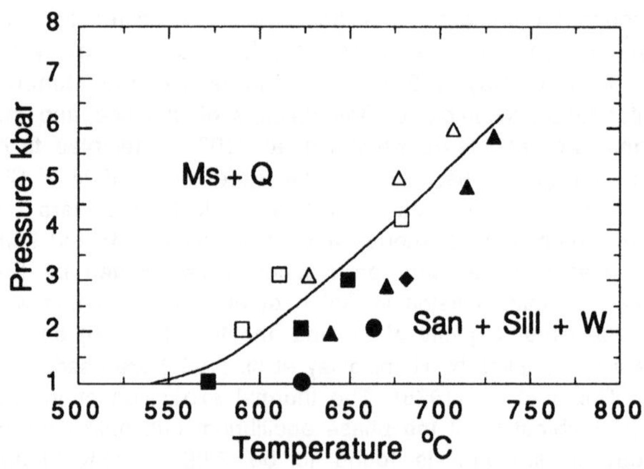

Fig.4.12a. Comparison of phase equilibrium computations with experimental data of the reaction Ms + Q = San + Sill + H_2O (4.q): circles, Day (1973); triangles, Chatterjee and Johannes (1974); squares, Evans (1965); diamonds, Althaus et al. (1970).

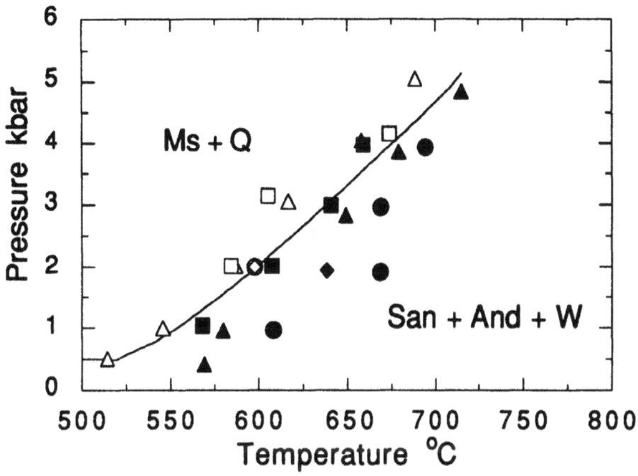

Fig. 4.12b. Comparison of phase equilibrium computations with experimental data of the reaction Ms + Q = San + And + H_2O (4.q): circles, Althaus et al. (1970); triangles, Chatterjee and Johannes (1974); squares, Evans (1965); diamonds, Kerrick (1972).

Jadeite, Albite and Paragonite

The following reaction relating jadeite and albite

$$Ab = Jd + Q \tag{4.r}$$

was studied by Holland (1979). Reactions involving jadeite, albite and paragonite are as follows:

$$Pg = Ab + Cor + H_2O \tag{4.s}$$

$$Pg + Q = Ab + Al_2SiO_5 + H_2O \tag{4.t}$$

$$Pg = Jd + Ky + Q \tag{4.u}$$

Experimental work has been done on reaction (4.s) by Chatterjee (1970), on reaction (4.t) by Chatterjee (1972) and on reaction (4.u) by Holland (1979). Entropy and Cp of jadeite are as listed in Robie et al. (1978). The evaluated enthalpy of jadeite from Hemingway et al. (1981) is adopted.

Albite and jadeite are related by the equilibrium (4.r). The free energies of albite and jadeite were evaluated by Hemingway et al. (1981). Holland (1979) also attempted to evaluate the free energy of albite on the basis of reaction (4.s). Hemingway et al.'s values on albite are based on a careful consideration of the low-albite ↔ albite transition. Their values on jadeite and albite are thus consistent with both the transition and the reaction (4.r). However, they did not consider the effect of thermal expansion and bulk modulus of these two phases. When these two parameters are considered, their second model, which assumes that the aluminum avoidance principle holds for albite, fits the experimental equilibria the best. Thus, the values recommended by Hemingway et al. (1981) are adopted for albite.

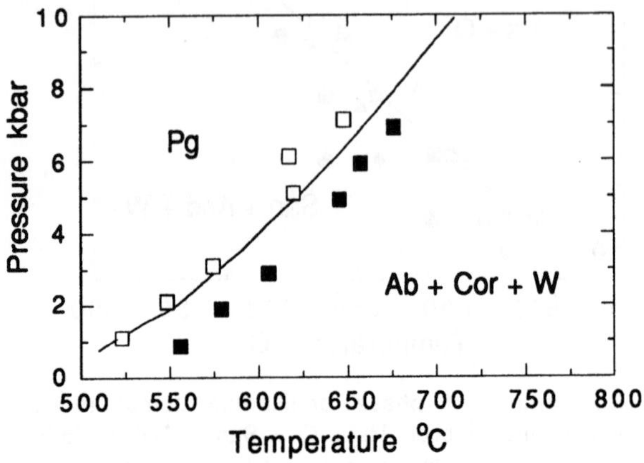

Fig.4.13. Comparison of phase equilibrium computations with experimental data of the reaction Pg = Ab + Cor + H_2O (4.s): experimental data from Chatterjee (1970).

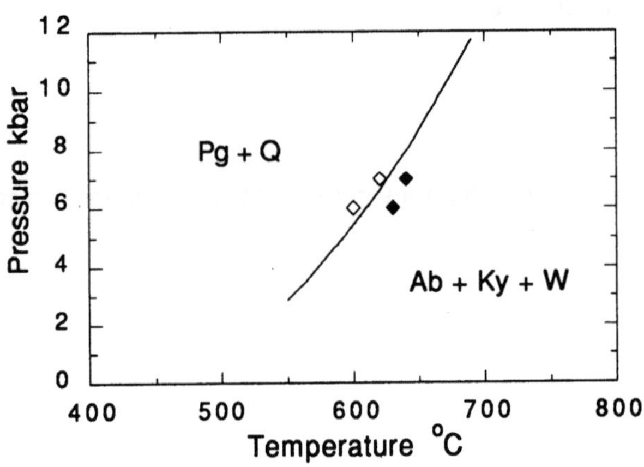

Fig. 4.14a. Comparison of phase equilibrium computations with experimental data of the reaction Pg + Q = Ab + Ky + H_2O (4.t): experimental data from Chatterjee (1972).

The stability limit of paragonite is defined by the experimentally investigated reactions (4.s), (4.t) and (4.u). In this work different approaches were made to calculate the free energy of paragonite. Assuming that the value for jadeite is correct, the enthalpy and entropy of paragonite were optimized from reaction (4.u) (heat capacity of paragonite being from the calometric measurement of Robie and Hemingway (1984a)). Enthalpy and entropy of albite were then optimized from reactions (4.s), (4.t) and (4.u). However, this gave way to a convergence problem. In the second approach, the free energies of albite and jadeite were adopted from Hemingway et al. (1981). Using the heat capacity of paragonite from Robie and Hemingway (1984a), the enthalpy and entropy of paragonite were calculated. The same problem of convergence was encountered again. In the final approach, the second approach is repeated only this time instead of the entropy (which is taken from Robie and Hemingway, 1984a), the heat capacity of paragonite is optimized. The thermal expansion and bulk modulus are also optimized simultaneously. The resulting value of the bulk modulus was not different from that of muscovite. The molar volume of paragonite is from Flux and Chatterjee (1986). The resulting data reproduce the reactions (4.s), (4.t) (for sillimanite and andalusite) and (4.u) Calculated curves for reactions (4.s), (4.t), and (4.u) and the experimental data are shown respectively in Figs. 4.13, 4.14 and 4.15.

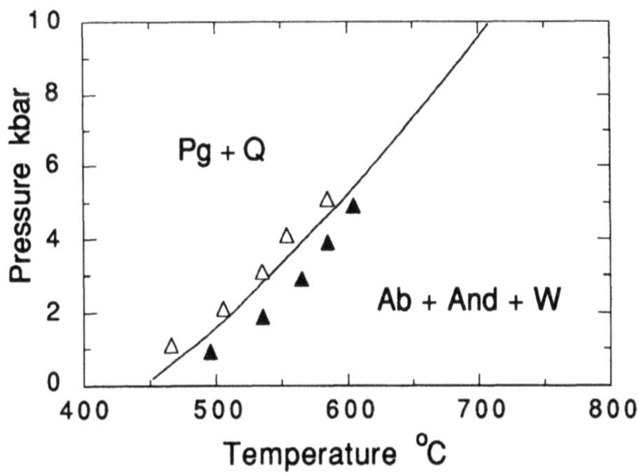

Fig. 4.14b. Comparison of phase equilibrium computations with experimental data of the reaction Pg + Q = Ab + And + H_2O (4.t): experimental data from Chatterjee (1972).

In the study above and in the work of Hemingway et al. (1981) and Holland (1979), it was assumed that the free energy of jadeite is correct. Hemingway et al. (1981) pointed out that heat capacity of jadeite has not been determined in the temperature range 5-50 K. This could have led to a spurious value of the entropy of jadeite. Enthalpy and entropy of jadeite can be determined from reaction (4.r) if the free energy of albite were known independently or from reaction (4.u) if the free energy of paragonite were known independently. Since this is not the case, the approach adopted here is the best. The difference between the calculated and the calorimetric values (enthalpy being the recommended value of Robie and Hemingway, 1984a) of the free energy of paragonite (0.23, 1.027 and 5.327 kJ/mol at 298, 500 and 800 K respectively) can be attributed to various types of energies associated with the paragonite structure, one of which could be the entropy of disorder. Paragonite probably has some zero-point entropy as indicated by the Schottky anomaly at lower temperatures (Robie and Hemingway, 1984a). However, more investigation is needed to characterize this energy from the theoretical point of view. The conclusion reached here can only be ascertained through a more accurate determination of the free energy of jadeite.

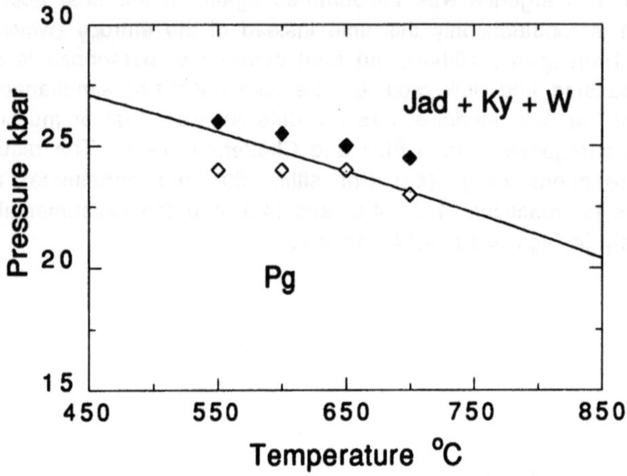

Fig. 4.15. Comparison of phase equilibrium computations with experimental data of the reaction Pg = Jad + Ky + Q (4.u): experimental data from Holland (1979).

Phlogopite

The following phlogopite breakdown reaction was studied by Wood (1976), Wones and Dodge (1977) and Bohlen et al. (1983c):

$$\text{Phl} + \text{Q} = \text{En} + \text{San} + H_2O \qquad (4.v)$$

Enthalpy of phlogopite was determined by Clemens et al. (1987) by solution calorimetry. Their value of -6215.33 kJ/mol was derived by considering a calorimetric cycle for the reaction:

$$\text{Per} + \text{Br} + \text{San} = \text{Phl} \qquad (4.w)$$

and using the free energies of periclase, brucite and sanidine from Robie et al. (1978) and the heat capacity of phlogopite from Robie and Hemingway (1984a). The same procedure is repeated here. A different value for the enthalpy of phlogopite is obtained for the following reason. Saxena (1989) and Fei and Chatterjee (1988) using the experimental data on the reaction (4.ee) from Irving et al. (1977), which are up to a temperature of 1400 K, obtained an optimized heat capacity expression for brucite. The heat capacities of brucite obtained from this expression differ from the original data of King et al. (1975) (used by Clemens et al.) by 0.8 and 1.35 J/mol at 500 and 700 K respectively. Also, the enthalpy of brucite obtained from optimization by Saxena (1989) and Fei and Chatterjee (1988) is 0.08 J/mol less than the value quoted in Robie et al. (1978). The enthalpy of phlogopite obtained using these enthalpy and heat capacities of brucite is -6203.9 kJ/mol. The entropy of phlogopite is also from Robie and Hemingway (1984a). This includes a configurational entropy of 18.7 J/mol.K assuming total Al-Si disorder.

Clemens et al. (1987) found that there is a considerable discrepancy between the experimental data of Bohlen et al. (1983c) and those of Wood (1976) and Wones and Dodge (1977) on reaction (4.v). The enthalpy of phlogopite calculated by Robie and Hemingway (1984a) on the basis of the data of Wood (1976) and Wones and Dodge (1977) is -6226 kJ/mol. However, the measured value of Clemens et al. (1987) (recalculated here) agrees well with the data of Bohlen et al. (1983c).

Annite and the Phlogopite-Annite Solution

Reversed distribution data of Fe and Mg between biotite and garnet are available from Ferry and Spear (1978) and Perchuk and Lavrent'eva (1983). These data, along with the free energy data of phlogopite as discussed above, are used to calculate the free-energy and volume data of annite and the solution properties of the biotite solution. The relevant reaction is as follows:

Alm + Phl = Pyr + Ann (4.x)

Free energy of almandine is from Chatterjee (1987) and of pyrope as discussed before. The enthalpy, entropy and volume of annite and the interaction parameters of the phlogopite-annite solution are calculated assuming ΔC_p of reaction (4.x) as zero and from the following equilibrium relation:

$$\Delta G^o_{(4.x)}(P,T) + RT\ln K_{D\,(4.x)} + RT\ln K_{\gamma\,(4.x)} = 0.0 \quad (4.5)$$

the only unknown quantities in the term ΔG^o (4.x)(P,T) being the enthalpy, entropy and volume of annite at P and T and where, K_D is from Ferry and Spear (1978) or Perchuk and Lavrent'eva (1983) and K_γ is as follows:

$$K_{\gamma\,(4.x)} = (\gamma_{Mggar} \cdot \gamma_{FeBi})/(\gamma_{Fegar} \cdot \gamma_{MgBi}) \quad (4.6)$$

The activity formulations for phlogopite and annite, following the Margules formulation for asymmetric solutions, are as follows:

$$RT\ln\gamma_{Ph} = X_{Ann}{}^2[W_{Ph\text{-}Ann} + 2X_{Ph}(W_{Ann\text{-}Ph} - W_{Ann\text{-}Ph})] \quad (4.7)$$

$$RT\ln\gamma_{Ann} = X_{Ph}{}^2[W_{Ann\text{-}Ph} + 2X_{Ann}(W_{Ph\text{-}Ann} - W_{Ph\text{-}Ann})] \quad (4.8)$$

where, W's are the interaction parameters.

When the data of Ferry and Spear (1978) and Perchuk and Lavrent'eva (1983) are used in conjunction, convergence is not achieved. It is concluded that the two sets of data are inconsistent. When only the data of Ferry and Spear (1978) are used, the fit obtained is not satisfactory. Also, the solution parameters obtained suggest an extremely asymmetric solution. This could be due to the fact that the data are only in the Fe-rich end of the binary. When only the data of Perchuk and Lavrent'eva (1983) are used, a much better fit is obtained. Also, the solution parameters show a closely symmetric, not very non-ideal solution. Since Perchuk and Lavrent'eva's data cover a wide range of the binary and the results obtained are much better, the free-energy of annite and the interaction parameters of the phlogopite-annite solution obtained from their data are recommended for use in calculations. An uncertainty of 700 joules is found in the enthalpy of annite. The values for W obtained are as follows:

$W_{Phl\text{-}Ann}$ = 5540.7 J/mol
$W_{Ann\text{-}Phl}$ = 5068.5 J/mol

The molar volume of annite is from Robie and Bethke (1963). The bulk modulus is optimized with a start value equal to that of phlogopite. The resulting value is the same as that of phlogopite. This is because the experiments were performed only up to a higher pressure limit of 6 kbar. Thermal expansion was not optimized as no data were found for the thermal expansion of phlogopite.

Talc and Anthophyllite

The stability limits of anthophyllite and talc were recently studied in detail by Chernosky et al. (1985) under silica saturated and silica undersaturated conditions. The following reactions were studied:

$$\text{Tc} = \text{En} + \text{Q} + H_2O \tag{4.y}$$

$$\text{Tc} + \text{Fo} = \text{En} + H_2O \tag{4.z}$$

$$\text{Anth} = \text{En} + \text{Q} + H_2O \tag{4.aa}$$

$$\text{Anth} + \text{Fo} = \text{En} + H_2O \tag{4.bb}$$

$$\text{Tc} + \text{En} = \text{Anth} \tag{4.cc}$$

$$\text{Tc} + \text{Fo} = \text{Anth} + H_2O \tag{4.dd}$$

$$\text{Tc} = \text{Anth} + \text{Q} + H_2O \tag{4.ee}$$

The three silica undersaturated reactions (4.z), (4.bb) and (4.dd) along with the reaction (4.cc) intersect to give an invariant point at 7.7 kbar and 955 K. The three silica saturated reactions (4.y), (4.aa) and (4.ee) along with the reaction (4.cc) intersect to give an invariant point at 10.5 kbar and 1067 K. Greenwood (1963) studied the same reactions. However, his results gave different locations of the invariant points.

Experimental data on reactions (4.y) through (4.ee) are used to determine the free energy data of talc and anthophyllite. These seven reactions involve talc and two of them, (4.y) and (4.z), contain no other hydrous phases. The optimization of thermodynamic data for talc is started with these two reactions since the other three anhydrous phases, quartz, enstatite and forsterite, are well evaluated (Saxena and Chatterjee, 1986 and Chatterjee, 1987 see above). It is found that the phase equilibrium data for reactions (4.y) and (4.z) are not consistent with each other. There is no way to adjust thermodynamic data on talc to fit the experimental

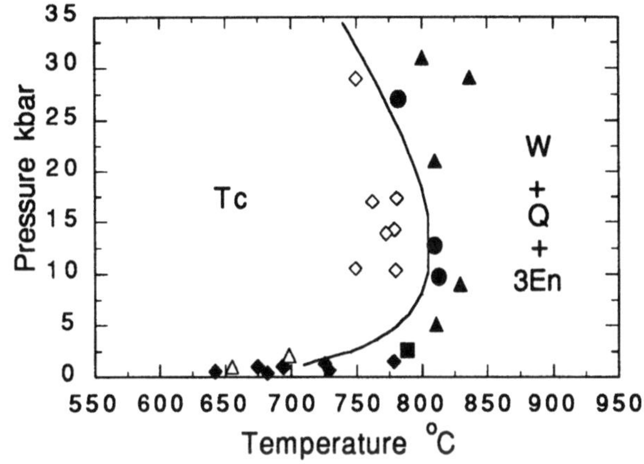

Fig. 4.16. Comparison of phase equilibrium computations with experimental data of the reaction Tc = En + Q + H_2O (4.y): circles and diamonds, Chernosky et al. (1985); Chernosky (1976); triangles, Kitihara et al. (1966); squares, Skippen (1971).

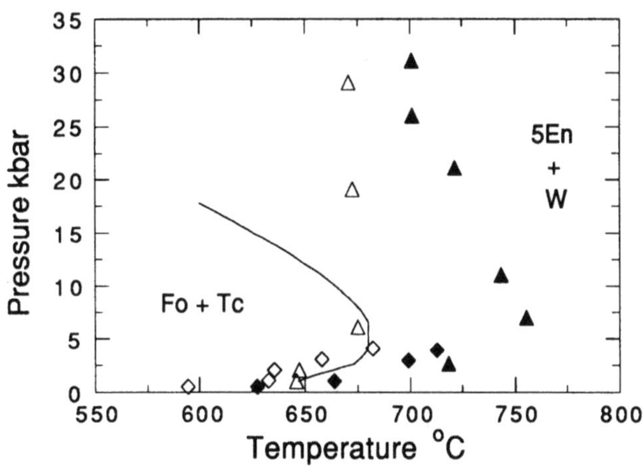

Fig. 4.17. Comparison of phase equilibrium computations with experimental data of the reaction Fo + Tc = En + H_2O (4.z): triangles, Kitihara et al. (1966); diamonds, Chernosky (1976).

data on both reactions simultaneously. Optimization of thermodynamic data for anthophyllite from the five anthophyllite related reactions, (4.aa) through (4.ee), indicates that the data on talc optimized from reaction (4.y) are more consistent with other talc related equilibrium data than that from reaction (4.z). Therefore, the final optimized results on talc and anthophyllite are obtained from reactions (4.y) and (4.aa) through (4.ee). Berman et al. (1986) faced the same problem in their work. They ignored the data of Kitihara et al. (1966) on reactions (4.y) and (4.z) in their analysis. In this work also, the data of Kitihara et al. (1966) are found to be inconsistent with the other experimental data on reaction (4.z). The reason for this

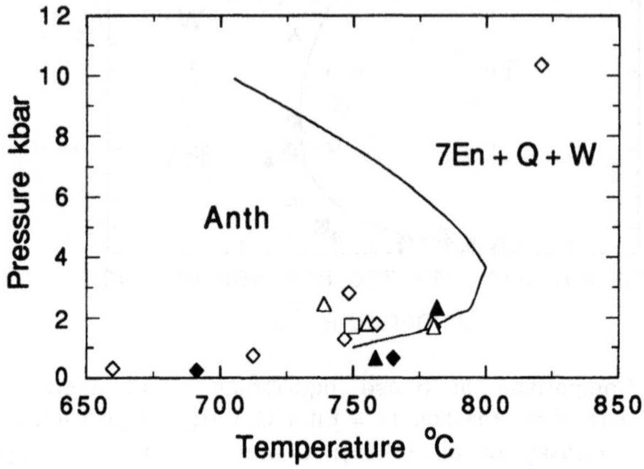

Fig.4.18. Comparison of phase equilibrium computations with experimental data of the reaction Anth = En + Q + H_2O (4.aa): triangles, Greenwood (1963); diamonds, Chernosky et al. (1985); squares, Fyfe (1962).

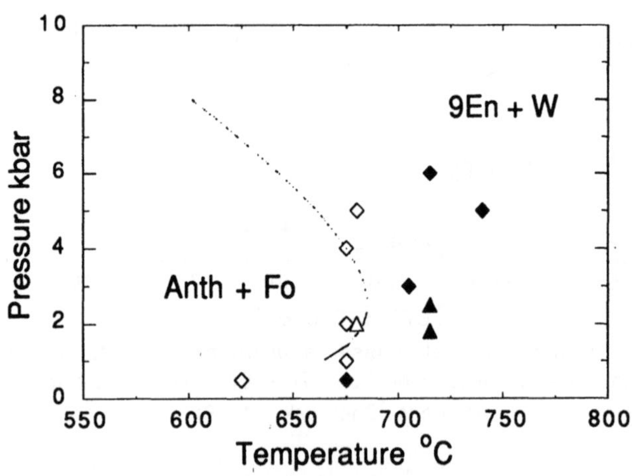

Fig. 4.19. Comparison of phase equilibrium computations with experimental data of the reaction Anth + Fo = En + H_2O (4.bb): triangles, Greenwood (1963); diamonds, Chernosky et al. (1985).

may be that reaction (4.y) takes place at higher temperatures than reaction (4.z) and so is less influenced by kinetic factors. The heat capacity data of talc and anthophyllite in the temperature range 300-700 K are from Krupka et al. (1985a,b). These measured data are also included in the optimization for thermal expansion and bulk modulus.

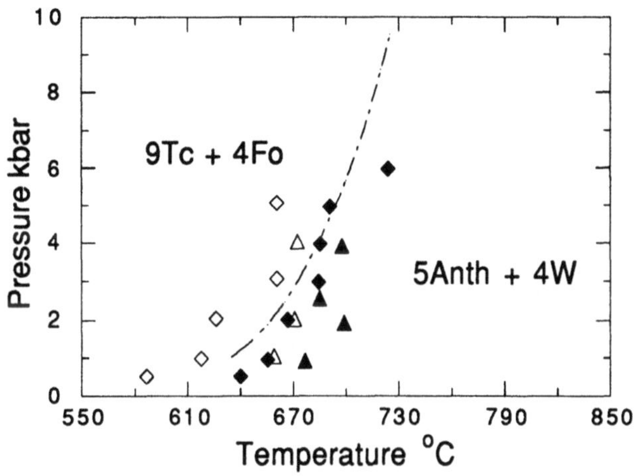

Fig. 4.20. Comparison of phase equilibrium computations with experimental data of the reaction Tc + Fo = Anth + H_2O (4.dd): triangles, Greenwood (1963); diamonds, Chernosky et al. (1985).

Figures 4.16 to 4.21 show equilibrium curves for some of the reactions discussed above. The curve for reaction (4.z) lies outside the experimental error limits in the higher pressure region (Kitihara et al., 1966) for the reasons mentioned above. The curves in Figs. 4.13 [reactions (4.aa), (4.bb) and (4.cc)] and 4.14 [reactions (4.dd) and (4.ee)] are calculated on the basis of the optimized parameters for anthophyllite. The curves agree well with the experimental data on reactions (4.aa) through (4.ee). The two triple points are consistent with the experiments of Greenwood (1963) and Chernosky et al. (1985) and the solubility experiments of Hemley et al. (1977a,b).

Brucite, Chrysotile and Antigorite

The equilibrium data on the following reaction were determined by Barnes and Ernst (1963), Irving et al. (1977) and Schramke et al. (1982):

Br = Per + H_2O (4.kk)

The following reactions involve chrysotile:

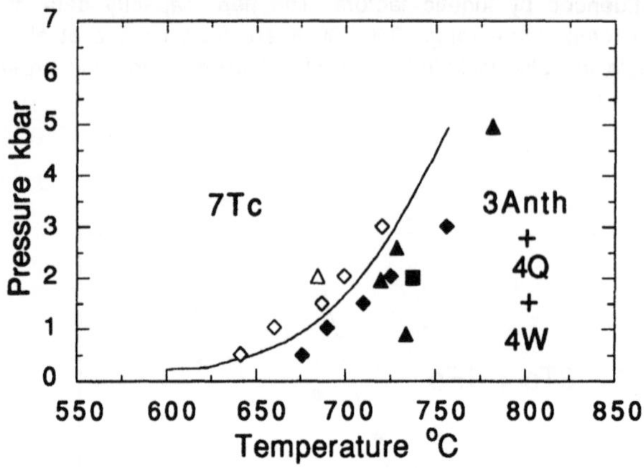

Fig.4.21. Comparison of phase equilibrium computations with experimental data of the reaction Tc = Anth + Q + H_2O (4.ee): triangles, Greenwood (1963); diamonds, Chernosky and Autio (1979); squares, Fyfe (1962).

$$\text{Chr} + \text{Br} = 2\text{Fo} + 3\text{H}_2\text{O} \tag{4.ll}$$

$$5\text{Chr} = 6\text{Fo} + \text{Tc} + 9\text{H}_2\text{O} \tag{4.mm}$$

The experimental data on these two reactions are from Kitihara et al. (1966), Johannes (1968) and Chernosky (1982).

The stability of antigorite is defined by the following reaction:

$$\text{Ant} = 18\text{Fo} + 4\text{Tc} + 27\text{H}_2\text{O} \tag{4.nn}$$

The equilibria for this reaction were experimentally determined by Evans et al. (1976).

Experimental data on reaction (4.kk) are used to determine the free energy data of brucite. The heat capacity data of brucite are available up to temperatures of 700 K (King et al., 1975). The equilibrium data of Irving et al. are up to a temperature of approximately 1400 K. Thus it is necessary to extrapolate the heat capacity using the Fei-Saxena formulation. The heat capacities obtained from the new expression (optimized by Saxena, 1989) deviate in small amounts from the original data of King et al. (-0.68, 0.8 and 1.35 J/mol at 298, 500 and 700 K respectively). Enthalpy, entropy, thermal expansion and bulk modulus of brucite are optimized from the experimental data. A detailed discussion of data evaluation on brucite is given in Saxena (1989). Figure 4.22 shows the calculated curve for the reaction (4.kk) along with the experimental data. Experimental data on reactions (4.ll) and (4.mm) are used to optimize the free energy of chrysotile. The data of some other workers at low pressures on reaction (4.mm) (see figure caption for details) were also included in the analysis. All these data are well reproduced from the results obtained. No heat capacity data for chrysotile are available. Therefore, the optimized parameters may not be very well constrained. Experimental data on reaction (4.nn) are used to optimize the free energy of antigorite. The heat capacity data for antigorite were measured up to a temperature of 850 K by King et al. (1967). The equilibria for reaction (4.nn) were experimentally determined by Evans et al. (1976). The optimized parameters for antigorite include enthalpy, entropy, thermal expansion and bulk modulus.

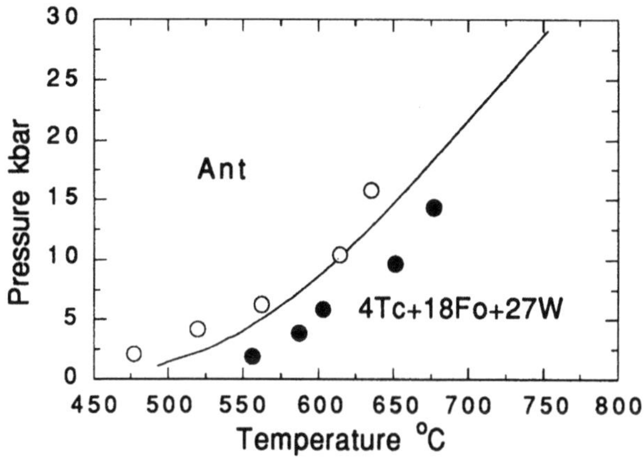

Fig. 4.22. Comparison of phase equilibrium computations with experimental data of the reaction Ant = Tc + Fo + H_2O (4.kk): experimental data from Evans et al. (1976).

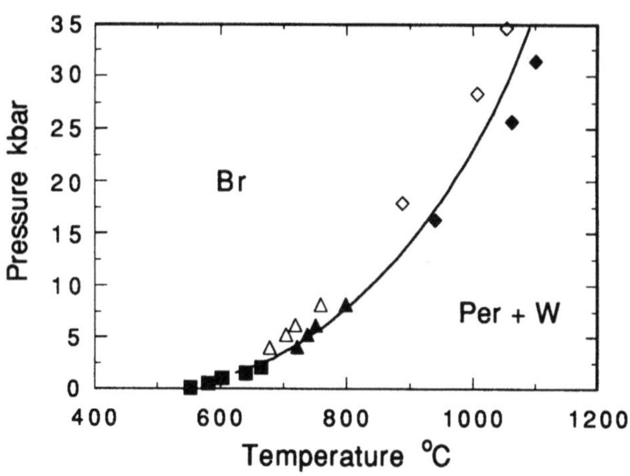

Fig. 4.23. Comparison of phase equilibrium computations with experimental data of the reaction Br = Per + H_2O (4.ll): triangles, Schramke et al. (1982); diamonds, Iriving et al. (1977); squares, Barnes and Ernst (1963).

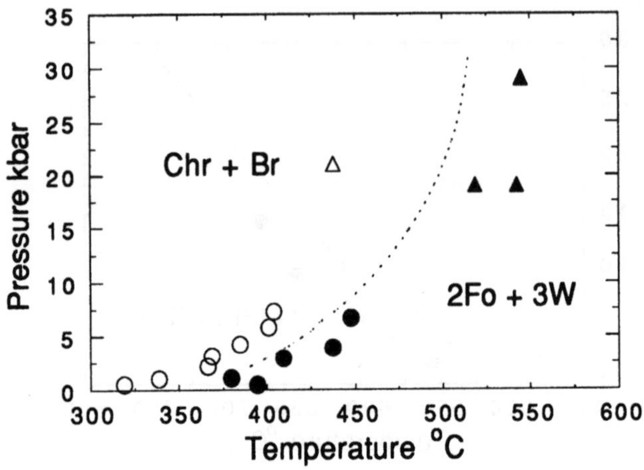

Fig. 4.24. Comparison of phase equilibrium computations with experimental data of the reaction Chr + Br = Fo + H_2O (4.mm): circles, Johannes (1968); triangles, Kitihara et al. (1966).

The curves in Figs. 4.22 and 4.23 involving brucite, chrysotile and antigorite are calculated from the optimized values. They agree very well with the experimental data of Johannes (1968), Kitihara et al. (1966), Chernosky (1982), Scarfe and Wyllie (1967), Kalinin and Zubkov (1981), Evans et al. (1976), Irving et al. (1977), Schramke et al. (1982) and Barnes and Ernst (1963). Berman et al. (1986) put an artificial constraint on the basis of the solubility data of Hemley et al. (1977a,b) and some field observations (Evans et al., 1976 Evans, 1977) on the equilibrium at 523 K and 2 kbar. In doing so they could not reproduce the higher pressure equilibrium data of Kitihara et al. (1966) for the reactions (4.ll) and (4.mm). Since the results reproduce the low pressure-temperature points of the above mentioned workers except Hemley et al. (1977a,b) just as well, the data of Hemley et al. were not considered. There is clearly a major inconsistency between the data of Hemley et al. and the data of the other workers. Berman et al. (1986) suggest that the hydrothermal data of Chernosky (1982) and Evans et al. (1976) are more reliable than the solubility data of Hemley et al. which were probably affected by kinetic factors. Figures 4.24 and 4.25 are for the reactions (4.mm) and (4.nn), respectively.

Grunerite

The grunerite breakdown reaction was studied by Mel'nik and Radchuk (1977) and Forbes (1977). Their results agree at the WM buffer. The reaction is as follows:

Grun = Fay + Q + H_2O (4.oo)

Experimental data on the reaction (4.oo) are used to optimize the free energy of grunerite. Miyano and Klein (1983) determined the free energy of grunerite by the same method. However, since they used different free energy values of fayalite and quartz than the values used in this study, the procedure had to be repeated. Their heat capacity equation is accepted in this work while calculating the enthalpy and entropy of grunerite. The molar volume of grunerite is from Klein and Waldbaum (1967).

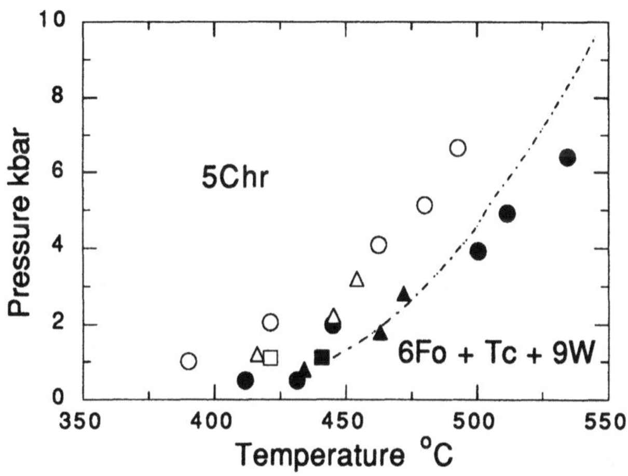

Fig. 4.25. Comparison of phase equilibrium computations with experimental data of the reaction Chr = Fo + Tc + H_2O (4.nn): circles, Chernosky et al. (1982); triangles, Scarfe and Wyllie (1967); squares, Kalinin and Zubkov (1981).

Ferrocordierite and the Cordierite-Ferrocordierite Solution

Experimental data on the distribution of Fe and Mg between garnet and cordierite were obtained by Hensen and Green (1973), Holdaway and Lee (1977) and Perchuk and Lavrent'eva (1983). The reaction is as follows:

Fcord + Pyr = Cord + Alm (4.pp)

The equation given below represents the average of all available experimental data:

$$\ln K_D \text{ (4.pp)} = 3039.06/T\text{(K)} - 1.242 \quad (4.9)$$

where,

$$K_D \text{ (4.pp)} = (X_{Mgcord} \cdot X_{Fegar})/(X_{Fecord} \cdot X_{Mggar}) \quad (4.10)$$

The free energy and volume of ferrocordierite and the interaction parameters of the cordierite-ferrocordierite solution are calculated assuming WCp of reaction (4.pp) as zero and from the following equilibrium relation:

$$\Delta G^{o}(4.qq)(P,T) + RT \ln K_D (4.qq) + RT \ln K_\gamma(4.qq) = 0 \quad (4.11)$$

the only unknown quantity in the term $\Delta G^{o}(P,T)$ (4.qq) being the free energy of ferrocordierite at P and T and where $\ln K_D$ is given by equation (4.10) and K_γ is as follows:

$$K_\gamma (4.qq) = (\gamma_{Mgcord}.\gamma_{Fegar})/(\gamma_{Fecord}.\gamma_{Mggar}) \quad (4.12)$$

The activity formulations for cordierite and ferrocordierite, following the Guggenheim formulation for simple mixtures, are as follows:

$$RT \ln\gamma_{Mg.cord} = X_{Fe.cord}{}^2 W \quad (4.13)$$

$$RT \ln\gamma_{Fe.cord} = X_{Mg.cord}{}^2 W \quad (4.14)$$

where W is the interaction parameter. The solution shows a near-ideal behavior (W = 1830.6 J/mol) in the temperature range considered (600-1100 °C). The thermodynamic properties obtained for ferrocordierite are consistent with the results of Richardson (1968) and Holdaway and Lee (1977), who studied the upper stability limit of ferrocordierite given by the following reaction:

Fecord = Alm + Sill + Q (4.qq)

Hydrous Cordierite

For the reaction

Cord + H_2O = Hydrous-cord (4.rr)

Bhattacharya and Sen (1985) give the following relations in the range 773-1073 K:

$$\Delta G_{solids}(1,T) = -271018.6 + 134.976T \text{ J/mol} \quad (4.15)$$

$$\Delta V_{solids}(1,T) = -3.7656\text{x}10\text{-}3T - 2.1514 \text{ J/bar/mol} \quad (4.16)$$

Using these relations, the enthalpy, entropy, molar volume and thermal expansion of hydrous cordierite are calculated. The heat capacity is assumed to be the same as that of cordierite.

Chapter 5

Assessment of Data on Some Oxides and Silicates: Calorimetry and High-Pressure Phase Equilibrium Experiments

The high pressure phase equilibrium calculations require a special consideration of both the *P-V-T* properties of the solids and fluids and the high-temperature extrapolation of the heat capacity data. The appropriate thermodynamic formulations were discussed in Chapter 2. Here we continue with systematics of the high *P* and *T* properties and our arguments for the choice of the data. It may be noted that the heat capacity data for all phases discussed here have been refitted to follow the $Cp/Cv/\alpha/\beta$ optimization discussed earlier. The data on enthalpy and entropy have accordingly been reassessed to fit the new data. Therefore the tabulated data may differ from the data of previous studies (e.g. Fei and Saxena, 1986; Chatterjee, 1989; Fei, 1990; Fei et al. 1990).

System Fe-O

Important minerals in this system are: iron polymorphs (bcc-Fe, fcc-Fe and hcp-Fe), wüstite (variable Fe:O), magnetite polymorphs (Fe_3O_4 and high-pressure-Fe_3O_4) and hematite polymorphs (Fe_2O_3 and high-pressure-Fe_2O_3).

Polymorphism of Fe

Phase relations of Fe at high pressure and high temperature have been extensively studied (e.g. Bundy, 1965; Strong et al., 1973; Fukizawa, 1982; Manghnani et al., 1985; Mao et al., 1987; Akimoto et al., 1987; Manghnani et al., 1987; Huang et al., 1987). Guillermet and Gustafson (1985) critically evaluated the thermochemical and thermophysical data for iron. Their data are adopted in this study to construct the phase diagram of iron.

Wustite

Several phase equilibrium studies are available for this system (e.g. Goel et al., 1980, Bjorkman, 1984) and Kubaschewski (1982) has summarized the phase relations. Hazen and Jeanloz (1984) have discussed the crystal chemistry of wüstite. As described by these authors, the structure of wüstite is complex and difficult to determine. The variables which may affect the thermodynamic properties of wüstite are a) degree of nonstoichiometry, b) the ratio of octahedral vacancies to tetrahedral ferric iron, and c) the size and shape of defect clusters, their periodicity and the extent of magnetite or iron exsolution. Each wüstite sample may have its own structural peculiarities depending on the conditions of synthesis and thermal history; this makes measurements of thermochemical and physical properties a very difficult task. From the numerous phase equilibrium studies in the system Fe-O cited in Kubaschewski (1982), it appears that certain phase boundary relationships are commonly accepted. However, details on various allotropic modifications of wüstite are not well known (see Hazen and Jeanloz, 1984). Such modifications are the reflections of discontinuities in the thermochemical properties of the wüstite solid solution.

Wüstite has a temperature-composition field within which its composition varies mainly between $Fe_{0.947}O$ and $Fe_{0.85}O$. The system may be modeled by considering wüstite as a solid solution of two fictive, isostructural components FeO and $FeO_{1.5}$. Through such modeling one can hope to treat the thermodynamics of the hetereogeneous reactions (i.e. wüstite reacting with other phases) generally within the errors in the thermochemical data of the reacting phases. A precise model for the wüstite solid solution particularly for modeling the crystal chemical behavior must include the effect of additional components (other than FeO and $FeO_{1.5}$) and other parameters discussed by Hazen and Jeanloz (1984). The reaction within the field of wüstite solution is

$$\{FeO\} + 0.25O_2 = \{FeO_{1.5}\} \tag{5.a}$$

and in the field towards magnetite, the reaction may be conceived as

$$\{5FeO\} + O_2 = \{2FeO_{1.5}\} + Fe_3O_4 \tag{5.b}$$

where the brackets denote the components in the wüstite solution. The computed W_{ij} values for wüstite are (FeO = 1, $FeO_{1.5}$ = 2): W_{12} = -7229, W_{21} = -2142 (J/mol) which are the same as in Bjorkman (1984). However, note that the thermochemical data on fictive end member components used here are different. The enthalpy and entropy data on the two fictive end member components are determined simultaneously through these calculations. The C_P coefficients for fictive FeO and $FeO_{1.5}$ are the same as for the stoichiometric FeO and hematite, respectively. A ternary solution model may be more suitable for the wüstite (Fe - FeO - $FeO_{1.5}$) as done by Goel et al. (1980) and Bjorkman (1984). However, in view of the uncertainties introduced at high pressures (see later) we have refrained from using the ternary model at present. The phase diagram of the Fe-O system used here is based on several experimental results reviewed by Kubaschewski (1982), Goel et al. (1980) and Bjorkman (1984). References to the literature may be found in the above mentioned articles.

The errors in the W_{12} and W_{21} and in the ΔH^o_f and ΔS^o_f are related and cannot be independently evaluated. However, for fixed value of W_{ij}, a change of a few hundred joules in ΔH^o_f is enough to make the results on phase boundaries change beyond the quoted experimental uncertainty.

For calculations at high pressure, we need data on volume of mixing and on compressibility of the solid solution components FeO and $FeO_{1.5}$. Volume of mixing is assumed to be ideal with volume of the solution given by:

$$V_{Fe_xO}=X_{FeO}V_{FeO}+X_{FeO_{1.5}}V_{FeO_{1.5}}=(3-2/X)V_{FeO}+(2/X-2)V_{FeO} \quad (5.1)$$

Using the molar volume data of FeO as 12.25 cm^3 (extrapolated value from Rosenhauer et al., 1976) and the V_{Fe_xO} data from Simons (1980) and Hentschel (1970), the molar volume of the fictive $FeO_{1.5}$ is determined to be 15.97 cm^3. For estimating the bulk modulus of Fe_xO (change in bulk modulus as a function of composition of wüstite), there are two sets of data available (see review, Jeanloz and Hazen, 1983). The data from dynamic measurements indicate a $K_{T,0}$ of 180 GPa for FeO and a similar value for $FeO_{1.5}$. The static measurements of $K_{T,0}$ lead to values of 155 GPa and 137 GPa for FeO and $FeO_{1.5}$, respectively. Both sets of data with their indicated limits have been used in the phase equilibrium calculations.

Magnetite and Hematite

Magnetite transforms to its high-pressure polymorph at pressure of about 25 GPa at room temperature (Mao et al., 1974). The magnetite and high-pressure magnetite boundary was determined to have a negative slope ($\delta T/\delta P$ = -45 K/GPa, Huang and Bassett, 1986). Shock-wave data indicate that hematite transforms to its high-pressure polymorph below pressure of 60 GPa. The transition also was not observed in the diamond-anvil cell below 30 GPa (Mao and Bell, 1977). Some physical properties of these phases are available (e.g. Ahrens et al., 1969; Mao et al., 1974; Wilburn et al., 1978, Sato and Akimoto, 1979; Finger and Hazen, 1980; Finger et al., 1986). Thermochemical data for magnetite and hematite are from Haas (see Fei and Saxena, 1986).

There are several buffer reactions, wüstite-iron (WI), magnetite-iron (MI), magnetite-wüstite (MW) and hematite-magnetite (HM) buffers, in the system Fe-O. The phase equilibria in the system were experimentally determined by Eugster and Wones (1962). The computed phase equilibria, based on evaluated data for phases in the system, are in good agreement with the experimental data (Fei and Saxena, 1986).

System MgO-SiO_2

Phases in the MgO-SiO_2 system include all the mineral structures proposed as likely in the mantle. These minerals are MgO (rock salt), the polymorphs of SiO_2 (quartz, coesite and stishovite), Mg_2SiO_4 (olivine, β-phase and spinel) and $MgSiO_3$ (the pyroxene polymorphs, ilmenite, garnet and perovskite). The thermochemical and thermophysical data for periclase are well known (Suzuki, 1975; Robie et al., 1978; Jackson and Niesler, 1982; Anderson and Zou, 1989). Data selections on other phases are discussed below. The results on the $MgO-SiO_2$ system included here are from Fei et al. (1990).

Polymorphism of SiO_2

The polymorphism of SiO_2 is quite complicated. Here we only focus on three high pressure polymorphs (quartz, coesite and stishovite). Weaver et al. (1979) and recently Kuskov and Fabrichnaya (1987) compared the thermochemical-thermophysical data with phase equilibrium data for the quartz-coesite-stishovite transformations. Akaogi and Navrotsky (1984), using new calorimetric data, calculated phase relations in the three polymorphs. Fei and Saxena (1986) started with the thermochemical data on quartz from Robie et al. (1978), and evaluated the thermochemical data on coesite and stishovite; these data were then combined with the phase equilibrium data for the quartz-coesite transformation (Bohlen and Boettcher, 1982), for the coesite-stishovite transformation (Yagi and Akimoto, 1976) and with the calorimetric data from Akaogi and Navrotsky (1984). There are several high pressure reactions involving stishovite (e.g. pyroxene = β-phase + stishovite in the Mg-Si-O system, pyroxene = spinel + stishovite and spinel = wüstite + stishovite in the Fe-Si-O system, and perovskite = magnesiowustite + stishovite in the Mg-Fe-Si-O system). The physical properties of stishovite play an important role in those phase equilibrium relations. The thermal expansion of stishovite was determined by Ito et al. (1974) in a temperature range of 291 - 873 K. The compression of stishovite has been extensively studied by various methods, namely, the static compression, the ultrasonic, the shock wave, the brillouin scattering and theoretical calculation. The reported values of bulk modulus for stishovite range from 250 (Liebermann et al., 1976) to 343 GPa (Akimoto, 1972). The choice of bulk modulus of 316 GPa for stishovite (Weidner et al., 1982) in this study is supported by a global optimization in which all high pressure reactions involving stishovite are included. The data on stishovite were evaluated by simultaneously fitting several phase boundaries involving stishovite (Fei and Saxena, 1986). Figure 4.1 shows the calculated reaction curves in the system.

Polymorphism of Mg_2SiO_4

The phase relations for the three Mg_2SiO_4 polymorphs (olivine, β-phase and spinel) have been experimentally investigated by Ringwood and Major (1970), Suito (1972), Kawada (1977), Suito (1977), Ohtani (1979), Fukizawa (1982), Sawamoto (1986b), Katsura and Ito (1989), and Ito and Takahashi (1989). For pressure calibration at high temperatures, Suito (1977) used the data of Yagi and Akimoto (1976) who determined the coesite-stishovite boundary by the in-situ X-ray diffraction method. The pressures for the α-β transition determined by Suito (1977) were higher than that determined by Ringwood and Major (1970), Suito (1972) and Kawada (1977). Suito argued that the discrepancies were due to different starting materials, pressure media, apparatus and pressure scale (Suito, 1977). Fukizawa (1982) determined the α-β transition pressure using the in-situ X-ray diffraction method based on a gold internal pressure standard. The calculated result is about 1 GPa lower in pressure than that determined by Suito (1977). Very recently, Katsura and Ito (1989) have precisely determined the α-β transition pressures at 1473 and 1873

K to be 14 and 15 GPa, respectively. Their result is close to Suito's at 1273 K and to Fukizawa's at 1873 K. On the other hand, the calorimetric data for the α-β transition obtained by Akaogi et al. (1984) are consistent with Suito's results when the entropy change by the value of -10.5 J/mol.K for the α-β transition is used. The calorimetric data are also consistent with Katsura and Ito's results if the entropy change by the value of -7.7 J/mol.K is chosen (Akaogi et al., 1989). The optimized data in this study favor Katsura and Ito's results (see discussion below) as shown in Fig. 5.1.

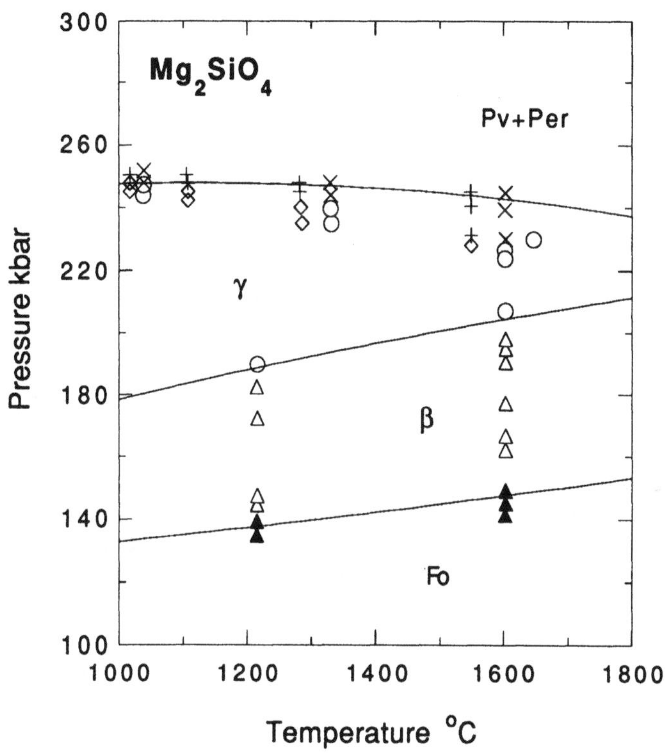

Fig. 5.1. Comparison of calculated phase boundaries in Mg_2SiO_4 with experimental phase equilibrium data. Solid lines are calculated results. Experimental data: Katsura and Ito (1989) (solid-triangles-Ol; open triangles-β; circle-γ); Ito and Takahashi (1989) (circles-γ; crosses-Pv+Per); Ito et al. (1990) (diamonds-γ; plusses-Pv+Per).

The pressures for the β-γ transition determined by Ohtani (1979) are about 0.8 GPa lower than those determined by Suito (1977). The disagreement could be totally due to the pressure calibration. Ohtani determined the phase boundary of the pyroxene-ilmenite transition in $ZnSiO_3$ as P(GPa) = 8.0(±0.5) + 0.0018T(K) and used it for comparing the pressure calibration. Such pressures are about 0.6 GPa lower than those determined with the in-situ X-ray diffraction experiments [P(GPa) = 8.6 + 0.002T(K), Akimoto et al., 1977]. When the pressures for the β-γ transition determined by Ohtani (1979) are calibrated using the experimental results of Akimoto et al. (1977), they agree with Suito's data. Recently, additional data on the β-γ transition at high temperatures were reported by Sawamoto (1986b). The data provided an important constraint on the slope of the β-γ transition boundary. Precise determination of the β-γ transition boundary has been made by Katsura and Ito (1989). Their results are generally consistent with those of Suito (1977) and Sawamoto (1986b). The pressures for decomposition of spinel into perovskite and periclase were determined by Ito and Yamada (1982). The pressure calibration at high temperature is based on the pyroxene-ilmenite transition in $ZnSiO_3$ (Akimoto et al., 1977) and decomposition of $MgAl_2O_4$ (Liu, 1980) and γ-Ni_2SiO_4 (Navrotsky et al., 1979) into the constituent oxide mixtures. Recently, Ito and Takahashi (1989) reported a new result on the spinel = perovskite + periclase transformation [P(GPa) = 28.4 - 0.0028T(K)].

After examining the experimental equilibrium data in the system Mg_2SiO_4, the following data were selected for evaluating an internally consistent data set: the α-β transition by Katsura and Ito (1989) and β-γ transition by Katsura and Ito (1989) (1473 - 1873 K) and by Sawamoto (1986b) (1873 - 2373 K) and, the decomposition of spinel into perovskite and periclase by Ito and Takahashi (1989). This choice of phase equilibrium data in the system Mg_2SiO_4 is also supported by the calorimetric data and the phase equilibrium data in the system $MgSiO_3$ (see discussion below). Figure 5.1 shows the final results.

Sumino et al. (1977), Sawamoto et al. (1984) and Weidner et al. (1984) studied the physical properties of the three Mg_2SiO_4 polymorphs. The heat capacity and the thermophysical data of the three Mg_2SiO_4 polymorphs are listed in the Appendix. The enthalpy and entropy of forsaterite are from Saxena and Chatterjee (1986) and Robie et al. (1982a). The enthalpies and entropies of β-phase and spinel have been evaluated by accepting the phase transition boundaries in the Mg_2SiO_4 system as determined by Katsura and Ito (1989) and Sawamoto (1986b). The calculated transition parameters (ΔH^o, ΔS^o and ΔV) at 1000 K are found to be consistent with the calorimetric data reported by Akaogi et al. (1984), and by a recent refinement of the enthalpy of the β-γ transition (Akaogi et al., 1989). The thermochemical and thermophysical data for periclase are well known (Suzuki, 1975; Robie et al., 1978; Jackson and Niesler, 1982; Anderson and Zou, 1989). The thermochemical data for perovskite are evaluated from transformation data on decomposition of spinel into perovskite and periclase and on the ilmenite-perovskite transition as determined by Ito and Takahashi (1989). They are consistent with the data set for the $MgSiO_3$ system (see below).

Polymorphism of $MgSiO_3$

There are six polymorphs of $MgSiO_3$: proto-, ortho- and clino-enstatite, garnet, ilmenite and perovskite. The status of clinopyroxene has been under some discussion (see Pacalo and Gasparik, 1990). These authors have considered low-clino ($P2_1/c$), high-P (C2/c) and high-T (C2/c) forms. These transitions are difficult to study because the enthalpy, entropy, and volume changes are extremely small and the transitions tend to be sluggish and susceptible to the influence of impurities and shear stresses. The data on orthoenstatite are constrained by several other sets of calorimetric and phase equilibrium relations (see Appendix). From 173 and 1073 K the difference between the heat capacities of the clino and the ortho forms is small (Haselton, 1979, Watanabe, 1982; Krupka et al., 1985a,b; Ashida et al., 1988). The data for pyroxenes as evaluated here are consistent with the data for the b-phase, stishovite and with the rest of the thermochemical data in the system. Finally, the calculated phase equilibrium diagram (Fig.5.2) shows that the data are consistent with most of the experimental information available thus far including the recent studies of Pacalo and Gasparik (1990) and Kanzaki (1991). The result agrees with the experimentally determined transformation boundary by Akaogi and Akimoto (1977), Ito and Navrotsky (1985) and Sawamoto (1987).

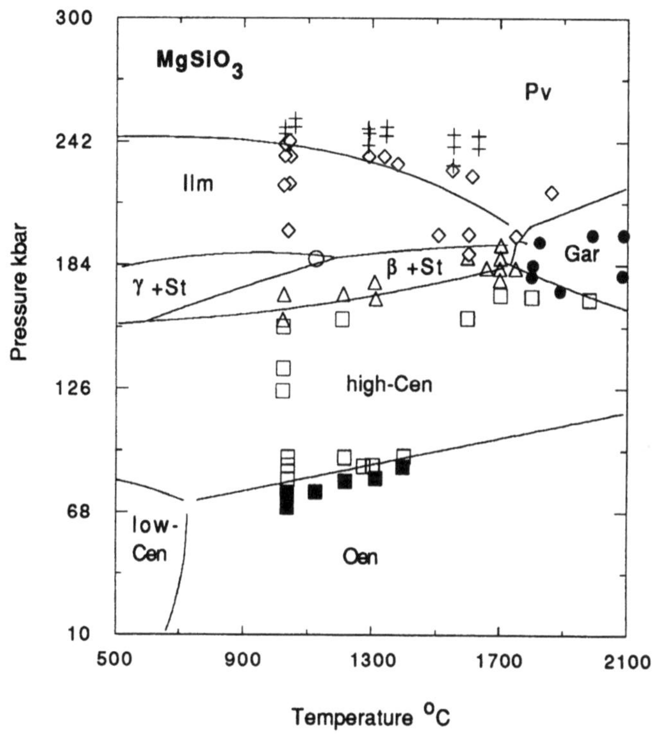

Fig. 5.2. Phase relations in $MgSiO_3$ in the pressure range between 10 and 300 kbar: a comparison between the calculated and the experimental results. Solid lines are calculated phase boundaries. Experimental data: Akaogi and Akimoto (1977) (open squares, highCen; triangles, β+St); Sawamoto (1987) (open squares, highCen; solid circles, Gar; triangles, β+St; diamond, Ilm); Ito and Navrotsky (1985) (open circles, γ+St; diamond, Ilm); Ito and Takahashi (1989) (diamonds, Ilm; plusses, Pv); Kanzaki (1991) (solid squares, Oen; open squares, highCen); Pacalo and Gasparik (1990) (solid squares, Oen; open squares, highCen).

The data for garnet and ilmenite should be considered as uncertain. The following discussion from Fei et al. (1990) apply here as well. Kato and Kumazawa (1985,1986) reported that garnet is stable to very high temperature. It is tetragonal possibly due to Mg-Si ordering on octahedral sites; a cubic form may also exist and a cubic (majorite) garnet solid solution between $Mg_4Si_4O_{12}$ ($4MgSiO_3$) and $Mg_3Al_2Si_3O_{12}$ (pyrope) is reasonably well documented (Ringwood, 1967; Akaogi and Akimoto, 1977; Liu, 1977; Akaogi et al, 1987). Akaogi et al. (1987) estimated the enthalpy of the pyroxene-cubic garnet transition as 36.5 kJ (per mole $MgSiO_3$) and the bulk modulus of cubic garnet as 154 GPa (assuming $K' = 4.0$) from enthalpies of solution and volume compression along the $Mg_4Si_4O_{12}$-$Mg_3Al_2Si_3O_{12}$ join. Their extrapolated molar volume for cubic garnet is 28.50 cm^3/mol ($MgSiO_3$), slightly larger than volumes near 28.1 to 28.2 cm^3/mol for tetragonal garnet (Kato and Kumazawa, 1985, 1986). Because the enthalpy of the cubic-tetragonal transition is unknown, we shall use data for the cubic phase in subsequent calculations. Lacking thermal expansion data, we assume that is the same as for pyrope (Leitner et al., 1980). Heat capacities for $MgSiO_3$ garnet are unavailable. One can either use values estimated from vibrational calculations (see below) or those for pyrope; both are similar. Kato and Kumazawa (1985) and Sawamoto (1986a) found that at temperatures above 1973 K at approximately 17 GPa, pyroxene transforms directly to garnet. From these data and their enthalpy, Akaogi et al. (1987) calculated ΔS^o (pyroxene = garnet) of -3.65±0.73 J/mol.K. This entropy change is quite different from our value but their calculated phase boundaries are not very different from ours. The unavailability of thermochemical data for tetragonal garnet adds an unknown (but presumably small) uncertainty to these calculations. Some data are available for $MgSiO_3$ ilmenite. The bulk modulus and elasticity have been reported by Weidner and Ito (1985) and heat capacity at 300 - 600 K has been measured by Watanabe (1982). Recently Ashida et al. (1988) have measured heat capacity over a somewhat wider range (160 - 550 K) and the thermal expansion at 295 - 875 K. McMillan and Ross (1987) calculated the entropy and heat capacity at 0 - 1800 K using vibrational models. Saxena and Shen (1992) calculated the heat capacity from the vibrational model and the available thermophysical data. All these estimates of C_P are in close agreement. We evaluated the enthalpy and entropy of ilmenite from phase equilibrium data for the reaction, spinel + stishovite = ilmenite (Ito and Navrotsky, 1985) and for the ilmenite-perovskite transition by Ito and Takahashi (1989), P(GPa) = 27.5 - 0.0025 T(K), a slightly lower curve than previously determined (Ito and Yamada, 1982). Since the data for ilmenite have to be consistent with both transformations, very little adjustment can be done to the ilmenite data. Calculated enthalpy of the pyroxene-ilmenite transition at 298 K (60.6 kJ/mol) is in good agreement with the new calorimetric value of 59.03±4.26 kJ/mol (Ashida et al., 1988) which revises the earlier preliminary calorimetric value of 71.8 kJ/mol (Ito and Navrotsky, 1985). Indeed, the calculations identified this discrepancy between earlier calorimetric and phase equilibrium data at the same time as further calorimetric studies independently revised the value of ΔH^o so as to remove the discrepancy.

Data for $MgSiO_3$ perovskite include estimates of compression by X-ray diffraction (Yagi et al., 1982; Kudoh et al., 1987) and from elasticity (Yeganeh-Haeri et al., 1989) and of thermal expansion (Knittle et al., 1986). The reactions ilmenite = perovskite and spinel = perovskite + periclase have been studied by Ito and Yamada (1982) and recently by Ito and Takahashi (1989). Both these transitions appear to have negative P-T slopes. No heat capacity measurements have been reported. An estimate of C_p from vibrational models has been made in this study. Recently Ito, Akaogi and Navrotsky (Navrotsky, pers. comm.) attempted to measure the enthalpy of the perovskite by drop solution calorimetry. They obtained a value of ΔH^o (ilmenite = perovskite) in the range 40 - 60 kJ/mol; the large uncertainty reflecting the presence of significant amounts of other phases in the sample. Vibrational calculations (see Fei et al., 1990) support that the entropy of the ilmenite-perovskite transition is positive, implying a negative $\delta P/\delta T$, but they suggest that ΔS^o may depend quite strongly on temperature. The parameters shown in tables generally satisfy all the above observations.

System FeO-SiO_2

Stable iron silcates in this system include fayalite, spinel and orthoferrosillite. β-phase and perovskite are not stable and will be discussed as fictive components in solid solutions.

Polymorphism of Fe_2SiO_4

Fe_2SiO_4, unlike Mg_2SiO_4, directly transforms from olivine (α) to spinel (γ) at high pressure. The transformation has been experimently studied using the quench technique (Akimoto et al., 1965, 1967) and using the in-situ X-ray diffraction method (Inoue, 1975; Sung and Burns, 1976; and Furnish and Bassett, 1983). Yagi et al. (1987) accurately determined the equilibrium phase boundary between 1073 and 1473 K using a cubic anvil type of high-pressure and high temperature X-ray diffraction apparatus combined with synchrotron radiation. The transition pressure P(GPa) = 3.43 + 0.0025 T(K). The in-situ observation also provides information on thermal expansion at high pressure. The thermophysical properties, such as molar volume, thermal expansion and bulk modulus for α-Fe_2SiO_4 and γ-Fe_2SiO_4 have been studied by many geophysicists (e.g. Akimoto et al., 1976; Marumo et al., 1977; Sato, 1977; Sumino, 1979; Suzuki et al., 1979, 1981). These data have been used to calculate the volumes at high pressure and high temperature using the equation of state discussed in Chapter 2. The calculated results are in good agreement with that observed in in-situ X-ray diffraction experiments (Yagi et al., 1987). The thermochemical data, such as enthalpy, entropy and heat capacity, for α-Fe_2SiO_4 are from Robie et al. (1982a). Heat capacity for γ-Fe_2SiO_4 was determined by Watanabe (1982). The enthalpy and entropy for γ-Fe_2SiO_4 are evaluated from the phase equilibrium data (Yagi et al., 1987), constrained by the calorimetric data on ΔH^o_{1000} and ΔS^o_{1000} of the α-γ transition as obtained by Akaogi et al. (1989). The evaluated data on γ-Fe_2SiO_4 along with the data on FeO (wüstite) and stishovite are consistent with the experimental data on reaction, γ-Fe_2SiO_4 = FeO (wüstite) + SiO_2 (stishovite), determined by Yagi et al. (1979) and Ohtani (1979). β-Fe_2SiO_4 is treated as a fictive component in the system. Its thermochemical and thermophysical properties are either extrapolated from limited solution data or estimated by using an analog (Fei and Saxena, 1986). The calorimetric data on ΔH^o_{1000} and ΔS^o_{1000} of the α-β transition as estimated by Akaogi et al. (1989) are used as additional constraints in the calculations.

Polymorphism of $FeSiO_3$

Perovskite, garnet (cubic and non-cubic) and ilmenite in $FeSiO_3$ are not stable and can be treated as fictive components. Garnet and ilmenite in $FeSiO_3$ are not considered in this study because of the small iron solubility in both garnet and ilmenite solutions and because of lack of thermochemical and thermophysical data on these fictive components. The data on Fe-perovskite are estimated from limited solution data and from other related phases (Fei and Saxena, 1986).

For ferrosillite ($FeSiO_3$), the data are based on the data of orthoferrosillite (Syono et al., 1971; Akimoto, 1972; Bohlen and Boettcher, 1981). These data along with the data on spinel (Fe_2SiO_4) and stishovite evaluated from the systems Fe_2SiO_4 and SiO_2 are consistent with the experimental data on reaction, $2FeSiO_3$ (ferrosillite) = Fe_2SiO_4 (spinel) + SiO_2 (stishovite), determined by Akimoto and Syono (1970).

Some discussion is appropriate regarding the effect of uncertainties in the calculated value of ΔG^o on the calculated phase equilibria. Because the pressure-temperature field is divided into many fields of stable assemblages, the margin for error in ΔG^o values for individual phases is small and the system is well constrained. If one changes the ΔG^o of a phase by more than 300 J/mol, the calculated stability fields of several related minerals can be completely upset. For example, if the enthalpy of garnet is decreased by 350 J/mol, the garnet phase will be calculated to form directly from pyroxene at 1273 K. Decrease of about 1000 J/mol in ΔH^o_{298} of ilmenite would eliminate the two-phase region between enstatite and ilmenite; and an increase of about 1000 J/mol would totally eliminate the ilmenite stability field between spinel plus stishovite and perovskite. Thus we conclude that the calculated free energies of transition are generally constrained to ±500 J/mol or better. The uncertainties in ΔG^o are difficult to attribute to uncertainties in individual thermochemical and thermophysical parameters unless one has compelling specific evidence. Data evaluated both from phase equilibrium reactions and from calorimetric measurements generally are within the limits of the calorimetric data. In the systems Mg_2SiO_4 and $MgSiO_3$ several phases, β-phase, spinel and perovskite, are common. Therefore, data on β-phase have to be evaluated by simultaneously fitting the phase equilibrium data of the α-β and β-γ transitions in Mg_2SiO_4 and the pyroxene = β + stishovite transition in $MgSiO_3$. Since data on stishovite, pyroxene and olivine (α) are well established from other systems (Fei and Saxena, 1986; Saxena and Chatterjee, 1986), evaluated data on β-phase are tightly constrained in such global optimization. Similar considerations apply to evaluation of data for perovskite and ilmenite. From the derived data for perovskite and ilmenite we may closely reproduce the phase equilibrium relations of the spinel + stishovite = ilmenite, ilmenite = perovskite and spinel = perovskite + periclase reactions. The Gibbs free energy for ilmenite evaluated from phase equilibrium data is more negative than that calculated from lattice vibrational model. If the entropy and heat capacity of ilmenite calculated from the vibrational model were used, the ilmenite stability field would be eliminated from the $MgSiO_3$ phase diagram. Therefore, the thermochemical and thermophysical data set derived here is generally adequate for calculating phase equilibria with uncertainties of 0.5 GPa or so in the location of phase boundaries.

Chapter 6

Assessment of Solid Solution Parameters for Some Silicates

Solid Solution Models

Various solid solution models were discussed by Fei et al. (1986). There are two different groups of solution models. The first group of models has their origin in the Flory-Huggins model (Flory, 1953), in which solutions are considered as athermal with zero excess enthalpy of mixing. The later refined versions, which are the Wilson model (Wilson, 1964), the quasi-chemical (Guggenheim, 1952) and the non-random two-liquids (NRTL) model (Renon and Prausnitz, 1968), do involve enthalpy of mixing and have been reviewed by Acree (1984). The second group of models simply expresses functions as power series in mole fraction. The Redlich-Kister and the two-constant Margules model fall in this category. We note that there are many studies of the application of the sub-lattice model for minerals with non-equivalent crystallographic sites (e.g. Davidson and Lindsley, 1989; Sack and Ghiorso, 1989). In this book, we are mostly concerned with single-site formulations. In certain cases, we have first used the two-site sub-lattice formulation and then converted the energies to a single-site ionic form (e.g. Shi et al., 1992).

The Margules model or the Redlich-Kister model for binary solution has been popularly used in geochemistry because of their simplicity of formulation, and are used in this study as well. For determining the properties of solid solutions from experimental data on compositions of coexisting phases, the distribution of a component between two coexisting solid solutions, e.g. (1,2)-M and (1,2)-N, where 1 and 2 are exchangeable components, may be considered as an ion-exchange reaction as discussed elsewhere (e.g. Ganguly and Saxena, 1988).

We have used an asymmetric non-ideal solution model for binary solid solutions, for which the excess free energy of mixing (Margules formulation, see Thompson and Waldbaum,1969) is defined as:

$$\Delta G^{ex} = X_1 X_2 (W_{12} X_2 + W_{21} X_1) \tag{6.1}$$

For the asymmetric model the activity coefficient of a component in a binary solution is given by

$$RT ln \gamma_1 = X_2^2 [W_{12} + 2X_1 (W_{21} - W_{12})] \tag{6.2}$$

Here W_{ij} parameters are adjustable constants and functions of both pressure and temperature. According to Thompson and Waldbaum (1969) for W_{ij}, we may write:

$$W_{ij} = W_{ij}^H - TW_{ij}^S + PW_{ij}^V \tag{6.3}$$

where W_{ij}^H, W_{ij}^S and W_{ij}^V represent the excess entropy and excess volume contributions to the interaction energy W_{ij}. The Kohler model (see Bertrand et al. 1983) is used for ternary and multicomponent solutions. According to the model of Bertrand et al. (1983), an excess property of a multicomponent solution is given by

$$\Delta Z^{ex}_{12...N} = \sum_{i=1}^{N} \sum_{j>i}^{N} (X_i + X_j)(f_i + f_j)(\Delta Z^{ex}_{ij})^* \tag{6.4}$$

in which $(\Delta Z_{ij}^{ex})^*$ is the molar excess property (enthalpy, entropy, volume, free energy etc.) of the binary system with components at the same molar ratio as the multicomponent system and f_i and f_j are weighted mole fractions using weighting factors based on the excess properties of the binary systems. X_i is used as the mole fraction in the multicomponent system.

For the activity coefficient of a component i, we have

$$\begin{aligned}(RTln\gamma_i)_{12...N} = &\sum_{j}^{?} (X_i + X_j)(f_i + f_j)(\Delta G^{ex}_{ij})^* \\ &- \sum_{j,k}^{?} (X_j + X_k)(f_j + f_k)(\Delta G^{ex}_{jk})^* \\ &+ \sum_{j}^{?} (f_j + f_k)(RTln\gamma_{ij})^* \qquad (i \neq j \neq k)\end{aligned} \tag{6.5}$$

where the binary functions are denoted by an asterisk. As discussed by Fei et al. (1986), the Kohler method of predicting the ternary solution properties is as good as the Wohl's formulation (see Saxena, 1973; Ganguly and Saxena, 1984) with the additional advantage of simplicity in extension to multicomponent solutions. In equation

$$RTlnK = -\Delta G^0_{P,T} + RTln\gamma_i \tag{6.6}$$

we may substitute appropriately for the activity coefficients from the binary or multicomponent models noted above and depending on the nature of the observational data, equation (6.6) may be used to determine the unknowns, which may be $\Delta G_{P,T}$ or W_{ij} or a combination of these. Details of such calculations are discussed amply in the literature (e.g. Saxena,1973; Ganguly and Saxena,1984,1987).

In this chapter, the interaction parameters for some of the common silicate solutions following the ionic model are described. These parameters are either assessed in this work or are adopted from the assessment of other workers. The assessments are mainly based on the distribution of major element data between two coexisting solution phases. In some cases, in which such data are not available, the interaction parameters are directly taken from calorimetric sources. When an assessment is done, the parameters are adjusted in such a way that they are constrained from calorimetric determinations, if any.

Gehlenite - Akermanite

This solution has a positive deviation from ideality. The asymmetric model of Charlu et al. (1981) determined from calorimetric experiments is recommended. They measured the enthalpy of solution in molten lead borate of synthetic melilites along the gehlenite - akermanite join. They found that the execess enthalpy of the join is given by:

$$W_H^{ex}(\text{KJ, 0.38 KJ}) = 24.2881\ X_1 X_2^2 + 0.5021\ X_2 X_1^2 \qquad (6.7)$$

where X_1 and X_2 are the mole fractions of gehlenite and akermanite respectively.

Spinel - Hercynite

Fe^{2+}-Mg^{2+} distribution data for the exchange reaction

$$\text{Fay} + \text{Sp} = \text{Fo} + \text{Her} \qquad (6.a)$$

were experimentally determined by Engi (1983) and Jamieson and Roeder (1984). A mixing model for spinel is obtained using the distribution data of Engi (1983). This is done through the relation

$$\Delta G^o(P,T) + RT\ln K_D(8.a) + RT\ln Kq(6.a) = 0 \qquad (6.8)$$

which can be rewritten as

$$\Delta G^o + RT\ln K_D(8.a) + (2X_{FeOl}-1)W^{Ol}$$
$$= W_{MgFe}^{Sp}[X_{FeSp}(3X_{FeSp}-2)] + W_{FeMg}^{Sp}[-1+4X_{FeSp}-3(X_{FeSp})^2] \qquad (6.9)$$

The W values obtained from equation (6.8) are as follows:

W_{MgFe}^{Sp} = 9124 + 1145 J/mol
W_{FeMg}^{Sp} = 0.0 J/mol

with a standard deviation of 1283.9. Jamieson and Roeder (1984) estimated an interaction parameter of -3100.34 + 627.6 J/mol at 1300 °C. Ganguly and Saxena (1984) found a positive deviation (W = 7698.6 + 2584.1 J/mol) through an estimation using Fujii's (1977) experimental data on coexisting olivine and spinel between 1200 and 1350 °C (P = 1 to 5 kbar). According to Jamieson and Roeder (1984), $\ln K_D$(6.a) at X_{FeSp} = 0.5 is zero. In such a case equation (6.9) reduces to

$$\Delta G^O = W_{MgFe}^{Sp}(-0.25) + W_{FeMg}^{Sp}(0.25) = -2281\ \text{J}.$$

Using the free energy data from this study, $\Delta G^o(1,1573)$ = -7537.5 J. Hence, their data are not reproduced. This discrepancy of -5.2 KJ points to an inconsistency between the data of Engi (1983) and Jamieson and Roeder (1984).

Forsterite - Fayalite

Fe^{2+}-Mg^{2+} distribution data for the exchange reaction

$$\text{Fay} + \text{En} = \text{Fo} + \text{Fer} \tag{6.b}$$

were determined by Matsui and Nishizawa (1974). Experiments involving Fe-rich olivine and quartz in equilibrium with Fe-rich orthopyroxene were done by Bohlen and Boettcher (1981).

A symmetric model for the Fe^{2+}-Mg^{2+} mixing in olivine with a W value of 4500 J/mol of cation fits best with the experiments of Bohlen and Boettcher (1981). Wood and Kleppa (1981) measured the enthalpies of solution of synthetic forsterite-fayalite olivines and determined the following interaction parameters at 970 K:

$W_{\text{Mg-Fe}}$ = 8368 J/mol
$W_{\text{Fe-Mg}}$ = 4184 J/mol

If the measurements of Wood and Kleppa (1981) are fitted to a symmetric solution model, the value of the interaction parameter will agree closely with the one recommended here.

Forsterite - Monticellite

Adams and Bishop (1985) conducted equilibrium experiments on this join and found an asymmetric miscibility gap in a pressure range 5 to 10 kbar and a temperature range 1100 to 1490 °C. They calculated the interaction parameters for the join on the basis of their experiments. Their values are recommended here.

Fayalite - Monticellite

Mukhopadhyay and Lindsley (1983) conducted reversed experiments on the fayalite - kirschsteinite ($[CaFeSi]_{0.5}O_2$) join and found a miscibility gap under 1041 °C at P_{H2O} = 1 kbar and Co-CoO or WM buffer. They calculated the interaction parameter for the join. The value obtained by them is 10933 (72) J/mol. As a first approximation, an average of this value and the interaction parameter of the forsterite - fayalite solution is recommended as the interaction parameter of the fayalite - monticellite solution.

Enstatite - Ferrosilite (Ortho- and Clino-)

The enthalpy of solution measurements of Chatillon-Colinet et al. (1983b) indicates a slight positive deviation (1987.4 J/mol at 1023 K) from an ideal behavior of the orthopyroxene solution. Using the site-occupancy data of Saxena and Ghose (1971), Ganguly and Saxena (1984) calculated the excess free energy at 1273 K and found a considerable negative deviation. They concluded that if the measured W^H of Chatillon-Colinet et al. is combined with a probable enthalpy of disorder and entropy, the solution becomes practically ideal and intrinsically stable over a large temperature range. Here, the orthopyroxene solution is recommended as ideal above 873 K. The solution shows a negative deviation as found by Saxena et al. (1986) from consideration of phase equilibrium data of Lindsley (1983) on quadrilateral pyroxenes. Their result is recommended below 873 K for enstatite-ferrosilite and 673 K for ferrosilite-enstatite. The clinoenstatite-clinoferrosilite solution is assumed to be ideal.

Enstatite - Diopside (Ortho- and Clino-)

The binary enstatite-diopside solid solutions have been discussed by Saxena (1973), Holland et al. (1979), Lindsley et al. (1981) and Nickel and Brey (1984). Although Nickel and Brey fitted a one constant symmetric model to clinopyroxene, the model of Lindsley et al. (1981) is recommended. The difference between the two models does not significantly affect the calculated phase equilibria, which reproduce the experimentally determined equilibrium data as reviewed by Lindsley et al. (1981) very well.

Clinoenstatite - Jadeite

Windom and Boettcher (1981) worked on the phase relations for the join clinoenstatite-jadeite and found a large miscibility gap at low temperatures. The interaction parameter for this solution is estimated from their result.

Diopside - Jadeite and Hedenbergite - Jadeite

The model recommended here is that of Gasparik (1985) who did phase equilibrium experiments on the diopside-jadeite join and found a large asymmetric non-ideal behavior. For the hedenbergite-jadeite solution, the same model may be assumed.

Muscovite - Paragonite

The muscovite-paragonite solution was studied extensively in the past by Eugster et al. (1972) and Chatterjee and Froese (1975) and recently by Flux and Chatterjee (1986) and Chatterjee and Flux (1986). Flux and Chatterjee (1986) determined the mixing properties of this binary by studying the exchange of K and Na between muscovite-paragonite and KCl-NaCl salts. They proposed a Redlich-Kister model expressed as follows:

$$G^{ex} = (1-X)\,X\,[A + B(1-2X) + C(1-2X)^2] \qquad (6.10)$$

where X is the mole fraction of muscovite in the muscovite-paragonite solution and the constants are a function of temperature and pressure given by:

$$A = 11222 + 1.389T + 0.2359P$$
$$B = -1134 + 6.806T - 0.0840P$$
$$C = -7305 + 9.043T$$

where the units for temperature, pressure, and excess free energy are kelvin, bar and J/mol, respectively.

Phlogopite - Annite

Wones and Eugster (1965) considered biotite a ternary solution with phlogopite, annite and "oxybiotite" [KFe_3+$3AlSi_3O_{12}$(H-1)] as the end members. Oxybiotite is present in natural biotites in significant amounts. They also calculated the activity of annite in the phlogopite-annite solution from the reaction

$$\mathrm{Ann = San + Mag + H_2} \qquad (6.c)$$

They used the fugacity of hydrogen determined by Eugster and Wones (1962) and Shaw and Wones (1964). They calculated a symmetric interaction parameter of -4211 J/mol. Wones (1972) later revised their work and concluded that the phlogopite-annite solution is ideal (see Hewitt and Wones, 1984). However, the fugacity of hydrogen has been revised many times since then (Saxena and Fei, 1988a). Thus, their calculations need to be revised.

In this work, the interaction parameters for this binary are determined in conjunction with the free energy of annite. This is discussed in Chapter 4 under the section "annite and the phlogopite-annite solution".

Cordierite - Ferrocordierite

This is discussed in Chapter 4 under the section "ferrocordierite and the cordierite-ferrocordierite solution".

Cordierite - Hydrous Cordierite and Ferrocordierite - Hydrous Cordierite

The cordierite - hydrous cordierite solution is considered ideal. The ferrocordierite - hydrous cordierite solution is approximated as the same as the ferrocordierite - cordierite solution.

Anorthite - Albite, Anorthite - Sanidine and Albite - Sanidine

The interaction parameters on these solutions are as evaluated by Nekvasil (1986) on the albite-anorthite-orthoclase solution.

Interaction Parameters

The recommended interaction parameters are listed in Table 6.1.

Table 6.1. Recommended interaction parameters.

MELILITE

Gehlenite-Akermanite $Ca_2Al_2SiO_7$-$Ca_2MgSi_2O_7$
W_{12} = 24288.1
W_{21} = 502.1
Charlu et al. (1981) (calorimetric determination)

SPINEL

Spinel-Hercynite $MgAl_2O_4$-$FeAl_2O_4$
W_{12} = 9124.0
W_{21} = 0.0
Chatterjee (1987) [evaluated from Engi, 1983 and valid in the range 923-1176 K and X_{Mg}(Sp)=0.4-0.9]

OLIVINE

Forsterite-Fayalite 1/2(Mg_2SiO_4-Fe_2SiO_4)
W = 4500.0
Chatterjee (1987) (evaluated from Bohlen and Boettcher, 1981)

Forsterite-Monticellite 1/2(Mg_2SiO_4-$CaMgSiO_4$)
$W_{12} = 21550.0 - 3.005\ T$
$W_{21} = 30600.0 - 7.6\ T$
Adams and Bishop (1985) (phase equilibrium experiment)

Fayalite-Monticellite 1/2(Fe_2SiO_4-$CaMgSiO_4$)
W = 7716.5
estimated

ORTHOPYROXENE

Enstatite-Ferrosilite $MgSiO_3$-$FeSiO_3$
$W_{12} = 13100.0 - 15.0\ T$ ($T \leq 873$ K)
= 0.0 (> 873 K)

W_{21} = 3370.0 - 5.0 T ($T \leq 673$ K)
= 0.0 (> 673 K)
Saxena et al. (1986) (when non-ideal) and Chatterjee (1987) (when ideal: approximately equal to the calorimetric values of Chatillon-Colinet et al., 1983b)

Enstatite-Orthodiopside $MgSiO∂$-1/2($CaMgSi_2O_6$)
W = 25000.0
Lindsley et al. (1981) (evaluated from phase equilibria)
Enstatite-Orthohedenbergite $MgSiO_3$-1/2($CaFeSi_2O_6$)
W_{12} = 60000.0
W_{21} = 30000.0 - 8.34 T
Saxena et al. (1986) (evaluated from Lindsley, 1983)

CLINOPYROXENE

Clinoenstatite-Diopside $MgSiO_3$-1/2($CaMgSi_2O_6$)
W_{12} = 25484.0 + 0.0812 P
W_{21} = 31216.0 - 0.0061 P
Lindsley et al. (1981) (evaluated from phase equilibria)

Clinoenstatite-Jadeite $MgSiO_3$-1/2($NaAlSi_2O_6$)
W = 30000.0
estimated from Windom & Boettcher (1981)

Diopside-Jadeite 1/2($CaMgSi_2O_6$-$NaAlSi_2O_6$)
Redlich-Kister model
A = 6300 - 4.725 T
B = 6300 - 3.8 T
C = -10700 + 8.1 T
Gasparik (1985) (phase equilibrium experiment)

MUSCOVITE

Muscovite-Paragonite $KAl_2AlSi_3O_{10}(OH)_2$-$NaAl_2AlSi_3O_{10}(OH)_2$
A = 11222 + 1.389 T + 0.2359 P
B = -1134 + 6.806 T - 0.0840 P
C = -7305 + 9.043 T
Flux and Chatterjee (1986) (calorimetric determination)

BIOTITE

Phlogopite-Annite $KMg_3AlSi_3O_{10}(OH)_2$-$KFe_3AlSi_3O_{10}(OH)_2$
W_{12} = 5540.7
W_{21} = 5068.5
This study (evaluated from experimental data of Perchuk and Lavrent'eva, 1983)

CORDIERITE

Cordierite-Ferrocordierite $Mg_2Al_4Si_5O_{18}$-$Fe_2Al_4Si_5O_{18}$
W = 1830.6

This study (evaluated from experimental data in Perchuk and Lavrent'eva, 1983)

Cordierite-Hydrous Cordierite $Mg_2Al_4Si_5O_{18}$- $Mg_2Al_4Si_5O_{18}.H_2O$
$W = 0.0$
(assumed)

Ferrocordierite-Hydrous Cordierite $Fe_2Al_4Si_5O_{18}$- $Mg_2Al_4Si_5O_{18}.H_2O$
$W = 1830.6$
same as Cordierite-Ferrocordierite

FELDSPAR

Anorthite-Albite $CaAl_2Si_2O_8$-$NaAlSi_3O_8$
$W_{12} = 11221.5 - 7.891\ T$
$W_{21} = 14129.4 - 6.18\ T$
Nekvasil (1986) (evaluated)

Anorthite-Sanidine $CaAl_2Si_2O_8$-$KAlSi_3O_8$
$W_{12} = 25020.3 + 10.795\ T$
$W_{21} = 75023.3 - 23\ T$
Nekvasil (1986) (evaluated)

Albite-Sanidine $NaAlSi_3O_8$-$KAlSi_3O_8$
$W_{12} = 30978.3 - 21.422\ T$
$W_{21} = 17062.35$
Nekvasil (1986) (evaluated)

References

Acree, W. E. Jr. (1984) *Thermodynamics Properties of Nonelectrolyte Solutions*, Academic Press, New York, 308.

Adams, G.E. and Bishop, F.C. (1985) An experimental investigation of thermodynamic mixing properties and unit-cell parameters of forsterite-monticellite solid solution, *Am. Mineral.*, **70**, 714-722.

Ahrens, T. J., Anderson, D. L. and Ringwood, A. E. (1969) Equations of state and crystal structure of high-pressure phases of shocked silicates and oxides, *Rev. Geophys.*, **7**, 667-707.

Akaogi, M. and Akimoto, S. (1977) Pyroxene-garnet solid solution equilibria in the system $Mg_4Si_4O_{12}$-$Mg_3Al_2Si_3O_{12}$ and $Fe_4Si_4O_{12}$ at high pressures and temperatures, *Phys. Earth Planet. Int.*, **15**, 90-106.

Akaogi, M. and Navrotsky A. (1984) The quartz-coesite-stishovite transformations: new calorimetric measurements and calculation of phase diagrams, *Phys. Earth Planet. Inter.*, **36**, 124-134.

Akaogi, M., Ross, N.L., McMillan P. and Navrotsky A. (1984) The Mg_2SiO_4 polymorphs(olivine, modified spinel and spinel)- thermodynamic properties from oxide-melt solution calorimetry, phase relations, and model of lattice vibrations, *Am. Mineral.*, **69**, 499-512

Akaogi, M., Navrotsky, A., Yagi, T. and Akimoto, S. (1987) Pyroxene-garnet transition: thermochemistry and elasticity of garnet solid solutions, and application to a pyrolite mantle, in *High Pressure Research in Mineral Physics*, edited by M. H. Manghnani and Y. Syono, pp. 251-260, Terra Publications, Tokyo, Japan.

Akaogi, M., Ito, E. and Navrotsky A. (1989) Olivine modified spinel-spinel transitions in the system Mg_2SiO_4-Fe_2SiO_4: calorimetric measurements, thermochemical calculation, and geophysical application, *J. Geophys. Res.*, **94**, 15671-15686.

Akimoto, S. (1972) The system MgO-FeO-SiO_2 at high pressures and temperatures - phase equilibria and elastic properties, *Tectonophysics*, **13**, 167-187.

Akimoto, S. (1987) High-pressure research in geophysics: past, present and future, in *High Pressure Research in Mineral Physics*, edited by M. H. Manghnani and Y. Syono, pp. 1-13, Terra Publications, Tokyo, Japan.

Akimoto, S. and Fujisawa, H. (1968) Olivine-spinel solid solution equilibria in the system Mg_2SiO_4-Fe_2SiO_4, *J. Geophys. Res.*, **73**, 1467-1473.

Akimoto, S. and Syono, Y. (1970) High-pressure decomposition of the system $FeSiO_3$-$MgSiO_3$, *Phys. Earth Planet. Int.*, **3**, 186-188.

Akimoto, S., Katsura, T., Syono, Y., Fujisawa, H. and Komada, E. (1965) Polymorphic transition of pyroxenes $FeSiO_3$ and $CoSiO_3$ at high pressures and temperatures, *J. Geophys. Res.*, **70**, 5269-5278.

Akimoto, S., Komada, E. and Kushiro, I. (1967) Effect of pressure on the melting of olivine and spinel polymorphs of Fe_2SiO_4, *J. Geophys. Res.*, **68**, 679-686.

Akimoto, S., Matsui, Y. and Syono, Y. (1976) High pressure crystal chemistry in orthosilicates and formation of the mantle transition zone, in *Physics and Chemistry of Minerals and Rocks*, edited by R. J. Strens, pp. 327-363, John Wiley, London.

Akimoto, S., Yagi, T. and Inoue, K. (1977) High temperature-high pressure phase boundaries in silicate systems using in-situ X-ray diffraction, in *High-Pressure Research: Application to Geophysics*, edited by M. H. Manghnani and S. Akimoto, pp. 585-602, Academic Press, New York.

Akimoto, S., Suzuki, T., Yagi, T. and Shimomura, O. (1987) Phase diagram of iron determined by high-pressure/temperature X-ray diffraction using syncrotron radiation, in *High Pressure Research in Mineral Physics*, edited by M.H. Manghnani and Y. Syono, pp. 149-154, Terra Publications, Tokyo, Japan.

Althaus, E. (1967) The triple point andalusite-sillimanite-kyanite, *Contrib. Mineral. Petrol.*, **16**, 29-44.

Althaus, E., Karotke, E., Nitsch, K.H. and Winkler, H.G.F. (1970) An experimental re-examination of the upper stability limit of muscovite plus quartz, *Neues. Jahrb. Mineral. Monatsh.*, **1970**, 325-336.

Anderson, O.L. (1989) The relationship between the adiabatic bulk modulus and enthalpy for matle-related minerals, *Phys. Chem. Min.*, **16**, 559-562.

Anderson, O.L. and Goto, T. (1989) Measurement of elastic constants of mantle-related minerals at temperatures up to 1800 K, *Phys. Earth Planet. Int.*, **55**, 241-253.

Anderson, O.L. and Zou, K. (1989) Formulation of the thermadynamic functions for mantle minerals: MgO as an example, *Phys. Chem. Min.*, **16**, 642-648.

Anderson, O.L., Boehler, R. and Sumino, Y. (1982) Anharmonicity in equation of state at high temperature for some geophysically important solids, in *High - Pressure Research in Geophysics*, edited by S. Akimoto and M. H. Manghnani, pp.273 - 283, Center for Academic Publishing, Tokyo, Japan.

Anderson, O.L., Isaak, D.L. and Oda, H. (1991) Thermoelastic parameters for six minerals at high temperature, *J. Geophys. Res.*, **96**, 18037-18046.

Anovitz, L.M., Treiman, A.H., Essene, E.J., Hemingway, B.S., Westrum, E.F., Wall, V.J., Burriel, R. and Bohlen, S.R. (1985) The heat capacity of ilmenite and phase equilibria in the system Fe-Ti-O, *Geochim. Cosmochim. Acta*, **49**, 2027-2040.

Anovitz, L.M., Essene, E.J., Hemingway, B.S., Komada, N.L. and Westrum, E.F. (1988) The heat-capacities of grunerite and deerite: phase equilibria in the system Fe-Si-C-O-H and implications for the metamorphism of banded iron formations, *EOS Trans.*, AGU, **69**, no. 16, pp. 515.

Ansara, I. ans Sundman, B. (1986) Scientific Group Thermodata Europe. Proceedings Conf. CODATA, Ottawa Canada

Ashida, T., Kume, S., Ito, E. and Navrotsky, A. (1988) $MgSiO_3$ ilmenite: heat capacity, thermal expansivity, and enthalpy of transformation, *Phys. Chem. Min.*, **16**, 239-245.

Barnes, H.L. and Ernst, W.G. (1963) Ideality and ionization in hydrothermal fluids: the system MgO-H_2O-$NaOH$, *Am. J. Sci.*, **261**, 129-150.

Barron, T.H.K. and Munn, R.W. (1968) Heat capacities of non-cubic solids, *Proc. Phys. Soc. Lond., Sect. C*, **2**, 1-10.

Barron, T.H.K., Berg, K.W.T. and Morrison, J.A. (1959) On the heat capacity of crystalline magnesium oxide, *Proc. R. Soc. London, Ser.A.*, **250**, 70-83.

Bass, J.D. and Weidner, D.J. (1984) Elasticity of single-crystal orthoferrosilite, *J. Geophys. Res.*, **89**, 4359-4371.

Berman, R.G. (1988) Internally consistent thermodynamic data for minerals in the system Na_2O-K_2O-CaO-MgO-FeO-Fe_2O_3-Al_2O_3-SiO_2-TiO_2-H_2O-CO_2, *J. Petrol.*, **29**, 445-552.

Berman, R.G. and Brown, T.H. (1984) A thermodynamic model for multicomponent melts, with application to the system CaO-Al_2O_3-SiO_2, *Geochim. Cosmochim. Acta*, **48**, 661-678.

Berman, R.G., Engi, M., Greenwood, H.J. and Brown, T.H. (1986) Derivation of internally-consistent thermodynamic data by the technique of mathematical programming: a review with application to the system MgO-SiO_2-H_2O, *J. Petrol.*, **27**,1331-1364.

Bertrand, G.L., Acree, W.E. and Burchfield, T. (1983) Thermochemical excess properties of multicomponent systems: representation and estimation from binary mixing data, *J.Sol.Chem.*, **12**, 327-340.

Bhattacharya, A. and Sen, S.K. (1985) Energetics of hydration of cordierite and water barometry in cordierite-granulites, *Contrib. Mineral. Petrol.*, **89**, 370-378.

Birch, F. (1966) Compressibility, elastic constants, in *Handbook of Physical Constants*, edited by S.P. Clark, *Geol. Soc. Am. Mem.*, **97**, 77-173.

Bjorkman, B. (1984) Quantitative equilibrium calculations on systems with relevance to copper melting and converting. *Ph.D. Thesis*, Univ. of Umea, Sweden.

Boettcher, A.L. (1970) The system CaO-Al_2O_3-SiO_2-H_2O at high pressures and temperatures., *J. Petrol.*, **11**, 337-379.

Bohlen, S.R. and Boettcher, A.L. (1981) Experimental investigations and geological applications of orthopyroxene geobarometry, *Am. Mineral.*, **66**, 951-964.

Bohlen, S.R. and Boettcher, A.L. (1982) The quartz=coesite transformation: a precise determination and the effects of other components, *J. Geophys. Res.*, **87**(B8), 7073-7078.

Bohlen, S.R. and Liotta, J.J. (1986) A barometer for garnet amphibolites and garnet granulites, *J. Petrol.*, **27**, 1025-1034.

Bohlen, S.R., Essene, E.J. and Boettcher, A.L. (1980) Reinvestigation and application of olivine-quartz-orthopyroxene barometry, *Earth Planet. Sci. Lett.*, **47**, 1-10.

Bohlen, S.R., Wall, V.J. and Boettcher, A.L. (1983a) Experimental investigation and application of garnet granulite equilibria, *Contrib. Mineral. Petrol.*, **83**, 52-61.

Bohlen, S.R., Wall, V.J. and Boettcher, A.L. (1983b) Experimental investigations and geological applications of equilibria in the system FeO-TiO_2-Al_2O_3-SiO_2-H_2O, *Am. Mineral.*, **68**, 1049-1058.

Bohlen, S.R., Boettcher, A.L., Wall, V.J. and Clemens, J.D. (1983c) Stability of phlogopite-quartz and sanidine-quartz: a model for melting in the lower crust, *Contrib. Mineral. Petrol.*, **83**, 273-277.

Bohlen, S.R., Dollase, W.A. and Wall, V.J. (1986) Calibration and applications of spinel equilibria in the system FeO-Al_2O_3-SiO_2, *J. Petrol.*, **27**, 1143-1156.

Bokreta, M. (1989) Lattice energy calculations. *Ph.D. Thesis*, Univ. Penn., Philadelphia.

Bonzcar, L.J., Graham, E.K. and Wang, H. (1977) The pressure and temperature dependence of the elastic constants of pyrope garnet, *J. Geoph. Res.*, **82**, 2529-2534.

Boyd, A.R. (1973) Appendix of mineral analyses, in *Lesotho Kimberlites*, edited by P.H. Nixon, pp. 33-36, Lesotho National Development Corporation, Maseru.

Boyd, F. R., England, J. L. and Davis, B. T. C. (1964) Effect of pressure on the melting and polymorphism of enstatite, $MgSiO_3$, *J. Geophys. Res.*, **69**, 2101-2109.

Brace, W.F., Scholz, C.H. and LaMori, P.N. (1969) Isothermal compressibility of kyanite, andalusite and sillimanite from synthetic aggregates, *J. Geophys. Res.*, **74**, 2089-2098.

Brousse, C., Newton, R.C. and Kleppa, O.J. (1984) Enthalpy of formation of forsterite, enstatite, akermanite, monticellite and merwinite at 1073 K determined by alkali borate solution calorimetry, *Geochim. Cosmochim. Acta*, **48**, 1081-1088.

Bundy, F,P. (1965) Pressure-temperature phase diagram of iron to 200 kbar, 900 °C, *J.Appl. Phys.*, **36**, 616-620.

Burnham, C.W., Holloway, J.R. and Davis, N.F. (1969) Thermodynamic properties of water to 1000 °C and 10,000 bars, *Geol. Soc. Am. Spec Pap.*, **132**, Washington D. C.

Carlson, W.D. and Lindsley, D.H. (1988) Thermochemistry of pyroxenes on the join $Mg_2Si_2O_6$-$CaMgSi_2O_6$, *Am. Mineral.*, **73**, 242-252.

Carmichael, R.S. (1982) *Handbook of Physical Properties of Rocks.* CRC Press, Boca Raton.

Chang, Z.P. and Barsch, G.R. (1973) Pressure dependence of single-crystal elastic constants and anharmonic properties of spinel, *J. Geophy. Res.*, **78**, 2418.

Charlu, T.V., Newton, R.C. and Kleppa, O.J. (1975) Enthalpies of formation in the system MgO-Al_2O_3-SiO_2 from high temperature solution calorimetry, *Geochim. Cosmochim. Acta*, **39**, 1487-1497.

Charlu, T.V., Newton, R.C. and Kleppa, O.J. (1978) Enthalpy of formation of lime silicates by high temperature calorimetry, with discussion of high pressure phase equilibrium, *Geochim. Cosmochim. Acta*, **42**, 367-375.

Charlu, T.V., Newton, R.C. and Kleppa O.J. (1981) Thermochemistry of synthetic $Ca_2Al_2SiO_7$ (gehlenite) -$Ca_2MgSi_2O_7$ (akermanite) melilites, *Geochim. Cosmochim. Acta*, **45**, 1609-1617.

Chatillon-Colinet, C., Kleppa, O.J., Newton, R.C. and Perkins D. (1983a) Enthalpy of formation of $Fe_3Al_2Si_3O_{12}$ (almandine) by high temperature alkali borate solution calorimetry, *Geochim. Cosmochim. Acta*, **47**, 439-444.

Chatillon-Colinet, C., Newton, R.C., Perkins, D. and Kleppa, O.J.(1983b) Thermochemistry of $(Fe^{2+},Mg^{2+})SiO_3$ orthopyroxene, *Geochim. Cosmochim. Acta* **47**, 1597-1603.

Chatterjee, N. (1987) Evaluation of thermochemical data on Fe-Mg olivine, orthopyroxene, spinel and Ca-Fe-Mg-Al garnet, *Geochim. Cosmochim. Acta*, **51**, 2515-2525.

Chatterjee, N. (1989) An internally consistent thermodynamic data base on minerals: applications to the earth's crust and upper mantle, *Ph.D. Thesis*, City Univ. of New York.

Chatterjee, N.D. (1970) Synthesis and upper stability of paragonite, *Contrib. Mineral. Petrol.*, **27**, 244-257.

Chatterjee, N.D. (1972) The upper stability limit of the assemblage paragonite + quartz and its natural occurrences, *Contrib. Mineral. Petrol.*, **34**, 288-303.

Chatterjee, N.D. and Flux, S. (1986) Thermodynamic mixing properties of muscovite-paragonite crystalline solutions at high temperatures and pressures, and their geological applications, *J. Petrol.*, **27**, 677-693.

Chatterjee, N.D. and Froese, E. (1975) A thermodynamic study of the pseudobinary join muscovite-paragonite in the system $KAlSi_3O_8$-$NaAlSi_3O_8$-Al_2O_3-SiO_2-H_2O, *Am. Mineral.*, **60**, 985-993.

Chatterjee, N.D. and Johannes, W. (1974) Thermal stability and standard thermodynamic properties of synthetic $2M_1$ muscovite, $KAl_2[AlSi_3O_{10}](OH)_2$, *Contrib. Mineral. Petrol.*, **48**, 89-114.

Chernosky, J.V. (1976) The stability field of anthophyllite - a reevaluation based on new experimental data, *Amer. Mineral.*, **61**, 1145-1155.

Chernosky, J.V. (1982) The stability of clinochrysotile, *Can. Mineral.*, **20**, 19-27.

Chernosky, J.V. and Autio, L.K. (1979) The stability of anthophyllite in the presence of quartz, *Am. Mineral.*, **64**, 294-303.

Chernosky, J.V., Day, H.W. and Caruso, L.J. (1985) Equilibria in the system MgO-SiO_2-H_2O: experimental determination of the stability of Mg-anthophyllite, *Am. Mineral.*, **70**, 223-236.

Chopelas, A. (1990a) Thermal properties of forsterite at mantle pressures derived from vibrational spectroscopy, *Phys. Chem. Min.*, **17**, 149-156.

Chopelas, A. (1990b) Thermal expansion, heat capacity, and entropy of MgO at mantle pressures, *Phys. Chem. Mineral.*, **17**, 142-148.

Clemens, J.D., Circone, S., Navrotsky, A.,McMillan, P.F., Smith, B.K. and Wall, V.J. (1987) Phlogopite: high temperature solution calorimetry, thermodynamic properties, Al-Si and stacking disorder, and phase equilibria, *Geochim. Cosmochim. Acta*, **51**, 2569-2578.

Cohen, L.H. and Klement, W. (1967) High-low quartz inversion: determination to 35 kilobars, *J. Geophys. Res.*, **72**, 4245-4251.

Czank, M. and Schultz, H. (1971) Thermal expansion of anorthite, *Naturwissenschaften*, **58**, 94.

Danckwerth, P.A. and Newton, R.C. (1978) Experimental determination of the spinel peridotite to garnet peridotite reaction in the system MgO-Al_2O_3-SiO_2 in the range 900-1100 ^oC and Al_2O_3 isopleths of enstatite in the spinel field, *Contrib. Mineral. Petrol.*, **66**, 189-201.

Davidson, P.M. and Lindsley, D.H. (1989) Thermodynamic analysis of pyroxene-olivine-quartz equilibria in the system CaO-MgO-FeO-SiO_2, *Am. Mineral.*, **74**, 18-30.

Davis, B. T. C., and England, J. L. (1964) The melting of forsterite up to 50 kbar, *J. Geophys. Res.*, **69**, 1113-1116.

Day, H.W. (1973) The high temperature stability of muscovite plus quartz, *Am. Mineral.*, **58**, 255-262.

Day, H.W. and Kumin, H.J. (1980) Thermodynamic analysis of the aluminum silicate triple point, *Am. J. Sci.*, **280**, 265-287.

Day, H.W., Chernosky, J.V. and Kumin, H.J. (1985) Equilibria in the system MgO-SiO_2-H_2O; a thermodynamic analysis, *Am. Mineral.*, **70**, 237-248.

Debye, P. (1912) Zur Theorie der spezifischen Warmen, *Ann. Phys.*, **39**, 789-839.

Dixon, J.R. (1980) A spinel lherzolite barometer, *Ph.D. Thesis*, Univ. of Texas, Dallas.

Einstein, A. (1907) Die Plancksche Theorie der Strahlung und die Theorie der spezifischen Warmen, *Ann. Phys.*, **22**, 180-190.

Engi, M. (1980) The solid solution behavior of olivine in the temperature range from 500 K - 1500 K, *Geol. Soc. Am. Abst. Prog.*, **12**, 421.

Engi, M. (1983) Equilibria involving Al-Cr spinel: Mg-Fe exchange with olivine. Experiments, thermodynamic analysis, and consequences for geothermometry, *Am. J. Sci.* **283A**, 29-71.

Eriksson, G. (1975) Quantitative equilibrium calculations in multiphase systems at high temperatures, with special reference to the roasting of chalocopyrite ($CuFeS_2$), *Akademisk Avhandling*, Umea Univesitet.

Eriksson, G. and Rosen, E. (1973) Thermodynamic studies of high temperature equilibria. VIII. General equations for the calculation of equilibria in multiphase systems, *Chemica Scripta*, **4**, 193-194.

Eugster, H.P. and Wones, D.R. (1962) Stability relations of the ferruginous biotite, annite, *J. Petrol.*, **3**, 82-125.

Eugster, H.P., Albee, A.L., Bence, A.E., Thompson, J.B., Jr. Waldbaum, D.R. (1972) The two-phase region and excess mixing properties of paragonite-muscovite crystalline solutions, *J. Petrol.*, **13**, 147-179.

Evans, B.W. (1965) Application of a reaction-rate method to the breakdown equilibria of muscovite and muscovite plus quartz, *Am. J. Sci.*, **263**, 647-667.

Evans, B.W. (1977) Metamorphism of alpine peridotite and serpentinite, *Annu. Rev. Earth Planet. Sci.*, **5**, 397-447.

Evans, B.W., Johannes, W., Oterdoom, W.H. and Trommsdorff, V. (1976) Stability of chrysotile and antigorite in the serpentinite multisystem. Schweiz, *Mineral. Petrograph. Mitt.*, **56**, 79-93.

Fawcett, J.J. and Yoder, H.S. (1966) Phase relations of chlorites in the system $MgO-Al_2O_3-SiO_2-H_2O$ at 2 kbar water pressure, *Am. Mineral.*, **51**, 353-380.

Fei, Y. (1990) Thermochemical-thermophysical data on phases in the Mg-Fe-Si-O system: a synthesis of theory and experimental data and computation of phase equilibrium, *Ph.D. Thesis*, City Univ. of New York.

Fei, Y. and Chatterjee, N. (1988) Optimized thermodynamic data on hydrous phases in the system $MgO-SiO_2-H_2O$, *EOS Trans.*, AGU, **69**, p. 473.

Fei, Y. and Saxena, S.K. (1986) A thermochemical data base for phase equilibria in the system Fe-Mg-Si-O at high pressure and temperature, *Phys. Chem. Minerals*, **13**, 311-324.

Fei, Y. and Saxena, S.K. (1987) An equation for the heat capacity of solids, *Geochim. Cosmochim. Acta*, **51**, 251-254.

Fei, Y., Saxena, S. K. and Ericksson, G. (1986) Some binary and ternary silicate solution models, *Contrib. Mineral. Petrol.*, **94**, 221-229.

Fei, Y., Saxena, S.K. and Navrotsky, A. (1990) Internally consistent thermodynamic data and equilibrium phase relations for compounds in the system $MgO-SiO_2$ at high pressure and high temperature, *J. Geophys. Res.*, **95**, 6915-6928.

Fenner, C.L. (1913) The stability relations of the silica minerals, *Am. J. Sci.*, **36**, 331-384.

Ferry, J.M. and Spear, F.S. (1978) Experimental calibration of the partitioning of Fe and Mg between biotite and garnet, *Contrib. Mineral. Petrol.*, **66**, 113-117.

Finger, L. W. and Hazen, R. M. (1980) Crystal structure and isothermal compression of Fe_2O_3, Cr_2O_3, and V_2O_3 to 50 kbars, *J. Appl. Phys.*, **51**, 5362-5366.

Finger, L. W., Hazen, R. M. and Hofmeister, A. M. (1986) High-pressure crystal chemistry of spinel ($MgAl_2O_4$) and magnetite (Fe_3O_4): comparisons with silicate spinels, *Phys. Chem. Minerals*, **13**, 215-220.

Fisher, J.R. and Zen, E-an (1971) Thermodynamic calculations from hydrothermal phase equilibrium data and the free energy of H_2O, *Am. J. Sci.*, **270**, 297-314.

Fletcher, R. (1970) A new approach to variable matric algorithms, *Comput. J.*, **13**, 317-322.

Flory, P. J.(1953) Molecular configuration of polyelectrolytes, *J. Chem. Phys.*, **21**, 162-163.

Flux, S. and Chatterjee, N.D. (1986) Experimental reversal of the Na-K exchange reaction between muscovite-paragonite crystalline solutions and a 2 molal aqueous (Na,K)Cl fluid, *J. Petrol.*, **27**, 665-676.

Fonarev, V.I., Korol'kov, G.Ya., and Dokina, T.N. (1979) Laboratory data on the stability field of the cummingtonite+olivine+quartz association, *Geokhimiya*, **7**, 984-995.

Forbes, W.C. (1977) Stability relations of grunerite, $Fe_7Si_8O_{22}(OH)_2$. *Am. J. Sci.*, **277**, 735-749.

Franz, G. (1982) The brucite-periclase equilibrium at reduced H_2O activities: some information about the system H_2O-NaCl, *Am. J. Sci.*, **282**, 1325-1339.

Fritz, I.J. (1974) Pressure and temperature dependences of the elastic properties of rutile (TiO_2), *J. Phys. Chem. Solids*, **35**, 817.

Fujii, T. (1977) Pyrixene equilibria in spinel lherzilite, *Carnegie Inst. Wash. Yearb.*, **76**, 569-572.

Fukizawa, T. (1982) Direct determination of phase equilibria of geophysically important matearials under high pressure and high temperature usinf in situ x-ray difffraction method, *Ph.D. Thesis*, 125 pp., University of Tokyo.

Furnish, M. D., and Bassett, W. A. (1983) Investigation of the mechanism of the olivine-spinel transition in fayalite by synchrotron radiation, *J. Geophys. Res.*, **88**, 10333-10341.

Fyfe, W.S. (1962) On the relative stabilities of talc, anthophyllite and enstatite, *Am. J. Sci.*, **270**, 151-154.

Ganguly, J. and Ghose, S. (1979) Aluminous orthopyroxene, order-disorder, thermodynamic properties and petrological implications, *Contrib. Mineral. Petrol.*, **68**, 375-385.

Ganguly, J. and Saxena, S.K. (1984) Mixing properties of aluminosilicate garnets: constraints from natural and experimental data and application to geothermo-barometry, *Amer. Mineral.*, **69**,88-97.

Ganguly, J. and Saxena, S.K. (1988) *Mixtures and Mineral Reactions*, Springer-Verlag, New York, 291 pp.

Gasparik, T. (1981) Thermodynamic properties of pyroxenes in the NCMAS system saturated with silica, *Ph.D. Thesis,* State Univ. New York, Stony Brook.

Gasparik, T. (1984a) Two-pyroxene thermobarometry with new experimental data in the system CaO-MgO-Al_2O_3-SiO_2, *Contrib. Mineral. Petrol.*, **87**, 87-97.

Gasparik, T. (1984b) Experimental study of subsolidus phase relations and mineral properties of pyroxene in the system CaO-Al_2O_3-SiO_2, *Geochim. Cosmochim. Acta,* **48**, 2537-2545.

Gasparik, T. (1984c) Experimentally determined stability of clinopyroxene+garnet+corundum in the system CaO-MgO-Al_2O_3-SiO_2, *Am. Mineral.*, **69**, 1025-1035.

Gasparik, T. (1985) Experimentally determined compositions of diopside-jadeite pyroxene in equilibrium with albite and quartz at 1200-1350 ^oC and 15-34 kbar, *Geochim. Cosmochim. Acta,* **49**, 865-870.

Gasparik, T. (1990) Phase relation in the transition zone, *J. Geophys. Res.*, **95**, 15751-15769.

Gasparik, T. and Newton, R.C. (1984) The reversed alumina contents of orthopyroxene in equilbrium with spinel and forsterite in the system MgO-Al_2O_3-SiO_2, *Contr. Mineral. Petrol.*, **85**, 186-196.

Gillet, Ph., Reynard, B. and Tequi, C. (1989) Thermodynamic properties of Glaucophane new data from calorimetric and spectroscopic measurements, *Phys, Chem. Mineral.*, **16**, 659-667.

Gillet, Ph., Richet, P., Guyot F. and Fiquet, G. (1991) High-temperature thermodynamic properties of forsterite, *J. Geophys. Res.*, **96**, 11805-11816.

Goel, R.P., Kellogg, H.H. and Larrain, J. (1980) Mathematical description of the thermodynamic properties of the system iron-oxygen and iron-oxygensilica, *Metall. Trans. B*, **11**, 107-117.

Goldsmith, J.R. (1980) The melting and breakdown reactions of anorthite at high pressures and temperatures, *Am. Mineral.*, **65**, 272-284.

Gordon, T.M. (1973) Determination of internally consistent thermochemical data from phase equilibrium experiments, *J. Geol.*, **81**, 199-208.

Greenwood, H.J. (1963) The synthesis and stability of anthophyllite, *J. Petrol.*, **4**, 317-351.

Grimsditch, M.H. and Ramdas, A.K. (1976) Elastic and elasto-optic constants of rutile from a Brillouin scattering study, *Phys. Rev.*, **14**, 1670.

Grønvold, F., Stølen, S. and Svendsen, S.R. (1989) Heat capacity of α quartz from 298.15 to 847.3 K, and of β quartz from 847.3 to 1000 K – transition behaviour and reevaluation of the thermodynamic propertities, *Thermochimica Acta*, **139**, 225-243.

Guggenheim, E. A.(1952) *Mixtures*, Clarendon Press, Oxford, 365 pp.

Guillermet, A.F. and Gustafson, P. (1985) An assessment of the thermodynamic properties and the (p.T) phase diagram of iron, *High temperatures-high pressure*, **16**, 591-619.

Gustafson, P. (1986) An evaluation of the thermodynamic propertites and the P, T phase diagram of carbon, *Carbon*, 169-176.

Haar, C., Gallagher, J.S. and Kell, G.S. (1984) NBS/NRC steam tables. Thermodynamic and transport properties and computer programs for vapor and liquid states of water in SI units. Hemisphere Publishing Co., Washington.

Haas, J.L. (1988) Recommended standard electrochemical potentials and fugacities of oxygen for solid buffers and thermodynamic data in the system iron-silicon-oxygen, nickel-oxygen, and copper-oxygen, Preliminary report of January 17, 1988 to the CODATA Task Group on Chemical Thermodynamic Tables, U.S. Geological Survey, Reston, Virginia.

Haas, J.L. and Fisher, J.R. (1976) Simultaneous evaluation and correlation of thermochemical data, *Amer. J. Sci.*, **276**, 525-545.

Haas, J.L., Robinson, G.R.,Jr. and Hemingway, B.S. (1981) Thermodynamic tabulations for selected phases in the system $CaO-Al_2O_3-SiO_2-H_2O$ at 101.325 KPa (1 atm) between 273.15 and 1800 K, *J. Phys. Chem. Ref. Data*, **10**, 575-669.

Hackler, R.T. and Wood, B.J. (1988) Experimental determination of Fe, Mg, Mn exchange between garnet and olivine, *EOS Trans.*, AGU, **69**, no. 16, pp. 512.

Halbach, H. and Chatterjee, N.D. (1982) The use of linear parametric programming for determination of internally consistent thermodynamic data for minerals, in *High-Pressure Researches in Geoscience*, edited by : W. Schreyer, pp. 475-491, E. Schweizerbartsche Verlagsbuchhandlung, Stuttgart.

Halbach, H. and Chatterjee, N.D. (1984) An internally consistent set of thermodynamic data for twenty-one $CaO-Al_2O_3-SiO_2-H_2O$ phases by linear programming, *Contrib. Mineral. Petrol.*, **88**, 14-23.

Halleck, P.M. (1973) The compression and compressibility of grossular garnet, a comparison of X-ray and ultrasonic methods, *Ph.D. thesis*, Univ. of Chicago, Ill.

Hankey, R.E. and Schuele, D.E. (1970) Third order elastic constants of Al_2O_3, *J. Acoust. Soc. Am.*, **48**, 190-202.

Hariya, Y. and Kennedy, G.C. (1968) Equilibrium study of anorthite under high pressure and high temperature, *Am. J. Sci.*, **266**, 193-203.

Harley S.L. (1984) An experimental study of the partitioning of Fe and Mg between garnet and orthopyroxene, *Contrib. Mineral. Petrol.*, **86**, 359-373.

Haselton, H.T. (1979) Calorimetry of simple synthetic pyrope-grossular garnets and calculated stability relations. *Ph.D. Thesis*, Univ. of Chicago, Ill., 98 pp.

Haselton, H.T. and Westrum, E.F. (1980) Low-temperature heat capacities of synthetic pyrope, grossular and pyrope$_{60}$grossular$_{40}$, *Geochim. Cosmochim. Acta*, **44**, 701-710.

Haselton, H.T., Hemingway, B.S. and Robie, R.A. (1984) Low temperature heat capacities of $CaAl_2SiO_6$ pyroxene, *Am. Mineral.*, **69**, 481-489.

Haselton, H.T., Robie, R.A. and Hemingway, B.S. (1987) Heat capacities of synthetic hedenbergite, ferrobustamite, and $CaFeSi_2O_6$ glass, *Geochim. Cosmochim. Acta*, **51**, 2211-2218.

Hays, J.F. (1966) Lime-alumina-silica, *Carnegie Inst. Wash. Yearb.*, **65**, 234-239.

Hazen, R. M. and Jeanloz, R. (1984) Wustite ($Fe_{1-x}O$): a review of its defect structure and physical properties, *Rev. Geophys. Space Phys.*, **22**, 37-46.

Heinz, D.L. and Jeanloz, R. (1984) The equation of state of the gold calibration standard, *J. Appl. Phys.*, **55**, 885-893.

Heinz, D.L., Jeanloz, R. and O'Connell, R.J. (1982) Bulk attenuation in a polycrystalline earth, *J. Geophys. Res.*, **87**, 7772-7778.

Helgeson, H.C., Delany, J.M., Nesbitt, H.W. and Bird, D.K. (1978) Summary and critique of the thermodynamic properties of rock forming minerals, *Amer. J. Sci.*, **278 A**, 229.

Hemingway, B.S. and Robie, R.A. (1984) Heat capacity and thermodynamic functions for gehlenite and staurolite: with comments on the Schottky anomaly in the heat capacity of staurolite, *Am. Mineral.*, **69**, 307-318.

Hemingway, B.S., Krupka, K.M. and Robie, R.A. (1981) Heat capacities of the alkali feldspars between 350 and 1000 K from differential scanning calorimetry, the thermodynamic functions of the alkali feldspars from 298.15 to 1400 K, and the reaction quartz + jadeite = analbite, *Am. Mineral.*, **66**, 1202-1215.

Hemingway, B.S., Evans, H.T., Nord, G.L.Jr., Haselton, H.T., Robie, R.A. and McGee, J.J. (1986) Akermanite: phase transitions in heat capacity and thermal expansion, and revised thermodynamic data, *Can. Mineral.*, **24**, 425-434.

Hemley, J.J., Montoya, J.W., Christ, C.L. and Hostetler, P.B. (1977a) Equilibria in the $MgO-SiO_2-H_2O$ system:I. Talc-chrysotile-forsterite-brucite stability relations, *Am. J. Sci.*, **277**, 322-351.

Hemley, J.J., Shaw, D.R. and Luce, R.W. (1977b) Mineral equilibria in the $MgO-SiO_2-H_2O$ system:II. Talc-antigorite-forsterite-anthophyllite-enstatite stability relations and some geologic implications in the system, *Am. J. Sci.*, **277**, 353-383.

Hensen, B.J. (1972) Phase relations involving pyrope, enstatite and sapphirine in the system $MgO-Al_2O_3-SiO_2$, *Carnegie Instn. Wash.*,Yearb. , **71**, 421-427.

Hensen, B.J. and Green, D.H. (1973) Experimental study of the stability of cordierite and garnet in pelitic compositions at high pressures and temperatures. III. Synthesis of experimental data and geological applications, *Contrib. Mineral. Petrol.*, **38**, 151-166.

Hentschel, B. (1970) Stoichiometric FeO as metastable intermediate of the decomposition of wustite at 225 ^{o}C, *Z. Naturforsch.*, **A25**, 1996-1997.

Herzberg, C.T. (1976) The plagioclase spinel-lherzolite facies boundary; its bearing on corona structure formation and tectonics of the Norwegian caledonides, in *Progress in Experimental Petrology*, D 3233-3235, NERC Publications, London.

Herzberg, C.T. (1983) The reaction forsterite+cordierite = aluminous orthopyroxene+spinel in the system $MgO-Al_2O_3-SiO_2$, *Contr. Mineral. Petrol.*, **84**, 84-90.

Hewitt, D.A. and Wones, D.R. (1984) Experimental phase relations of the micas, in *Reviews in Mineralogy*, **13**, edited by S.W. Bailey, pp. 201-256, Mineralogical Society of America, Washington D. C.

Hofmeister, A.M. (1987) Single-crystal absorption and reflection infrared spectroscopy of forsterite and fayalite, *Phys. Chem. Mineral.*, **14**, 499-513.

Hofmeister, A.M. and Chopelas, A. (1991) Thermodynamic properties of pyrope and grossular from vibrational spectroscopy, *Am. Mineral.*, **76**, 880-891.

Holdaway, M.J. (1971) Stability of andalusite and the aluminosilicate phase diagram, *Am. J. Sci.*, **271**, 97-131.

Holdaway, M.J. and Lee, S.M. (1977) Fe-Mg cordierite stability in high-grade pelitic rocks based on experimental, theoretical, and natural observations, *Contrib. Mineral. Petrol.*, **63**, 175-198.

Holland, T.J.B. (1979) Experimental determination of the reaction paragonite = jadeite + kyanite + H_2O, and internally consistent thermodynamic data for part of the system Na_2O-Al_2O_3-SiO_2-H_2O, with applications to eclogites and blueschists, *Contrib. Mineral. Petrol.*, **68**, 293-301.

Holland, T.J.B. (1980) The reaction albite = jadeite + quartz determined experimentally in the range 600-1200°C, *Am. Mineral.*, **65**, 129-134.

Holland, T.J.B. (1981) Thermodynamic analysis of simple mineral systems, in *Advances in Physical Geochemistry I*, edited by R.C. Newton, A. Navrotsky and B.J. Wood, pp 19-34, Springer-Verlag, New York.

Holland, T. and Powell, R. (1990) An enlarged and updated internally consistent thermodynamic dataset with uncertainties and correlations: the system K_2O-Na_2O-CaO-MgO-MnO-FeO-Fe_2O_3-Al_2O_3-TiO_2-SiO_2-C-H_2-O_2. *J. Metam. Geol.*, **8**, 89-124.

Holland, T.J.B., Navrotsky, A. and Newton, R.C. (1979) Thermodynamic parameters of $CaMgSi_2O_6$-$Mg_2Si_2O_6$ pyroxenes, based on regular solutions and cooperative disordering models, *Contrib. Mineral. Petrol.*, **75**, 305-306.

Huang, E. and Bassett, W. A. (1986) Rapid determination of Fe_3O_4 phase diagram by synchrotron radiation, *J. Geophys. Res.*, **91**, 4697-4703.

Huang, W.L. and Wyllie, P.J. (1974) Melting relations of muscovite with quartz and sanidine in the K_2O-Al_2O_3-SiO_2-H_2O system to 30 kilobars and an outline of paragonite melting relations, *Am. J. Sci.*, **274**, 378-395.

Huang, E., Bassett, W.A. and Tao, P. (1987) Study of bcc-hcp iron transition by syncrotron radiation, in *High Pressure Research in Mineral Physics*, edited by M.H. Manghnani and Y. Syono, pp. 165-172, Terra Publications, Tokyo, Japan.

Huckenholz, H.G., Holzel, E. and Lindhuber, W. (1975) Grossularite, its solidus and liquidus relations in the CaO-Al_2O_3-SiO_2-H_2O system upto 10 kbar, *Neueslahr. Jb. Mineral. Abh.*, **124**, 1-46.

Inoue, K.(1975) Development of high temperature and high pressure X-ray diffraction apparatus with energy dispersive technique and its geophysical applications, *Ph.D. Thesis*, Tokyo Univ., Tokyo.

Irving, A.J., Huang, W.L. and Wyllie, J. (1977) Phase relation of portlandite, $Ca(OH)_2$ and brucite, $Mg(OH)_2$ to 33 kilobars, *Am. J. Sci.*, **277**, 313-321.

Isaak, D.G., Anderson, O.L. and Goto, T. (1989) Elasticity of single-crystal forsterite measured to 1700 K, *J. Geophys. Res.*, **94**, 5895-5906.

Isaak, D.G., Cohen, R.E. and Mehl, M.J. (1990) Calculated elastic and thermal properties of MgO at high pressures and temperatures, *J. Geophys. Res.*, **95**, 7055-7067.

Ito, E., (1984) Ultra-high pressure phase relations of the system MgO-FeO-SiO_2 and their geophysical implications, in *Materials Science of the Earth's Interior*, edited by I. Sunagawa, pp.387-394, Terra Publications, Tokyo, Japan.

Ito, E. and Matsui, Y. (1977) Silicate ilmenite and post-spinel transformations, in *High-Pressure Research: Application to Geophysics*, edited by M. H. Manghnani and S. Akimoto, pp. 193-208, Academic Press, New York.

Ito, E. and Navrotsky, A. (1985) $MgSiO_3$ ilmenite: calorimetry, phase equilibria, and decomposition at room pressure, *Amer. Mineral.*, **70**, 1020-1026.

Ito, E. and Takahashi, E. (1989) Post-spinel transformations in the system Mg_2SiO_4-Fe_2SiO_4 and some geophysical implications, *J. Geophys. Res.*, **94**, 10637-10646.

Ito, E. and Yamada, H. (1982) Stability relations of silicate spinel ilmenites and perovskites, in *High-Pressure Research in Geophysics*. edited by S. Akimoto and M. H. Manghnani, pp. 405-419, Center for Academic Publications., Tokyo, Japan.

Ito, H., Kawada, K. and Akimoto, S. (1974) Thermal expansion of stishovite, *Phys. Earth Planet. Int.*, **8**, 277-281.

Ito, E., Akaogi, M., Topor, L. and Navrotsky, A. (1990) Negtive P-T slopes for reactions forming $MgSiO_3$ perovskite in the mantle confirmed by calorimetry, *Science*, **249**, 1275-1278.

Ivanov, I.P. and Gusynin, V.F. (1970) Stability of paragonite in the system SiO_2-$NaAlSi_3O_8$-Al_2O_3-H_2O, *Geochem. Intern.*, **7**, 578-587, Trans. from *Geokhimiya*, **7**, 801-811.

Ivanov I.P., Potekhin, V.Y., Dmitriyenko, L.T. and Beloborodov, S.M. (1973) An experimental study of T and P conditions of equilibrium of reaction: muscovite - K-feldspar + corundum + H_2O at $P(H_2O) < P(total)$, Trans. from *Geokhimiya*, **9**, 1300-1310.

Jackson, I. and Niesler, H. (1982) The elasticity of periclase to 3 GPa and some geophysical implications, in *High-Pressure Research in Geophysics*, edited by S. Akimoto and M. H. Manghnani, pp. 93-113, Center for Academic Publications, Tokyo, Japan.

James, F. and Roots, M. (1975) MINUIT, *Comput. Phys. Commun.*, **10**, 343-367.

Jamieson, H.E. and Roeder, P.L. (1984) The distribution of Mg and Fe^{2+} between olivine and spinel at 1300 ^oC, *Am. Mineral.*, **69**, 283-291.

Jeanloz, R. and Hazen, R. M. (1983) Compression, nonstoichiometry and bulk viscosity of wustite, *Nature*, **304**, 620-622.

Jeanloz, R. and Knittle, E. (1986) Reduction of mantle and core properties to a standard state by adiabatic decompression, in *Chemistry and Physics of Terrestrial Planets*, edited by S.K. Saxena, pp. 275-309, Springer-Verlag, New York.

Jenkins, D.M. and Newton, R.C. (1979) Experimental determination of the spinel peridotite to garnet peridotite inversion at 900 ^oC and 1000 ^oC in the system CaO-MgO-Al_2O_3-SiO_2 and at 900 ^oC with natural garnet and olivine, *Contrib. Mineral. Petrol.*, **68**, 407-419.

Johannes, W. (1968) Experimental investigation of the reaction forsterite + H_2O = serpentine + brucite, *Contrib. Mineral. Petrol.*, **19**, 309-315.

Kajiyoshi, K. (1986) High-temperature equation of state for mantle minerals and their anharmonic properties, *M.S. Thesis*, Okayama Univ., Japan.

Kalinin, D.V. and Zubkov, M.Y. (1981) Kinetic investigation of the MgO-SiO_2-H_2O system, reaction: serpentine = forsterite + talc + water, *Sov. Geol. Geophys.*, **22**, 61-68.

Kanzaki, M. (1991) Ortho/clinoenstatite transition, *Phys. Chem. Min.*, **17**, 726-730.

Kato, T. and Kumazawa, M. (1983) Melting curves of Mg_2SiO_4 and $MgSiO_3$ up to 9 GPa, in Program and Abstract, the High Pressure Conference of Japan, pp. 198-199.

Kato, T. and Kumazawa, M. (1985) Garnet phase of $MgSiO_3$ filling the pyroxene-ilmenite gap at very high temperatures, *Nature*, **316**, 803-804.

Kato, T. and Kumazawa, M. (1986) Melting and phase relations in the system Mg_2SiO_4-$MgSiO_3$ at 20 GPa under hydrous conditions, *J. Geophys. Res.*, **91**, 9351-9355.

Katsura, T. and Ito, E. (1989) The system Mg_2SiO_4-Fe_2SiO_4 at high pressures and temperatures: precise determination of stabilities of olivine, modified spinel and spinel, *J. Geophy. Res.*, **94**, 15663-15670.

Kawada, K. (1977) System Mg_2SiO_4-Fe_2SiO_4 at high pressures and temperatures and the earth's interior, *Ph.D. Thesis*, Univ. of Tokyo, Tokyo.

Kawasaki, T. (1979) Thermodynamic analyses on the Fe-Mg exchange equilibrium between olivine and garnet: an application to the estimation of P-T relations of ultramafic rocks, *J. Jpn. Assoc. Mineral. Petrol. Econ. Geol.*, **74**, 395-405.

Kawasaki, T. and Matsui, Y. (1977) Partioning of Fe^{+2} and Mg^{+2} between olivine and garnet, *Earth Planet. Sci. Lett.*, **37**, 159-166.

Kawasaki, T. and Matsui, Y. (1983) Thermodynamic analysis of equilibria involving olivine, orthopyroxene and garnet, *Geochim. Cosmochim. Acta*, **47**, 1661-1680.

Kerrick, D.M. (1972) Experimental determination of muscovite + quartz stability with P(H2O) < P(total), *Am. J. Sci.*, **272**, 946-958.

Kerrick, D.M. and Ghent, E.D. (1979) P-T-XCO_2 relations of equilibria in the system: CaO-Al_2O_3-SiO_2-CO_2-H_2O. in *Problems in Physico-Chemical Petrology*, **2**, edited by V.A. Zharikov, V.I. Fonarev, V.A. Korikovshii, pp.32-52, Academy of Sciences of the USSR, Moscow.

Kerrick, D.M. and Henninger, S. G. (1984) The andalusite-sillimanite equilibrium revisited, *Geol. Soc. Am. Abstr. Prog.*, **16**, 558.

Kieffer, S.W. (1979a) Thermodynamics and lattice vibrations of minerals: 1. Mineral heat capacities and their relationships to simple lattice vibrational models, *Rev. Geophys. Space Phys.*, **17**, 1-19.

Kieffer, S.W. (1979b) Thermodynamics and lattice vibrations of minerals: 2. Vibrational characteristics of silicates, *Rev. Geophys. Space Phys.*, **17**, 20-34.

Kieffer, S.W. (1979c) Thermodynamics and lattice vibrations of minerals: 3. Lattice dynamics and an approximation for minerals with application to simple substances and framework silicates, *Rev. Geophys. Space Phys.*, **17**, 35-59.

Kieffer, S.W. (1980) Thermodynamics and lattice vibrations of minerals: 4. Application to chain and sheet silicates and orthosilicates, *Rev. Geophys. Space Phys.*, **18**, 862-886.

King, E.G., Barany, R., Weller, W.W. and Pankratz, L.B. (1967) Thermodynamic properties of forsterite and serpentine, *U.S. Bur. Mines Rep. Invest.*, **6962**, 19 pp.

King, E.G., Ferrante, M.J. and Pankratz, L.B. (1975) Thermodynamic data for $Mg(OH)_2$ (brucite), *U.S. Bur. Mines Rep. Invest.*, **8041**, 13 pp.

Kiseleva, I.A., Ogorodova, L.P., Topor, N,D, and Chigareva, O.G. (1979) A thermochemical study of the CaO-MgO-SiO_2 system, *Geokhimiya*, **12**, 1811-1825 (in Russian).

Kitihara, S., Takenouchi, S. and Kennedy, G.C. (1966) Phase relations in the system MgO-SiO_2-H2O at high temperatures and pressures, *Am. J. Sci.*, **264**, 223-233.

Klein, C. and Waldbaum, D.R. (1967) X-ray crystallographic properties of the cummingtonite-grunerite series, *J. Geol.*, **75**, 379-392.

Knittle, E., Jeanloz, R. and Smith, G.L. (1986) Thermal expansion of silicate perovskite and stratification of the earth's mantle, *Nature*, **319**, 214-216.

Koziol, A.M. and Newton, R.C. (1986) Definition of anorthite = grossular + kyanite + quartz in the range 650-1250 C, *Geol. Soc. Am. Abstr. Prog.*, **99**, 661.

Koziol, A.M. and Newton, R.C. (1988a) Redetermination of the anorthite breakdown reaction and improvement of the plagioclase-garnet-Al_2SiO_5-quartz geobarometer, *Am. Mineral.*, **73**, 216-223.

Koziol, A.M. and Newton, R.C. (1988b) Experimental determination of almandine-grossular and (Ca, Fe, Mg) ternary garnet solid solution relations, *EOS Trans.*, AGU, **69**, no. 16, pp. 498.

Krupka, K.M., Robie, R.A. and Hemingway, B.S. (1979) High-temperature heat capacities of corundum, periclase, anorthite, $CaAl_2Si_2O_8$ glass, muscovite, pyrophyllite, $KAlSi_3O_8$ glass, grossular and $NaAlSi_3O_8$ glass, *Am. Mineral.*, **64**, 86-101.

Krupka, K.M., Hemingway, B.S., Robie, R.A. and Kerrick, D.M. (1985a) Low temperature heat capacities and derived thermodynamic properties of anthophyllite, diopside, enstatite, bronzite and wollastonite, *Am. Mineral.*, **70**, 249-260.

Krupka, K.M., Hemingway, B.S., Robie, R.A. and Kerrick, D.M. (1985b) High temperature heat capacities and derived thermodynamic properties of anthophyllite, diopside, dolomite, enstatite, bronzite, talc, tremolite and wollastonite, *Am. Mineral.*, **70**, 261-271.

Kubaschewski, O. (1982) Iron-Binary Phase Diagrams, 185 pp., Springer-Verlag, New York.

Kudoh, Y., Ito, E. and Takeda, H. (1987) Effect of pressure on the crystal structure of perovskite-type $MgSiO_3$, *Phys. Chem. Minerals*, **14**, 350-354.

Kunze, G. (1961) Antigorit: Strukturtheoretische Grundlagen und ihre praktische Bedeutung fur die weitere Serpentin-Forschung, *Fortschr. Mineral.*, **39**, 206-324.

Kushiro, I. and Yoder, H.S. (1966) Anorthite-forsterite and anorthite-enstatite reactions and their bearing on the basalt-eclogite transformation, *J. Petrol.*, **7**, 337-362.

Kuskov, O. L. and Fabrichnaya, O. B. (1987) The SiO_2 polymorphs: the equations of state and thermodynamic properties of phase transformations, *Phys. Chem. Minerals*, **14**, 58-66.

Kuskov, O.L., Fabrichnaya, O.B., Galimzyanov, R.F. and Truskinovsky, L.M. (1989) Computer simulation of the phase diagram for the $MgO-SiO_2$ system at P-T parameters of the mantle transition zone, *Phys. Chem. Min.*, **16**, 442-454.

Lane, D. and Ganguly, J. (1980) Al_2O_3 solubility in orthopyroxene in the system $MgO-Al_2O_3-SiO_2$. A reevaluation of the mantle geotherm, *J. Geophys. Res.*, **85**, 6963-6972.

Lee, H.Y. and Ganguly, J. (1988) Fe^{2+}-Mg fractionation between garnet and orthopyroxene: experimental determination in the FMAS system and application to geothermometry, *J. Petrol.*, **29**, 93-113.

Leitner, B.J., Weidner, D.J. and Liebermann, R.C. (1980) Elasticity of single crystal pyrope and implications for garnet solid solution series, *Phys. Earth Planet. Int.*, **22**, 111.

Levien, L. and Prewitt, C.T. (1981) High pressure crystal structure and compressibility of coesite, *Am. Mineral.*, **66**, 324-333.

Liebermann, R.C. (1974) Compressional velocities of polycrystalline olivine, spinel and rutile minerals, *Earth Planet. Sci. Lett.*, **17**, 263-268.

Liebermann, R.C. and Ringwood, A.E. (1976) Elastic properties of anorthite and the nature of the lunar crust, *Earth Planet. Sci. Lett.*, **31**, 69-74.

Liebermann, R.C., Ringwood, A.E. and Major, A. (1976) Elasticity of polycrystalline stishovite, *Earth Planet. Sci. Lett.*, **32**, 127.

Liebermann, R.C. and Schreiber, E. (1968) Elastic constants of polycrystalline hematite as a function of pressure to 3 kbar, *J. Geophys. Res.*, **73**, 65-86.

Lindsley, D.H. (1983) Pyroxene thermometry, *Am. Mineral.*, **68**, 477-493.

Lindsley, D.H., Davis, B. T. C. and MacGregor, I. D. (1964) Ferrosilite ($FeSiO_3$): synthesis at high pressures and temperatures, *Science*, **144**, 73-74.

Lindsley, D.H., Grover, J.E. and Davidson, P.M. (1981) The thermodynamics of the $Mg_2Si_2O_6$-$CaMgSi_2O_6$ join: a review and an improved model. in *Advances in Physical Geochemistry*, **1**, edited by R.C. Newton, A. Navrotsky, and B.J. Wood, pp. 149-175, Springer-Verlag, New York.

Liu, L. G. (1977) The system enstatite-pyrope at high pressures and temperatures and the earth's mantle, *Earth Planet. Sci. Lett.*, **36**, 237-245.

Liu, L. G. (1980) The equilibrium boundary of spinel = corundum + periclase: a calibration curve for pressure above 100 kbar, *High Temperature-High Pressures*, **12**, 217-220.

Manghnani, M.H. and Matsui, T. (1981) Temperature dependence of pressure derivatives of single-crystal elastic constants of pure forsterite (abstract), *ISPEI Symposium on Properties of Materials at High Pressures and High Temperatures*, Toronto, Canada.

Manghnani, M.H., Ming, L.C., Balogh, J., Qadri, S.B., Skelton, E.F. and Schiferl, D. (1985) Equation of state and phase transition studies under in situ high P-T conditions using synchrotron radiation, in*Solid State Physics-Recent Advances with Anvil Devices*, edited by S. Minomura, pp. 343-350, Scientific Publisher, Tokyo.

Manghnani, M.H., Ming, L.C. and Nakagiri, N. (1987) Investigation of the α-Fe--ε-Fe phase transition by synchrotron radiation, in *High Pressure Research in Mineral Physics*, edited by M.H. Manghnani and Y. Syono, pp. 155-163, Terra Publications, Tokyo, Japan.

Mao, H. K. and Bell, P. M. (1977) Disproportionation equilibrium in iron-bearing systems at pressures above 100 kbar with applications to chemistry of the earth's mantle, in *Energetics of Geological Processes*, edited by S.K. Saxena and S. Bhattacharji, pp. 236-249, Springer-Verlag, New York.

Mao, H. K., Takahashi, T., Bassett, W.A., Kinsland, G. L. and Merrill, L. (1974) Isothermal compression of magnetite to 320 kbar and pressure-induced phase transformation, *J. Geophys. Res.*, **79**, 1165-1170.

Mao, H.K., Bell, P.M. and Hadidiacos C. (1987) Experimantal phase relations of iron to 360 kbar, 1400 °C, determined in an internally heated diamond-anvil apparatus, *High Pressure Research in Mineral Physics*, edited by M.H. Manghnani and Y. Syono, 135-138, Terra Scientific Publishing Company, Tokyo.

Mao, H.K., Hemely, R.J.,Fei, Y., Shu, J.F., Chen, L.C., Jephcoat, A.P.,Wu,Y. and Basset, W.A. (1991) Effect of pressure, temperature and composition of lattice parameters and density of (Fe,Mg)SiO_3-perovskite to 30 GPa, *J.Geophys. Res.*, **69**, 8069-8079.

Marumo, F., Isobe, M. and Akimoto, S. (1977) Electron density distributions in crystals for γ-Fe_2SiO_4 and γ-Co_2SiO_4, *Acta Crystallogr. Sect.*, **B33**, 713-716.

Matsui, Y. and Manghnani, M.H. (1985) Thermal expansion of single-crystal forsterite to 1023 K by Fizeau interferometry, *Phys. Chem. Min.*, **12**, 201-210.

Matsui, Y. and Nishizawa, O. (1974) Iron (II)-magnesium exchange equilibrium between olivine and calcium-free pyroxene over a temperature range 800 to 1300 ^{o}C, *Bull. Soc. Fr. Mineral. Crystallogr.*, **97**, 122-130.

Matsui, Y. and Syono, Y. (1968) Unit cell dimensions of some synthetic olivine group solid solutions, *Geochem. J.*, **2**, 51-59.

McMillan, P.F. and Ross, N.L. (1987) Heat capacity calculations for Al_2O_3 corundum and $MgSiO_3$ ilmenite, *Phys. Chem. Min.*, **14**, 225-234.

Mel'nik, Yu.P. and Radchuk, V.V. (1977) The stability of grunerite-$Fe_7Si_8O_{22}(OH)_2$, *Geokhimiya*, **5**, 693-704 (in Russian).

Metz, G.W., Anovitz, L.M. and Essene, E.J. (1983) The heat capacity and phase equilibria of almandine, *EOS Trans.*, AGU, **64**, no. 18, pp. 346.

Mirwald, P.W. and Massonne, H.J. (1980) The low-high quartz and quartz-coesite transition to 40 kbar between 600 and 1600°C and some reconnaissance data on the effect of $NaAlO_2$ component on the low quartz-coesite transition, *J. Geophys. Res.*, **85**, 6983-6990.

Miyano, T. and Klein, C. (1983) Phase relations of orthopyroxene, olivine and grunerite in high-grade metamorphic iron-formation, *Am. Mineral.*, **68**, 699-716.

Miyano, T. and Klein, C. (1986) Fluid behavior and phase relations in the system Fe-Mg-Si-C-O-H: Application to high grade metamorphism of iron-formations, *Am. J. Sci.*, **286**, 540-575.

Mukhopadhyay, D.K. and Lindsley, D.H. (1983) Phase relations in the join kirschsteinite ($CaFeSiO_4$)-fayalite (Fe_2O_4), *Am. Mineral.*, **68**, 1089-1094.

Nafziger, R. H., and Muan, A. (1967) Equilibrium phase compositions and thermodynamic properties of olivines and pyroxenes in the system MgO-"FeO"-SiO_2, *Am. Mineral.*, **52**, 1364-1385.

Navrotsky, A.(1987) High pressure transitions in silicates and related compounds. *Prog. Solid State Chemistry*, **17**, 53-86.

Navrotsky, A.(1989) Thermochemistry of perovskites, in *Perovskite, A Structure of Great Interest to Geophysics and Materials Science*, edited by A. Navrotsky and D. J. Weidner, pp. 67-80, Am. Geophys. Union, Washington D. C.

Navrotsky, A., Pintchovski, F. S. and Akimoto, S. (1979) Calorimetric study of the stability of high pressure phase in the system CoO-SiO_2 and "FeO"-SiO_2 and calculation of phase diagrams in MO-SiO_2 systems, *Phys. Earth Planet. Int.*, **19**, 275-292.

Navrotsky, A., Ziegler, A., D., Oestrike, R. and Maniar, P. (1988) Calorimetry of silicate melts at 1773 K: measurement of enthalpies of fusion and of mixing in the systems diopside-anorthite-albite and anorthite-forsterite, *Contrib. Mineral. Petrol.*, **101**, 122-130.

Nekvasil, H. (1986) A theoretical thermodynamic investigation Ab-Or-An-Qz-H_2O and implications for melt speciation, *Ph.D. Thesis*, Penn. State Univ., University Park, 268 pp.

Newton, R.C. (1966) Some calc-silicate equilibrium relations. *Am. J. Sci.*, **264**, 204-222.

Newton, R.C. (1969) Some high-pressure hydrothermal experiments on severely ground kyanite and sillimanite. *Am. J. Sci.*, **267**, 278-294.

Newton, R.C., Charlu, T.V. and Kleppa, O.J. (1977) Thermochemistry of high pressure garnets and clinopyroxenes in the system CaO-MgO-Al_2O_3-SiO_2, *Geochim. Cosmochim. Acta*, **41**, 369-377.

Newton, R.C., Geiger, C.A., Kleppa, O.J. and Brousse, C. (1986) Thermochemistry of binary and ternary garnet solid solutions. *IMA Abstr. Prog.*, p.186.

Nickel, K.G. and Brey, G. (1984) Subsolidus orthopyroxene-clinopyroxene systematics in the system CaO-MgO-SiO_2 to 60 kb, a reevaluation of the regular solution model, *Contrib. Mineral. Petrol.*, **87**, 35-42.

Ohtani, E. (1979) Melting relation of Fe_2SiO_4 up to 200 kbar, *J. Phys. Earth*, **27**, 189-208.

Ohtani, E. and Kumazawa, M. (1981) Melting of forsterite Mg_2SiO_4 up to 15 GPa, *Phys. Earth Planet. Int.*, **27**, 32-38.

O'Neill, H.S.C. and Wood, B.J. (1979) An experimental study of Fe-Mg partitioning between garnet and olivine and its calibration as a geothermometer, *Contrib. Mineral. Petrol.*, **70**, 59-70.

Openshaw, R.E., Hemingway, B.S., Robie, R.A., Waldbaum, D.R. and Krupka, K.M. (1976) The heat capacities at low temperature and entropies at 298.15 K of low albite, analbite, microcline, and sanidine. *J. Res. U.S. Geol. Surv.*, **4**, 195-204.

Pacalo, R.E.G. and Gasparik, T. (1990) Reversals of the orthoenstatite-clinoenstatite transition at high pressures and high temperature, *J.Geophys. Res.*, **95**, 15853-15858.

Patera, E.S. (1982) Phase equilibria in the upper Martian mantle, calculation and experiments, *Ph.D. Thesis*, Arizona State Univ., Tempe.

Perchuk, L.L. and Lavrent'eva, I.V. (1983) Experimental investigation of exchange equilibria in the system cordierite-garnet-biotite, in *Advances in Physical Geochemistry*, **3**, edited by S.K. Saxena, pp. 199-239, Springer-Verlag, New York.

Perkins, D. (1983) The stability of Mg-rich garnet in the system CaO-MgO-Al_2O_3-SiO_2. at 1000 to 1300^oC and high pressure, *Am. Mineral.*, **68**, 355-364.

Perkins, D., Holland, T.J.B. and Newton, R.C. (1981) The Al_2O_3 content of enstatite in equilibrium with garnet in the system MgO-Al_2O_3-SiO_2 at 15-40 kbar and 900-1600 ^oC, *Contrib. Mineral. Petrol.*, **78**, 99-109.

Presnall, D.C. (1976) Alumina content of enstatite as a geobarometer for plagioclase and spinel lherzolites, *Am. Mineral.*, **61**, 582-588.

Renon, H. and Prausnitz, J. M. (1968) Local compositions in thermodynamic excess functions for liquid mixtures, *Am. Inst. Chem. Eng. J.*, **14**, 135-144.

Richardson, S.W. (1968) Staurolite stability in a part of the system Fe-Al-Si-O-H, *J. Petrol.*, **9**, 468-488.

Richardson, S.W., Bell, P.M. and Gilbert, M.C. (1968) Kyanite-sillimanite equilibrium between 700 and 1500 K, *Am. J. Sci.*, **266**, 513-541.

Richardson, S.W., Gilbert, M.C. and Bell, P.M. (1969) Experimental determination of kyanite-andalusite and andalusite-sillimanite equilibria; the aluminum silicate triple point, *Am. J. Sci.*, **267**, 249.

Richet, P. and Fiquet, G. (1991) High-temperature heat capacity and premelting of minerals in the system MgO-CaO-Al2O3-SiO2, *J. Geophys. Res.*, **96**, 445-456.

Richet, P., Gillet, P. and Fiquet, G. (1992) Thermodynamic properties of minerals: macroscopic and microscopic approaches, in *Adv. Physical Geochem.*, **10**, *Thermodynamic data: Systematics and Estimation*, edited by S.K. Saxena, pp. 98-131, Springer-Verlag, New York.

Ringwood, A. E.(1967) The proxene-garnet transition in the earth's mantle, *Earth Planet. Sci. Lett.*, **3**, 255-263.

Ringwood, A. E. and Major, A. (1970) The system Mg_2SiO_4-Fe_2SiO_4 at pressures and temperatures, *Phys. Earth Planet. Int.*, **3**, 89-108.

Robie, R.A. and Bethke, P.M. (1963) Molar volumes and densities of minerals, *U.S. Geol. Surv. Rep.*, **TEI-822**, Open-File Report.

Robie, R.A. and Hemingway, B.S. (1984a) Heat capacities and entropies of phlogopite ($KMg_3[AlSi_3O_{10}](OH)_2$) and paragonite ($NaAl_2[AlSi_3O_{10}](OH)_2$) between 5 and 900 K and estimates of the enthalpies and Gibbs free energies of formation, *Am. Mineral.*, **69**, 858-868.

Robie, R.A. and Hemingway, B.S. (1984b) Entropies of kyanite, andalusite, and sillimanite: additional constraints on the pressure and temperature of the Al_2SiO_5 triple point, *Am. Mineral.*, **69**, 298-306.

Robie, R.A. and Stout, J.W. (1963) Heat capacity from 12 to 305 K and entropy of talc and tremolite, *J. Phys. Chem.*, **67**, 2252-2256.

Robie, R.A., Hemingway, B.S. and Fisher, J.R. (1978) Thermodynamic properties of minerals and related substances at 298.15 K and 1 bar (105 Pascals) pressure and at higher temperatures, *US Geol. Survey Bull.*, **1452**, 456 pp.

Robie, R.A., Finch, C.B. and Hemingway, B.S. (1982a) Heat capacity and entropy of fayalite (Fe_2SiO_4) between 5.1 and 383 K: comparison of calorimetric and equilibrium values for the QFM buffer reaction, *Am. Mineral.*, **67**, 463-469.

Robie, R.A., Hemingway, B.S. and Takei, H. (1982b) Heat capacities and entropies of Mg_2SiO_4, Mn_2SiO_4 and Co_2SiO_4 between 5 and 380 K, *Am. Mineral.*, **67**, 470-482.

Rosenhauer, M., Mao, H. K. and Woermann, E. (1976) Compressibility of $(Fe_{0.4}Mg_{0.6})O$ magnesiowustite to 264 kbar, *Carnegie Inst. Wash.*, Yearb. **75**, 513-515.

Ross, N.L. and Hazen, R.M. (1989) Single crystal X-ray diffraction study of $MgSiO_3$ perovskite from 77 to 400 K, *Phys. Chem. Min.*, **16**, 415-420.

Sack, R.O. (1982) Spinels as petrogenetic indicators: activity-composition relations at low pressures, *Contrib. Mineral. Petrol.*, **79**, 169-186.

Sack, R.O. and Ghiorso, M.S. (1989) Importance of considerations of mixing propertities in establishing an internally consistent thermodynamics database: thermochemistry of minerals in the system Mg_2SiO_4-Fe_2SiO_4-SiO_2, *Contrib. Mineral. Petrol.*, **102**, 41-68.

Sato, Y. (1977) Equation of state of mantle minerals determined through high-pressure X-ray study, in *High-pressure Research: Applications in Geophysics*, edited by M. H. Manghnani and S. Akimoto, pp. 307-323, Academic Pess, New York.

Sato, Y. and Akimoto, S. (1979) Hydrostatic compression of four corundum-type compounds: a-Al_2O_3, V_2O_3, Cr_2O_3, and α-Fe_2O_3, *J. Appl. Phys.*, **50**, 5285-5291.

Sawamoto, H. (1986a) Phase equilibrium of $MgSiO_3$ under high pressures and high temperatures: garnet-perovskite transformation (abstract), U.S.-Japan Seminar, *High-Pressure Research: Application in Geophysics and Geochemistry*, Corsponsored by Japan Society for the Promotion of Science and U. S. National Science Foundation, Honolulu, Hawaii, pp. 50-51.

Sawamoto, H. (1986b) Single crystal growth of the modified spinel (β) and spinel (γ) phases of $(Mg,Fe)_2SiO_4$ and some geophysical implications, *Phys. Chem. Min.*, **13**, 1-10.

Sawamoto, H. (1987) Phase diagram of $MgSiO_3$ at pressures up to 24 GPa and temperatures up to 2200 K: phase stability and properties of tetragonal garnet, in *High-Pressure Research in Mineral Physics*, edited by M. H. Manghnani and Y. Syono, pp. 209-219, Terra Publications, Tokyo, Japan.

Sawamoto, H., Weidner, D. J., Sasaki, S. and Kumazawa, M. (1984) Single-crystal elastic properties of the modified spinel (beta) phase of Mg_2SiO_4, *Science*, **224**, 749-751.

Saxena, S. K. (1973) *Thermodynamics of Rock-Forming Crystalline Solutions*, Springer-Verlag, Heidelberg, 189 pp.

Saxena, S.K. (1982) Computation of multicomponent phase equilibria, in: *Advances in Physical Geochemistry*, **2**, edited by S. K. Saxena, pp. 353-370, Springer-Verlag, New York.

Saxena, S. K. (1988) Assessment of thermal expansion, bulk modulus, and heat capacity of enstatite and forsterite, *J. Phys. Chem. Solids*, **49**, 1233-1235.

Saxena, S.K., (1989) Assessment of bulk modulus, thermal expansion and heat capacity of minerals, *Geochim. Cosmochim. Acta*, **53**, 785-789.

Saxena, S.K. and Chatterjee, N. (1986) Thermochemical data on mineral phases: the system CaO-MgO-Al_2O_3-SiO_2, *J. Petrol.*, **27**, 827-842.

Saxena, S.K. and Eriksson, G. (1983a) Theoretical computation of mineral assembalges in pyrolite and lherzolite, *J. Petrol.*, **24**, 538-555.

Saxena, S.K. and Eriksson, G. (1983b) High temperature phase equilibria in a solar composition gas, *Geochim. Cosmochim. Acta*, **47**, 1865-1874.

Saxena, S.K. and Eriksson, G. (1985) Anhydrous phase equilibria in the earth's upper mantle. *J. Petrol.*, **26**, 378-390.

Saxena, S.K. and Fei, Y. (1987a) Fluids at crustal pressures and temperatures I. Pure species, *Contrib. Mineral. Petrol.*, **95**, 370-375.

Saxena, S.K. and Fei, Y. (1987b) High pressure and high temperature fluid fugacities, *Geochim. Cosmochim. Acta*, **15**, 783-791.

Saxena, S.K. and Fei, Y. (1988a) The pressure-volume-temperature equation of hydrogen, *Geochim. Cosmochim. Acta*, **52**, 1195-1196.

Saxena, S.K. and Fei, Y. (1988b) Fluid mixtures in the C-H-O system at high pressure and temperature, *Geochim. Cosmochim. Acta*, **52**, 505-512.

Saxena, S.K. and Ghose, S. (1971) Mg^{2+}-Fe^{2+} order-disorder and the thermodynamics of the orthopyroxene crystalline solution, *Am. Mineral.*, **56**, 532-559.

Saxena, S.K. and Shen, G. (1992) Assessment of thermophysical and thermochemical data in some oxides and silicates, *J. Geophys. Res.*, 97, 19,831-19,825.

Saxena, S. K., and J. Zhang, (1989) Assessed high-temperature thermochemical data on some solids, *J. Phys. Chem. Solid*, **50**, 723-727.

Saxena, S. K., and J. Zhang, (1990) Themochemical and pressure-volume-temperature systematics of data on solids, examples: Tungsten and MgO, *Phys. Chem. Mineral.*, **17**, 45-51.

Saxena, S.K., Sykes, J. and Eriksson, G. (1986) Phase equilibria in the pyroxene quadrilateral, *J. Petrol.*, **27**, 843-852.

Scarfe, C.M. and Wyllie, P.J. (1967) Experimental redetermination of the upper stability limit of serpentine up to 3 kb pressure, *Trans. Am. Geophys. Union*, **48**, 255.

Schramke, J.K., Kerrick, D.M. and Blencoe, J.G. (1982) The experimental determination of the brucite = periclase + water equilibrium with a new volumetric technique, *Am. Mineral.*, **67**, 269-276.

Seifert, F. (1974) Stability of sapphirine: a study of the aluminous part of the system MgO-Al_2O_3-SiO_2-H_2O, *J. Geol.*, **82**, 173-204.

Sharp, A.D., Essene, L.M., Metz, G.W., Westrum, E.F., Hemingway, B.S. and Valley, J.W. (1986) The heat capacity of a natural monticellite and phase equilibria in the system CaO-MgO-SiO_2-CO_2, *Geochim. Cosmochim. Acta*, **50**, 1475-1484.

Shaw, H.R. and Wones, D.R. (1964) Fugacity coefficients for hydrogen gas between 0° and 1000^oC., for pressures to 3000 atm., *Am. J. Sci.*, **262**, 918-929.

Shi, P.F., Saxena, S.K. and Sundman, B. (1992) Sublattice solid solution model and its application to orthopyroxene $(Mg,Fe)_2Si_2O_6$, *Phys. Chem. Min.*, **18**, 393-405.

Shi, P., Saxena, S.K., Zhang, Z., and Sundman, B. (1993) Thermodynamics of the Ca-Mg-Fe-Al-Si-O pyroxenes: I. theretical model and assessment of the Ca-Mg-Si-O system, submittied to *Calphad.*

Shmulovich, K.I. (1974) Phase equilibria in the CaO-Al_2O_3-SiO_2-CO_2 system, *Geochem. Int.*, **11**, 883-887.

Simons, B.(1980) Composition-lattice parameter relationship of the magnesiowustite solid solution series, *Carnegie Inst. Wash.* Yearb. **79**, 376-380.

Skinner, B.J. (1956) Physical properties of end-members of the garnet group, *Am. Mineral.*, **41**, 428-436.

Skinner, B.J. (1962) Thermal expansion of ten minerals, *U.S. Geol. Surv. Prof. Pap.*, **450D**, 109-112.

Skinner, B.J. (1966) Thermal expansion, in *Handbook of Physical Constants*, edited by S.P. Clark, *Geol. Soc. Am. Memoirs*, **97**, 75-96.

Skippen, G. B. (1971) Experimental data for reactions in siliceous marbles, *J. Geol.*, **79**, 457-481.

Smith, D. (1971) Stability of the assemblage iron-rich orthopyroxene-olivine-quartz, *Am. J. Sci.*, **271**, 370-382.

Smith, W.R. and Missen, R.W. (1982) Chemical reaction equilibrium analysis, Wiley-Interscience, New York.

Soga, N. (1967) Elastic constants of garnet under pressure and temperature, *J. Geophys. Res.*, **72**, 4227-4234.

Soga, N. (1968) The temperature and pressure derivatives of isotropic sound velocities of a-quartz, *J. Geophys. Res.*, **73**, 827-829.

Staudigel, H. and Schreyer, W. (1977) The upper stability limit of clinochlore, $Mg_5Al[Si_3Al_{10}](OH)_8$, at 10-35 kbar $P(H_2O)$, *Contrib. Mineral. Petrol.*, **61**, 187-198.

Strong, H.M., Tuft, R.E. and Hanneman, R.E. (1973) The iron fusion curve and γ-δ-l triple point, *Met. Trans.*, **4**, 2657-2661.

Suito, K. (1972) Phase transitions of pure Mg_2SiO_4 into a spinel structure under high pressures and temperatures, *J. Phys. Earth*, **20**, 225-243.

Suito, K. (1977) Phase relations of pure Mg_2SiO_4 up to 200 kilobar, in *High-Pressure Research: Applications in Geophysics*, edited by M.H. Manghnani and S. Akimoto, pp. 255-266, Academic Press, New York.

Sumino, Y. (1979) The elastic constants of Mn_2SiO_4, Fe_2SiO_4 and Co_2SiO_4 and the elastic properties of olivine group minerals at high temperature, *J. Phys. Earth*, **27**, 209-238.

Sumino, Y., Nishizawa, O., Goto, T., Ohno, I. and Ozima, M. (1977) Temperature variation of elastic constants of single-crystal forsterite between 190 and 400^oC, *J. Phys. Earth*, **28**, 273-280.

Sung, C. M. and Burns, R. G. (1976) Kinetics of high-pressure phase transformations: implications to the evolution of the olivine-spinel transition in the downgoing lithosphere and its consequences on the dynamics of the mantle, *Tectonophysics*, **31**, 1-32.

Suzuki, I. (1975) Thermal expansion of periclase and olivine and their anharmonic properties, *J. Phys. Earth*, **23**, 145-159.

Suzuki, I. and Anderson, O.L. (1983) Elasticity and thermal expansion of a natural garnet up to 1,000 K, *J. Phys. Earth*, **31**, 125-138, 1983.

Suzuki, I., Ohtani, E. and Kumazawa, M. (1979) Thermal expansion of γ-Mg_2SiO_4, *J. Phys. Earth*, **27**, 53-61.

Suzuki,I., Ohtani, E. and Kumuzawa, M. (1980) Thermal expansion of modified spinel, b-Mg_2SiO_4, *J. Phys. Earth*, **28**, 273-280.

Suzuki, I., Seya, K., Takei, H. and Sumino, Y. (1981) Thermal expansion of fayalite, Fe_2SiO_4, *Phys. Chem. Mineral*, **7**, 60-63.

Swamy, V., Saxena, S.K., Sundman, B., Zhang, Z., and Liebermann, R.C. (1993) A thermodynamic assessment of silica phase diagram, submitted to *J. Geophys. Res.*

Syono, Y., Akimoto, S. and Matsui, Y. (1971) High pressure transformations in zinc silicates, *J. Solid State Chem.*, **3**, 369-380.

Thompson, J. B. Jr, and Waldbaum, D. R. (1969) Mixing properties of sanidine crystalline solutions. III Calculation based on two-phase data, *Am. Mineral.*, **54**, 811-838.

Trommsdorff, V. and Connolly, J.A.D. (1990) Constrains on phase diagram topology for the system CaO-MgO-SiO_2-CO_2-H_2O, *Contrib. Mineral. Petrol.*, **104**, 1-7.

Velde, B. (1966) Upper stability of muscovite, *Am. Mineral.*, **51**, 921-929.

Wagner, V.H. (1932) Zur Thermochemie der Metasilikate des Calciums und Magnesiums und des Diopsids, *Z. Anorg. Allg. Chemie*, **208**, 1-22.

Walter, L.S. (1963a) Experimental studies on Bowen's decarbonation series: I: P-T univariant equilibria of the 'monticellite' and 'akermanite' reactions, *Am. J. Sci.*, **261**, 488-500.

Walter, L.S. (1963b) Experimental studies on Bowen's decarbonation series: II: P-T univariant equilibria of the reaction: frosterite + calcite = monticellite + periclase + CO_2, *Am. J. Sci.*, **261**, 773-779.

Wang, H. and Simmons, G. (1972) Elasticity of some mantle crystal structures. I. Pleonaste and hercynite spinel, *J. Geophys. Res.*, **77**, 4379.

Watanabe, H. (1982) Thermochemical properties of synthetic high-pressure compounds relevant to the earth's mantle, in *High Pressure Research in Geophysics*, edited by H. Akimoto and M.H. Manghani, pp. 441-464, Center for Academic Publications, Tokyo, Japan.

Weaver, J. S., Chipman, D. W. and Takahashi, T. (1979) Comparison between thermochemical and phase stability data for the quartz-coesite-stishovite transformations, *Am. Miner.*, **64**, 604-614.

Weeks, W.F. (1955) Heats of formation of metamorphic minerals in the system CaO-MgO-SiO_2-H_2O and their petrological significance, *J. Geol.*, **64**, 456-473.

Weidner, D.J. (1986) Mantle model based on measured physical properties of minerals, In *Advances in Physical Geochemistry*, edited by S. K. Saxena, pp. 251-274, Springer-Verlag, New York.

Weidner, D.J. and Ito, E. (1985) Elasticity of $MgSiO_3$ in the ilmenite phase, *Phys. Earth Planet. Int.*, **40**, 65-70.

Weidner, D.J., Bass, J.D., Ringwood, A.E. and Sinclair, W. (1982) The single - crystal elastic moduli of stishovite, *J. Geophys. Res.*, **87**, 4740-4746.

Weidner, D. J., Sawamoto, H., Sasaki, S. and Kumazawa, M. (1984) Single-crystal elastic properties of the spinel phase of Mg_2SiO_4, *J. Geophys. Res.*, **89**, 7852-7860.

White, G.K., Roberts, R.B. and Collins, J.N. (1985) Thermal properties and Grüneisen functions of forsterite, Mg_2SiO_4, *High Temperatures-High Pressures*, **17**, 61-66.

White, W.P. (1919) Silicate specific heats, *Am. J. Sci.*, **47**, 1-21.

Wilburn, D. R., Bassett, W. A., Sato, Y. and Akimoto, S. (1978) X-ray diffraction compression studies of hematite under hydrostatic, isothermal conditions, *J. Geophys. Res.*, **83**, 3509-3512.

Wilson, G. M.(1964) A new expression for the excess free energy of mixing, *J. Am. Chem. Soc.*, **86**, 127-130.

Windom, K.E. and Boettcher, A.L. (1976) The effect of reduced activity of anorthite on the reaction grossular + quartz = anorthite + wollastonite: a model for plagioclase in the earth's lower crust and upper mantle, *Am. Mineral.*, **61**, 889-896.

Windom, K.E. and Boettcher, A.L. (1981) Phase relations for the joins jadeite-enstatite and jadeite-forsterite at 28 kb, *Am. J. Sci.*, **281**, 335-351.

Wones, D.R. (1972) Stability of biotite: a reply, *Amer. Mineral.*, **57**, 316-317.

Wones, D.R. and Dodge, F.C.W. (1977) The stability of phlogopite in the presence of quartz and diopside, in *Thermodynamics in Geology,* Dedited by D.G. Fraser, pp. 229-247, D. Reidel, Dordrecht.

Wones, D.R. and Eugster, H.P. (1965) Stability of biotite: experiment, theory, and application, *Am. Mineral.*, **50**, 1228-1272.

Wood, B.J. (1976) The reaction phlogopite + quartz = enstatite + sanidine + H_2O, in *Progress in Experimental Petrology, Nat. Environ. Res. Counc.*, publ. ser. D, **6**, 17-19.

Wood, B.J. (1978) Reactions involving anorthite and $CaAlSi_2O_6$ pyroxene at high pressures and temperatures, *Am. J. Sci.*, **278**, 930-942.

Wood, B.J. and Holloway, J.R. (1984) A thermodynamic model for subsolidus equilibria in the system $CaO-MgO-Al_2O_3-SiO_2$, *Geochim. Cosmochim. Acta*, **48**, 159-176.

Wood, B.J. and Kleppa, O.J. (1981) Thermochemistry of forsterite-fayalite olivine solutions, *Geochim. Cosmochim. Acta*, **45**, 529-534.

Wood, B.J. and Nicholls, J.W. (1978) The thermodynamic properties of reciprocal solid solutions, *Contrib. Mineral. Petrol.*, **66**, 389-400.

Wood, B.J., Kirkpatrick, R.J. and Montez, B. (1986) Order-disorder phenomena in $MgAl_2O_4$ spinel, *Am. Mineral.*, **71**, 999-1006.

Woodhead, J.A. (1977) The crystallographic and calorimetric effects of Al-Si distribution of the tetrahedral sites of melilite, *Ph.D. Thesis*, Princeton Univ. Microfilms, Ann Arbor, Michigan.

Yagi, T. and Akimoto, S. (1976) Direct determination of coesite-stishovite transition by in situ X-ray measurements, *Tectonophysics*, **35**, 259-270.

Yagi, T., Bell, P. M. and Mao, H. K. (1979) Phase relations in the system MgO-FeO-SiO_2 between 150 and 700 kbar at 1000 ^oC, *Carnegie Inst. Wash. Yearb.* **78**, 614-618.

Yagi, T., Mao, H. K. and Bell, P. M. (1982) Hydrostatic compression of perovskite type $MgSiO_3$, in *Advances in Physical Geochemistry*, **2**, edited by S. K. Saxena, pp. 317-326, Springer-Verlag, New York.

Yagi, T., Akaogi, M., Shimomura, O., Suzuki, T. and Akimoto, S. (1987) In situ observation of the olivine-spinel phase transformation in Fe_2SiO_4 using synchrotron radiation, *J. Geophy. Res.*, **92**, 6207-6213.

Yeganeh-Haeri, A., Weidner, D. J. and Ito, E. (1989) Single crystal elastic moduli of magnesium metasilicate perovskite, in *Perovskite, A Structure of Great Interest to Geophysics and Materials Science*, edited by A. Navrotsky and D. J. Weidner, pp. 13-15, Am. Geophys. Union, Washington D.C.

Yin, H.A. and Greenwood, H.J. (1983) Displacement of equilibria of OH-tremolite and F-tremolite solid solution, I, determination of the equilibrium P-T curve of OH-tremolite, *Trans. Am. Geophys. Union*, **64**, 347.

Yoder, H.S. (1968) Akermanite and related melilite-bearing assemblages, *Carnegie Inst. Wash.* Yearb., **66**, 471-477.

Yusa, H., Akaogi, M. and Ito, E. (1993) Calorimetric study of $MgSiO_3$ garnet and pyroxene: heat capacity, transition enthalpies and equilibrium phase relations in $MgSiO_3$ at high pressures and temperatures, *J. Geophys. Res.*, in press.

Zen, E.A. (1969) Free energy of formation of pyrophyllite from hydrothermal data: values, discrepancies and implications, *Am. Mineral.*, **54**, 1592-1606.

Zhang, J. and Saxena, S.K. (1991) Thermodynamic properties of andradite and application to skarn with coexisting andradite and hedenbergite, *Contrib. Mineral. Petrol.*, **107**, 255-263.

Thermodynamic Data

Appendix I: Tables of Thermodynamic Data

Chemical Formula Species Name	Formula wt. (g)
Enthalpy (J/mol) =	Reference
Entropy (J/mol.K) =	Reference
Heat Capacity *Cp*:	
No. of *Cp* ranges = 1	Reference
$a + b\,T + c\,T^{-2} + d\,T^{2} + e\,T^{-3} + g\,T^{-0.5} + h\,T^{-1}$	
Volume (cc/mol) =	Reference
Thermal expn:	
$a_0 + a_1\,T + a_2\,T^{-1} + a_3\,T^{-2}$	Reference
Isot.compressibility:	
$bo + b_1\,T + b_2\,T^{2} + b_3\,T^{3}$	Reference
P derivative of the bulk modulus	Reference
T derivative of the *P* derivative	Reference
Melting pt. and enthalpy of melting:	

NOTE: If more than one C_p range, the data on ΔH of transition follow the first C_p equation followed by the second C_p equation. All *T* derivatives of the variation of the *P* derivative of the bulk modulus are estimated in this study.

HERCYNITE

$FeAl_2O_4$
HERCYNITE Formula wt. (g) = 173.81

Enthalpy (J/mol) =	-1958200.000	This study
Entropy (J/mol.K) =	113.500	This study
Heat Capacity:		
No. of *Cp* ranges = 1		This study
	1.7572000D+02 2.2200D-02 8.1615D+06 0.0000D+00 -1.7335D+09 0.0000D+00 -2.4517D+04	
Volume (cc/mol) =	40.750	Robie et al. (1978)
Thermal expn:		
	4.89524D-05 8.62000D-09 -2.05490D-02 3.45353D+0	This study
Isot.compressibility:		
	4.59048D-07 8.55240D-11 -1.42740D-15 9.31640D-18	This study
P deriv.	4.00000D+00	Wang and Simmons (1972)
T deriv.	7.80000D-04	This study
Melting pt. and enthalpy of melting:		
	- -	

NOTE: Free energy of hercynite is determined from the following reaction:Alm + Sill = Her + Q. The heat capacity expression for hercynite used by Saxena and Eriksson (1983b) is adjusted here. The entropy is from Robie et al. (1978). Enthalpy is adjusted within the error limits of Robie et al. (1978) (-1966.48 ± 8.5 KJ/mol).

Thermochemical Data

T (K)	C_p (J)	H (kJ)	S (J)	α 10^{-4}	K 10^4	γ	V (cc)	δ_T
298	126.5	-1958.20	113.5	0.2145	206.327	1.44	40.8	8.38
300	127.1	-1957.97	114.3	0.2141	206.259	1.43	40.8	8.40
400	147.2	-1944.08	154.1	0.2261	202.583	1.29	40.8	7.96
500	156.6	-1928.85	188.0	0.2598	198.958	1.38	40.9	6.98
600	162.8	-1912.87	217.2	0.2947	195.366	1.49	41.1	6.22
700	167.8	-1896.33	242.6	0.3268	191.787	1.59	41.2	5.71
800	172.2	-1879.32	265.4	0.3556	188.209	1.68	41.3	5.35
900	176.2	-1861.90	285.9	0.3814	184.619	1.76	41.5	5.11
1000	179.8	-1844.10	304.6	0.4048	181.008	1.82	41.6	4.95
1100	183.3	-1825.94	321.9	0.4261	177.369	1.88	41.8	4.84
1200	186.6	-1807.45	338.0	0.4457	173.696	1.92	42.0	4.77
1300	189.8	-1788.63	353.1	0.4639	169.986	1.96	42.2	4.73
1400	192.8	-1769.50	367.2	0.4810	166.238	1.99	42.4	4.71
1500	195.8	-1750.07	380.7	0.4972	162.452	2.02	42.6	4.71
1600	198.7	-1730.34	393.4	0.5125	158.631	2.05	42.8	4.72
1700	201.5	-1710.33	405.5	0.5271	154.778	2.06	43.0	4.74
1800	204.3	-1690.04	417.1	0.5412	150.898	2.08	43.3	4.77
1900	207.0	-1669.48	428.2	0.5547	146.996	2.09	43.5	4.80
2000	209.7	-1648.64	438.9	0.5678	143.078	2.10	43.7	4.83
2100	212.3	-1627.54	449.2	0.5805	139.153	2.10	44.0	4.86
2200	214.9	-1606.18	459.1	0.5929	135.228	2.10	44.3	4.89
2300	217.5	-1584.55	468.8	0.6050	131.310	2.10	44.5	4.92
2400	220.1	-1562.67	478.1	0.6168	127.409	2.10	44.8	4.95
2500	222.6	-1540.54	487.1	0.6284	123.532	2.09	45.1	4.98
2600	225.1	-1518.15	495.9	0.6397	119.688	2.08	45.4	5.00
2700	227.6	-1495.52	504.4	0.6509	115.884	2.06	45.7	5.01
2800	230.1	-1472.63	512.8	0.6619	112.128	2.04	46.0	5.03
2900	232.5	-1449.50	520.9	0.6728	108.427	2.02	46.3	5.03
3000	235.0	-1426.12	528.8	0.6835	104.787	2.00	46.6	5.04

$FeAl_2O_4$
HERCYNITE

	<INT VdP (J)>						
	PRESSURE (GPa)						
T(K)	1	5	10	20	30	40	50
298	40652	201375	398339	780699	1149750	1507420	1855164
300	40653	201382	398352	780720	1149775	1507447	1855191
400	40739	201777	399063	781886	1151226	1509064	1856888
500	40836	202226	399882	783280	1153052	1511233	1859347
600	40948	202745	400842	784964	1155332	1514040	1862649
700	41074	203334	401938	786922	1158034	1517434	1866715
800	41212	203985	403158	789125	1161112	1521343	1871448
900	41362	204692	404487	791545	1164521	1525702	1876758
1000	41523	205451	405916	794161	1168222	1530458	1882573
1100	41694	206257	407435	796951	1172184	1535564	1888830
1200	41874	207105	409037	799899	1176382	1540983	1895480
1300	42062	207995	410716	802994	1180793	1546685	1902480
1400	42259	208922	412466	806223	1185400	1552643	1909797
1500	42463	209885	414284	809577	1190188	1558836	1917402
1600	42675	210883	416166	813050	1195146	1565248	1925272
1700	42894	211914	418109	816634	1200262	1571863	1933386
1800	43120	212977	420111	820325	1205528	1578668	1941730
1900	43354	214072	422170	824117	1210937	1585655	1950289
2000	43594	215196	424284	828009	1216485	1592816	1959053
2100	43841	216351	426452	831995	1222165	1600142	1968014
2200	44094	217535	428673	836076	1227974	1607631	1977165
2300	44355	218747	430945	840247	1233911	1615277	1986501
2400	44622	219989	433269	844509	1239972	1623079	1996019
2500	44895	221258	435643	848861	1246157	1631035	2005717
2600	45175	222556	438068	853301	1252464	1639143	2015595
2700	45462	223882	440543	857831	1258895	1647405	2025651
2800	45755	225236	443069	862449	1265449	1655820	2035888
2900	46055	226618	445644	867157	1272127	1664391	2046308
3000	46361	228028	448270	871954	1278931	1673119	2056912

GEHLENITE

$Ca_2Al_2SiO_7$
GEHLENITE Formula wt. (g) = 274.21

Enthalpy (J/mol) =	-3977900.000	Saxena and Chatterjee (1986)
Entropy (J/mol.K) =	204.350	Woodhead (1977)
Heat Capacity:		
No. of *Cp* ranges = 1		Saxena and Chatterjee (1986)
3.4154000D+02 3.7950D-03 -7.6630D+05 0.0000D+00		
4.0740D+08 0.0000D+00 -4.0370D+04		
Volume (cc/mol) =	90.060	Charlu et al. (1981)
Thermal expn:		
2.55000D-05 0.00000D+00 0.0000D+00 0.0000D+00		Skinner (1966)
Isot.compressibility:		
1.00000D-06 0.00000D+00 0.00000D+00 0.00000D+00		
P deriv. 4.00000D+00		
T deriv. 0.00000D+00		
Melting pt. and enthalpy of melting:		
1863.00000	-	Robie et al. (1978)

NOTE: Calorimetric data: H: -3983.93±1.3, Charlu et al. (1981); S, Hemingway and Robie (1984). Wollastonite and gehlenite participate in reactions involving the breakdown of anorthite. These reactions, studied by various workers, are as follows:

An + Woll = Gr + Q
An + Ge = Gr + Cor
An + Ge + Woll = Gr

For the second reaction, Gasparik (1984b) evaluated the experimental data of Boettcher (1970) and Huckenholz et al. (1975) and found the equilibrium pressure as P = 23.8T(C) - 17440. Using these data, $-\Delta C'p$, the difference between the experimental and the calculated heat capacities, for gehlenite is found to be $-\Delta C'p$ = 7.5491 (±0.134) + 0.0027126 (±0.00019) * T and a corresponding $-\Delta H_f$ are listed above. This value is comparable to -3979046 J/mol derived from the $-\Delta H_f(970)$ value of Charlu et al. (1981) using the *Cp* expression from Robie et al. (1978). The entropy of gehlenite is selected as 204.35 J/mol.K following Woodhead (1977) (a progressive disorder with temperature.

$Ca_2Al_2SiO_7$
GEHLENITE

Thermochemical Data

T(K)	C_p(J)	H(kJ)	S(J)
298	214.0	-3977.90	204.3
300	214.7	-3977.50	205.7
400	243.7	-3954.48	271.7
500	262.9	-3929.09	328.3
600	276.3	-3902.09	377.5
700	286.1	-3873.95	420.8
800	293.7	-3844.94	459.6
900	299.7	-3815.26	494.5
1000	304.6	-3785.03	526.4
1100	308.7	-3754.36	555.6
1200	312.2	-3723.31	582.6
1300	315.2	-3691.95	607.7
1400	317.8	-3660.30	631.2
1500	320.1	-3628.40	653.2
1600	322.2	-3596.28	673.9
1700	324.1	-3563.97	693.5
1800	325.8	-3531.48	712.1
1900	327.4	-3498.82	729.7
2000	328.8	-3466.01	746.5
2100	330.2	-3433.06	762.6
2200	331.4	-3399.98	778.0
2300	332.6	-3366.78	792.8
2400	333.7	-3333.46	806.9
2500	334.8	-3300.04	820.6
2600	335.8	-3266.51	833.7
2700	336.8	-3232.88	846.4
2800	337.7	-3199.16	858.7
2900	338.6	-3165.35	870.6
3000	339.4	-3131.45	882.0

AKERMANITE

$Ca_2MgSi_2O_7$
AKERMANITE Formula wt. (g) = 272.64

Enthalpy (J/mol) = -3867500.000 Robie et al. (1978), Yoder (1968)
Entropy (J/mol.K) = 208.200 Robie et al. (1978)
Heat Capacity:
No. of *Cp* ranges = 1 Robie et al. (1978)
3.4884000D+02 6.2740D-03 9.5150D+06 0.0000D+00
-8.7810D+08 0.0000D+00 -6.5390D+04
Volume (cc/mol) = 92.810 Robie et al. (1978)
Thermal expn:
- - - -
Isot.compressibility:
1.00000D-06 0.00000D+00 0.00000D+00 0.00000D+00
P deriv. 4.00000D+00
T deriv. 0.00000D+00
Melting pt. and enthalpy of melting:
1727.00000 - Robie et al. (1978)

NOTE: Calorimetric data: H: -3865.6 ±1.13, Charlu et al. (1981); S: 209.33±2.09, Robie et al. (1978). The experimental data of Yoder (1968) involving wollastonite, monticellite and akermanite are used to calculate the H of akermanite. The S and C_p data are from Robie et al.(1978). The data are close to the calorimetric data of Haas et al. (1981) and Charlu et al. (1981).

Thermochemical Data

T(K)	C_p(J)	H(kJ)	S(J)
298	205.3	-3867.50	208.2
300	206.0	-3867.12	209.5
400	233.6	-3845.03	272.8
500	252.2	-3820.69	327.1
600	266.0	-3794.75	374.3
700	276.7	-3767.59	416.2
800	285.3	-3739.48	453.7
900	292.4	-3710.59	487.7
1000	298.4	-3681.04	518.8
1100	303.5	-3650.94	547.5
1200	308.0	-3620.36	574.1
1300	311.9	-3589.37	598.9
1400	315.5	-3557.99	622.2
1500	318.6	-3526.29	644.1
1600	321.5	-3494.28	664.7
1700	324.2	-3461.99	684.3
1800	326.6	-3429.45	702.9
1900	328.9	-3396.68	720.6
2000	331.0	-3363.69	737.5
2100	332.9	-3330.49	753.7
2200	334.8	-3297.10	769.3
2300	336.6	-3263.53	784.2
2400	338.2	-3229.79	798.5
2500	339.8	-3195.89	812.4
2600	341.4	-3161.83	825.7
2700	342.8	-3127.62	838.6
2800	344.2	-3093.27	851.1
2900	345.6	-3058.78	863.2
3000	346.9	-3024.15	875.0

FORSTERITE

Mg_2SiO_4
FORSTERITE   Formula wt. (g) = 140.70

Enthalpy (J/mol) = -2174140.000   Haselton (1979)
Entropy (J/mol.K) = 94.110   Robie et al. (1982b)
Heat Capacity
No. of *Cp* ranges = 1   Robie et al. (1982b)
1.6580000D+02 1.8550D-02 -3.9710D+06 0.0000D+00
2.8610D+08 0.0000D+00 -5.6100D+03
Volume (cc/mol) = 43.670   Matsui and Syono (1968)

Thermal expn.
2.01000D-05 1.39000D-08 1.62700D-03 -3.38000D-01   Saxena and Shen (1992)

Isot.bulk mod.
7.42700D-07 1.24000D-10 6.90000D-15 1.70200D-17   Saxena and Shen (1992)

P deriv., T deriv.:
5.20000D+00 2.00000D-04   Sumino et al. (1977)

Melting pt. and enthalpy of melting:
2163.00000 -   Robie et al. (1978)

NOTE: Calorimetric data: H: -2171.72+3.76, S: 94.12+1.0; H, Brousse et al. (1984)(-2174.14, Haselton 1979); S, Robie et al.(1982b).

Brousse et al. (1984) discussed the calorimetric data on enthalpy of forsterite which range from a high of -57.823 to a low of -63.262 KJ/mol [del-Hf(298)-oxides]. The entropy and *Cp* of forsterite are from Robie et al. (1982b); the heat capacity of forsterite is in the range 5-380 K. The *Cp* recommended here is closely similar to the *Cp* determined by them. Thermal expansion and compressibility are consistent with the data of Suzuki (1975) and Sumino et al. (1977), respectively.

Mg_2SiO_4

FORSTERITE

Thermochemical Data

T(K)	C_p(J)	H(kJ)	S(J)	α 10^{-4}	K 10^4	γ	V(cc)	δ_T
298	118.6	-2174.140	94.1	0.2590	28.084	1.23	43.7	6.56
300	119.1	-2173.920	94.8	0.2594	28.044	1.23	43.7	6.55
400	138.8	-2160.918	132.1	0.2761	25.866	1.11	43.8	6.28
500	150.3	-2146.419	164.4	0.2895	23.678	1.06	43.9	6.14
600	157.9	-2130.991	192.5	0.3021	21.468	1.04	44.0	6.06
700	163.5	-2114.910	217.3	0.3146	19.229	1.04	44.2	6.01
800	168.0	-2098.328	239.4	0.3273	16.955	1.04	44.3	5.99
900	171.8	-2081.337	259.5	0.3400	14.640	1.04	44.5	5.99
1000	175.1	-2063.993	277.7	0.3529	12.281	1.05	44.6	6.01
1100	178.0	-2046.336	294.5	0.3659	09.878	1.05	44.8	6.03
1200	180.8	-2028.393	310.2	0.3790	07.429	1.06	45.0	6.07
1300	183.4	-2010.183	324.7	0.3922	04.937	1.07	45.1	6.11
1400	185.8	-1991.721	338.4	0.4055	02.404	1.07	45.3	6.15
1500	188.2	-1973.018	351.3	0.4188	99.834	1.08	45.5	6.19
1600	190.5	-1954.083	363.5	0.4322	97.231	1.08	45.7	6.23
1700	192.7	-1934.922	375.2	0.4457	94.602	1.09	45.9	6.26
1800	194.9	-1915.541	386.2	0.4592	91.953	1.09	46.1	6.29
1900	197.0	-1895.944	396.8	0.4727	89.290	1.09	46.3	6.32
2000	199.1	-1876.135	407.0	0.4863	86.621	1.09	46.5	6.34
2100	201.2	-1856.117	416.7	0.4999	83.952	1.09	46.8	6.35
2200	203.3	-1835.893	426.2	0.5135	81.293	1.08	47.0	6.36
2300	205.3	-1815.465	435.2	0.5271	78.648	1.08	47.2	6.35
2400	207.3	-1794.834	444.0	0.5408	76.027	1.07	47.5	6.34
2500	209.3	-1774.002	452.5	0.5545	73.434	1.07	47.8	6.33
2600	211.3	-1752.972	460.8	0.5682	70.877	1.06	48.0	6.30
2700	213.3	-1731.743	468.8	0.5819	68.362	1.05	48.3	6.27
2800	215.2	-1710.317	476.6	0.5956	65.893	1.04	48.6	6.23
2900	217.2	-1688.694	484.2	0.6093	63.475	1.03	48.9	6.18
3000	219.1	-1666.877	491.6	0.6230	61.112	1.02	49.2	6.13

Mg_2SiO_4
FORSTERITE

				<INT VdP (J)> PRESSURE (GPa)			
T(K)	1	5	10	20	30	40	50
298	43502	214395	421884	820381	1201427	1568578	1924208
300	43504	214404	421900	820407	1201459	1568614	1924246
400	43618	214917	422810	821889	1203320	1570728	1926524
500	43738	215460	423779	823485	1205354	1573080	1929112
600	43865	216030	424797	825172	1207523	1575613	1931931
700	43997	216625	425862	826942	1209809	1578299	1934942
800	44135	217246	426971	828790	1212202	1581123	1938119
900	44279	217893	428125	830712	1214695	1584070	1941444
1000	44429	218564	429323	832704	1217280	1587129	1944902
1100	44584	219262	430564	834765	1219953	1590293	1948478
1200	44746	219985	431848	836893	1222709	1593552	1952161
1300	44914	220733	433175	839084	1225542	1596899	1955942
1400	45089	221507	434543	841337	1228449	1600329	1959811
1500	45270	222307	435953	843650	1231427	1603836	1963762
1600	45457	223133	437404	846021	1234472	1607416	1967789
1700	45650	223985	438896	848450	1237583	1611066	1971887
1800	45851	224862	440429	850936	1240758	1614783	1976055
1900	46058	225766	442002	853477	1243996	1618567	1980290
2000	46272	226695	443615	856073	1247296	1622416	1984592
2100	46492	227650	445269	858724	1250659	1626332	1988963
2200	46720	228632	446963	861431	1254085	1630316	1993404
2300	46954	229640	448697	864194	1257576	1634371	1997920
2400	47196	230674	450472	867013	1261134	1638499	2002515
2500	47445	231735	452287	869891	1264760	1642705	2007194
2600	47701	232823	454145	872827	1268459	1646993	2011964
2700	47964	233937	456044	875825	1272233	1651368	2016832
2800	48235	235080	457986	878886	1276086	1655837	2021807
2900	48514	236250	459971	882013	1280023	1660406	2026897
3000	48800	237448	462000	885207	1284048	1665082	2032113

TEPHROITE

Mn_2SiO_4
TEPHROITE Formula wt. (g) = 201.96

Enthalpy (J/mol) =	-1728070.000	Saxena and Chatterjee (1986)
Entropy (J/mol.K) =	156.400	Robie et al. (1982b)
Heat Capacity:		
No. of *Cp* ranges = 1		Robie et al. (1982b), refitted
1.3290000D+02 2.3990D-02 -2.4070D+07 0.0000D+00		
3.3270D+09 0.0000D+00 4.0060D+04		
Volume (cc/mol) =	48.620	Matsui and Syono (1968)
Thermal expn:		
2.34726D-05 9.92510D-09 5.38870D-03 -1.83749		This study
Isot.compressibility:		
7.45300D-07 1.2633D-10 -1.6001D-14 8.9915D-18		This study
P deriv. 5.20000D+00		Sumino et al. (1977)
T deriv. 2.00000D-04		This study
Melting pt. and enthalpy of melting:		
2163.00000	-	Robie et al. (1978)

NOTE:

Thermochemical Data

T(K)	C_p(J)	H(kJ)	S(J)	α 10^{-4}	K 10^4	γ	V(cc)	δ_T
298	129.2	-1728.070	156.4	0.2383	127.913	1.16	48.6	6.40
300	129.4	-1727.831	157.2	0.2400	127.877	1.16	48.6	6.35
400	144.2	-1714.148	196.5	0.2943	125.969	1.27	48.8	5.04
500	155.4	-1699.132	229.9	0.3186	124.133	1.27	48.9	4.56
600	162.6	-1683.209	258.9	0.3330	122.357	1.26	49.1	4.29
700	167.5	-1666.689	284.4	0.3437	120.631	1.26	49.2	4.11
800	171.1	-1649.753	307.0	0.3528	118.945	1.25	49.4	3.98
900	173.8	-1632.503	327.3	0.3612	117.290	1.26	49.6	3.88
1000	176.2	-1614.998	345.8	0.3695	115.658	1.26	49.8	3.80
1100	178.3	-1597.270	362.6	0.3777	114.042	1.27	49.9	3.74
1200	180.3	-1579.339	378.2	0.3860	112.436	1.28	50.1	3.69
1300	182.2	-1561.216	392.8	0.3943	110.835	1.29	50.3	3.66
1400	184.0	-1542.906	406.3	0.4028	109.233	1.30	50.5	3.64
1500	185.9	-1524.410	419.1	0.4114	107.627	1.31	50.7	3.64
1600	187.7	-1505.730	431.1	0.4200	106.011	1.32	51.0	3.64
1700	189.6	-1486.863	442.6	0.4288	104.385	1.32	51.2	3.65
1800	191.5	-1467.810	453.5	0.4376	102.744	1.33	51.4	3.67
1900	193.4	-1448.567	463.9	0.4466	101.088	1.34	51.6	3.69
2000	195.3	-1429.133	473.8	0.4556	99.415	1.35	51.9	3.71
2100	197.3	-1409.505	483.4	0.4646	97.723	1.36	52.1	3.75
2200	199.2	-1389.681	492.6	0.4738	96.013	1.37	52.3	3.78
2300	201.2	-1369.659	501.5	0.4830	94.285	1.37	52.6	3.82
2400	203.2	-1349.436	510.1	0.4922	92.539	1.38	52.8	3.85
2500	205.3	-1329.012	518.5	0.5015	90.776	1.38	53.1	3.89
2600	207.3	-1308.384	526.6	0.5108	88.998	1.39	53.4	3.93
2700	209.4	-1287.549	534.4	0.5201	87.205	1.39	53.7	3.97
2800	211.5	-1266.508	542.1	0.5295	85.400	1.39	53.9	4.00
2900	213.6	-1245.257	549.5	0.5390	83.585	1.39	54.2	4.04
3000	215.7	-1223.795	556.8	0.5484	81.763	1.39	54.5	4.07

Mn_2SiO_4
TEPHROITE

<INT VdP (J)>

PRESSURE(GPa)

T(K)	1	5	10	20	30	40	50
298	48432	238691	469685	913308	1337489	1746191	2142061
300	48435	238701	469702	913335	1337523	1746229	2142101
400	48563	239287	470761	915114	1339827	1748931	2145111
500	48710	239964	471999	917257	1342687	1752389	2149084
600	48867	240688	473333	919595	1345851	1756265	2153595
700	49030	241444	474731	922067	1349223	1760430	2158482
800	49198	242226	476181	924645	1352761	1764826	2163667
900	49371	243032	477678	927318	1356444	1769422	2169110
1000	49549	243860	479218	930077	1360261	1774199	2174785
1100	49731	244710	480802	932920	1364203	1779146	2180676
1200	49919	245583	482427	935844	1368265	1784253	2186769
1300	50111	246478	484094	938848	1372444	1789514	2193054
1400	50308	247395	485804	941929	1376735	1794923	2199521
1500	50510	248335	487555	945087	1381136	1800474	2206162
1600	50717	249298	489348	948320	1385644	1806161	2212971
1700	50929	250283	491184	951628	1390256	1811982	2219939
1800	51146	251293	493061	955010	1394970	1817931	2227061
1900	51369	252325	494981	958465	1399784	1824004	2234331
2000	51597	253382	496942	961992	1404696	1830199	2241744
2100	51831	254462	498946	965591	1409703	1836511	2249295
2200	52070	255567	500992	969260	1414805	1842938	2256979
2300	52315	256696	503081	973000	1420000	1849478	2264794
2400	52566	257849	505212	976809	1425286	1856128	2272736
2500	52822	259027	507385	980688	1430663	1862887	2280801
2600	53085	260230	509601	984636	1436130	1869752	2288989
2700	53353	261458	511860	988653	1441686	1876724	2297297
2800	53628	262712	514162	992738	1447330	1883800	2305724
2900	53909	263991	516507	996893	1453063	1890982	2314270
3000	54196	265296	518895	1001117	1458885	1898268	2322935

FAYALITE

Fe_2SiO_4
FAYALITE Formula wt. (g) = 203.78

Enthalpy (J/mol) =	-1478170.000	Robie et al. (1982a)
Entropy (J/mol.K) =	151.000	Robie et al. (1982a)
Heat Capacity:		
No. of *Cp* ranges = 1		Robie et al. (1982a), refitted
2.0890000D+02 8.0640D-03 5.8940D+06 0.0000D+00 -7.8820D+08 0.0000D+00 -3.4660D+04		
Volume (cc/mol) =	46.280	Akimoto et al. (1976)
Thermal expn: 5.6728D-05 1.6315D-09 -2.5186D-03 -1.61331D+0		This study
Isot.compressibility: 6.63158D-07 2.0846D-10 -1.0135D-14 1.1668D-17		This study
P deriv. 4.00000D+00		Sumino (1979)
T deriv. 2.20000D-04		This study
Melting pt. and enthalpy of melting: 1490.00000 92173.00000		

NOTE: Free energy of fayalite is adopted from Robie et al. (1982a) which is determined by calorimetry. Thermal expansion data are consistent with Suzuki et al.(1981).

Thermochemical Data

T (K)	C_p (J)	H (kJ)	S (J)	α 10^{-4}	K 10^4	γ	V (cc)	δ_T
298	131.6	-1478.17	151.0	0.3062	137.985	1.51	46.3	9.26
300	132.1	-1477.93	151.8	0.3090	137.912	1.51	46.3	9.17
400	150.0	-1463.73	192.5	0.4100	134.108	1.75	46.5	6.74
500	160.9	-1448.15	227.3	0.4605	130.495	1.82	46.7	5.87
600	168.7	-1431.65	257.3	0.4903	127.048	1.82	46.9	5.41
700	174.8	-1414.47	283.8	0.5098	123.745	1.81	47.1	5.13
800	179.7	-1396.74	307.5	0.5236	120.567	1.79	47.4	4.94
900	183.8	-1378.56	328.9	0.5341	117.499	1.76	47.6	4.81
1000	187.4	-1359.99	348.4	0.5423	114.528	1.74	47.9	4.71
1100	190.5	-1341.09	366.4	0.5490	111.641	1.71	48.1	4.65
1200	193.3	-1321.89	383.1	0.5547	108.828	1.68	48.4	4.60
1300	195.9	-1302.43	398.7	0.5596	106.082	1.65	48.7	4.57
1400	198.2	-1282.73	413.3	0.5639	103.396	1.62	48.9	4.56
1500	200.3	-1262.81	427.1	0.5678	100.763	1.60	49.2	4.56
1600	202.2	-1242.68	440.0	0.5713	98.180	1.57	49.5	4.56
1700	204.1	-1222.36	452.4	0.5746	95.641	1.54	49.8	4.58
1800	205.8	-1201.86	464.1	0.5777	93.145	1.52	50.1	4.60
1900	207.5	-1181.19	475.3	0.5806	90.688	1.49	50.4	4.63
2000	209.1	-1160.37	485.9	0.5833	88.270	1.46	50.7	4.66
2100	210.6	-1139.38	496.2	0.5859	85.889	1.43	51.0	4.69
2200	212.0	-1118.25	506.0	0.5884	83.545	1.40	51.3	4.73
2300	213.4	-1096.98	515.5	0.5908	81.237	1.38	51.6	4.77
2400	214.8	-1075.57	524.6	0.5931	78.965	1.35	51.9	4.81
2500	216.1	-1054.02	533.4	0.5954	76.729	1.32	52.2	4.85
2600	217.4	-1032.35	541.9	0.5976	74.531	1.29	52.5	4.89
2700	218.6	-1010.55	550.1	0.5998	72.370	1.26	52.8	4.93
2800	219.8	-988.63	558.1	0.6019	70.248	1.23	53.1	4.97
2900	221.0	-966.59	565.8	0.6040	68.165	1.21	53.4	5.01
3000	222.2	-944.43	573.3	0.6060	66.122	1.18	53.8	5.05

Fe_2SiO_4

FAYALITE

<INT VdP (J)>

T(K)	1	5	10	20	30	40	50
	PRESSURE(GPa)						
298	46114	227440	447759	870662	1274129	1661690	2035828
300	46116	227451	447777	870688	1274155	1661711	2035842
400	46281	228179	449036	872597	1276341	1663921	2037894
500	46479	229074	450625	875164	1279516	1667462	2041633
600	46696	230061	452398	878096	1283249	1671764	2046344
700	46926	231108	454287	881262	1287341	1676559	2051694
800	47164	232197	456258	884592	1291686	1681706	2057502
900	47410	233318	458290	888043	1296219	1687116	2063655
1000	47660	234465	460370	891587	1300897	1692730	2070078
1100	47916	235633	462490	895207	1305692	1698507	2076717
1200	48176	236819	464642	898888	1310579	1704415	2083530
1300	48439	238021	466822	902618	1315542	1710429	2090484
1400	48707	239237	469026	906390	1320567	1716530	2097553
1500	48977	240466	471250	910196	1325642	1722700	2104715
1600	49251	241707	473493	914030	1330757	1728925	2111950
1700	49527	242959	475752	917887	1335903	1735194	2119244
1800	49807	244221	478025	921763	1341075	1741496	2126580
1900	50089	245492	480311	925655	1346264	1747821	2133949
2000	50374	246773	482608	929557	1351467	1754163	2141339
2100	50662	248061	484915	933469	1356677	1760515	2148740
2200	50952	249358	487230	937386	1361892	1766870	2156147
2300	51245	250662	489554	941308	1367107	1773223	2163552
2400	51541	251973	491884	945231	1372320	1779572	2170949
2500	51839	253290	494220	949155	1377528	1785911	2178335
2600	52140	254614	496561	953078	1382729	1792239	2185705
2700	52443	255944	498907	956999	1387922	1798554	2193058
2800	52748	257281	501257	960916	1393106	1804854	2200391
2900	53057	258622	503611	964830	1398279	1811138	2207704
3000	53367	259969	505968	968740	1403441	1817406	2214995

MONTICELLITE

$(CaMg)SiO_4$

MONTICELLITE Formula wt. (g) = 156.47

Enthalpy (J/mol) =	-2246300.000	Robie et al. (1978), Yoder (1968)
Entropy (J/mol.K) =	112.050	Robie et al. (1978)
Heat Capacity:		
No. of *Cp* ranges = 1		Helgeson et al. (1978)
1.8761200D+02 7.9900D-03 1.9283D+06 0.0000D+00		
-3.6248D+08 0.0000D+00 -2.2822D+04		
Volume (cc/mol) =	51.470	Robie et al. (1978)
Thermal expn:		
5.5412D-05 1.50966D-09 -2.48638D-02 4.4113D+0		This study
Isot.compressibility:		
7.2388D-07 7.21789D-11 5.0000D-14 -1.4437D-18		This study
P deriv. 4.00000D+00		
T deriv. 0.00000D+00		This study
Melting pt. and enthalpy of melting:		
- -		

NOTE: Calorimetric data: H: -2248.9 ±1.72, Brousse et al. (1984); S: 108.1 ±0.2, Sharp et al. (1986). The enthalpy of monticellite is adjusted from Brousse et al. (1984) so that an improved fit to Yoder's (1968) experimental data involving wollastonite, monticellite and akermanite is obtained. The entropy is from Robie et al. (1978) and the heat capacity equation is as recommended by Helgeson et al. (1978).

$(CaMg)SiO_4$
MONTICELLITE

<INT VdP (J)>

	PRESSURE (GPa)						
T(K)	1	5	10	20	30	40	50
298	51279	252802	497452	966551	1413610	1842703	2256677
300	51281	252811	497468	966577	1413643	1842738	2256714
400	51387	253286	498302	967886	1415203	1844397	2258361
500	51501	253794	499192	969278	1416854	1846144	2260085
600	51630	254373	500212	970895	1418806	1848253	2262222
700	51774	255023	501364	972740	1421064	1850732	2264784
800	51932	255737	502633	974788	1423590	1853532	2267711
900	52103	256507	504003	977008	1426340	1856598	2270937
1000	52284	257326	505462	979373	1429277	1859880	2274403
1100	52475	258189	506996	981861	1432367	1863338	2278059
1200	52675	259089	508596	984452	1435585	1866939	2281868
1300	52882	260023	510254	987133	1438910	1870656	2285799
1400	53097	260987	511963	989891	1442325	1874469	2289825
1500	53317	261979	513718	992715	1445816	1878361	2293930
1600	53545	262996	515514	995598	1449372	1882318	2298097
1700	53777	264037	517348	998533	1452985	1886331	2302314
1800	54015	265099	519217	1001515	1456646	1890390	2306574
1900	54258	266181	521117	1004538	1460351	1894491	2310870
2000	54506	267282	523047	1007600	1464095	1898627	2315196
2100	54758	268401	525004	1010697	1467874	1902795	2319550
2200	55015	269537	526988	1013827	1471686	1906993	2323929
2300	55275	270690	528996	1016988	1475528	1911218	2328332
2400	55540	271858	531027	1020177	1479400	1915471	2332758
2500	55809	273041	533082	1023395	1483299	1919749	2337209
2600	56082	274239	535158	1026639	1487225	1924054	2341684
2700	56358	275451	537255	1029909	1491179	1928385	2346185
2800	56638	276676	539373	1033206	1495159	1932744	2350715
2900	56922	277916	541510	1036527	1499166	1937131	2355273
3000	57210	279169	543668	1039873	1503201	1941547	2359863

β-FORSTERITE

Mg_2SiO_4
β-FORSTERITE Formula wt. (g) = 140.70

Enthalpy (J/mol) =	-2142000.000	This study
Entropy (J/mol.K) =	93.000	This study
Heat Capacity:		
No. of *Cp* ranges = 1		Watanabe (1982), refitted
1.7290000D+02 1.1290D-02 -1.0771D+06 0.0000D+00		
-2.1875D+08 0.0000D+00 -1.3477D+04		
Volume (cc/mol) =	40.540	Akimoto et al. (1976)
Thermal expn:		
2.319D-05 9.040D-09 -3.966D-03 7.496D-01		This study
Isot.compressibility:		
5.51282D-07 9.2017D-11 8.849D-15 1.1529D-17		This study
P deriv. 4.30000D+00		
T deriv. 3.00000D-04		This study
Melting pt. and enthalpy of melting:		
- -		

NOTE: The phase relations for the three Mg_2SiO_4 polymorphs (olivine, beta-phase and spinel) have been experimentally investigated by Ringwood and Major (1970), Suito (1972, 1977), Kawada (1977), Ohtani (1979), Fukizawa (1982), Sawamoto (1986b), and Katsura and Ito (1989). The calorimetric data for the olivine/beta-phase transition were determined by Akaogi et al. (1984). The data on beta-phase and spinel were evaluated from the olivine/beta-phase transition by Katsura and Ito (1989) and from the beta/spinel transition by Katsura and Ito (1989) (1473-1873 K) and by Sawamoto (1986b) (1873-2373 K), combined with the calorimetric data by Akaogi et al. (1984) and Akaogi et al. (1989) (see Fei et al., 1990 for discussion).

Mg_2SiO_4

β-FORSTERITE

Thermochemical Data

T(K)	C_p(J)	H(kJ)	S(J)	α 10^{-4}	K 10^4	γ	V(cc)	δ_T
298	110.7	-2142.00	93.0	0.2102	172.471	1.34	40.5	8.24
300	111.3	-2141.79	93.7	0.2101	172.415	1.33	40.5	8.24
400	133.6	-2129.42	129.1	0.2158	169.422	1.12	40.6	8.22
500	145.5	-2115.41	160.4	0.2278	166.405	1.07	40.7	8.00
600	153.2	-2100.45	187.6	0.2409	163.354	1.06	40.8	7.81
700	158.7	-2084.84	211.7	0.2538	160.260	1.07	40.9	7.66
800	163.0	-2068.75	233.1	0.2664	157.119	1.08	41.0	7.57
900	166.5	-2052.27	252.5	0.2784	153.924	1.09	41.1	7.52
1000	169.4	-2035.47	270.2	0.2901	150.676	1.10	41.2	7.49
1100	172.0	-2018.40	286.5	0.3015	147.372	1.11	41.4	7.50
1200	174.3	-2001.08	301.6	0.3125	144.016	1.12	41.5	7.51
1300	176.5	-1983.54	315.6	0.3233	140.610	1.12	41.6	7.54
1400	178.5	-1965.79	328.8	0.3340	137.158	1.13	41.8	7.58
1500	180.3	-1947.85	341.1	0.3444	133.667	1.13	41.9	7.62
1600	182.1	-1929.73	352.8	0.3547	130.143	1.13	42.1	7.66
1700	183.7	-1911.44	363.9	0.3648	126.594	1.14	42.2	7.71
1800	185.4	-1892.98	374.5	0.3749	123.028	1.13	42.4	7.74
1900	186.9	-1874.37	384.5	0.3849	119.455	1.13	42.5	7.78
2000	188.4	-1855.60	394.2	0.3947	115.882	1.13	42.7	7.80
2100	189.9	-1836.68	403.4	0.4046	112.320	1.12	42.9	7.82
2200	191.4	-1817.62	412.3	0.4143	108.777	1.12	43.0	7.83
2300	192.8	-1798.41	420.8	0.4240	105.263	1.11	43.2	7.84
2400	194.2	-1779.06	429.0	0.4336	101.784	1.10	43.4	7.83
2500	195.5	-1759.57	437.0	0.4432	98.351	1.09	43.6	7.82
2600	196.9	-1739.95	444.7	0.4528	94.969	1.08	43.8	7.80
2700	198.2	-1720.19	452.1	0.4623	91.645	1.07	44.0	7.77
2800	199.6	-1700.31	459.4	0.4718	88.387	1.05	44.2	7.73
2900	200.9	-1680.28	466.4	0.4813	85.198	1.04	44.4	7.69
3000	202.1	-1660.13	473.2	0.4907	82.084	1.02	44.6	7.63

	<INT VdP (J)> PRESSURE(GPa)						
T(K)	1	5	10	20	30	40	50
298	40423	199901	394697	771273	1133280	1483130	1822574
300	40425	199908	394709	771293	1133302	1483154	1822597
400	40508	200285	395375	772344	1134560	1484493	1823927
500	40596	200681	396075	773464	1135927	1485993	1825480
600	40689	201101	396821	774670	1137423	1487670	1827268
700	40788	201546	397613	775960	1139041	1489512	1829266
800	40891	202014	398449	777327	1140768	1491497	1831444
900	41000	202506	399325	778764	1142593	1493607	1833777
1000	41114	203020	400241	780265	1144503	1495824	1836240
1100	41233	203555	401192	781825	1146490	1498135	1838814
1200	41357	204110	402178	783438	1148545	1500527	1841481
1300	41485	204685	403197	785101	1150661	1502988	1844228
1400	41618	205279	404247	786810	1152831	1505512	1847042
1500	41755	205892	405327	788561	1155050	1508089	1849914
1600	41898	206523	406436	790352	1157315	1510714	1852835
1700	42044	207172	407573	792181	1159621	1513382	1855801
1800	42195	207839	408737	794045	1161966	1516090	1858806
1900	42351	208523	409927	795944	1164347	1518835	1861847
2000	42511	209224	411143	797875	1166764	1521615	1864922
2100	42676	209942	412385	799839	1169215	1524430	1868032
2200	42845	210677	413651	801834	1171700	1527280	1871176
2300	43018	211428	414942	803861	1174219	1530165	1874356
2400	43197	212196	416257	805920	1176773	1533087	1877576
2500	43379	212981	417597	808010	1179364	1536050	1880838
2600	43567	213782	418961	810133	1181992	1539055	1884146
2700	43758	214599	420349	812289	1184661	1542105	1887506
2800	43955	215433	421762	814480	1187372	1545206	1890924
2900	44156	216283	423199	816706	1190128	1548362	1894404
3000	44362	217150	424662	818969	1192932	1551576	1897955

Mg_2SiO_4

β-FAYALITE

Fe_2SiO_4
β-FAYALITE Formula wt. (g) = 203.78

Enthalpy (J/mol) =	-1468000.000	This study
Entropy (J/mol.K) =	141.800	This study
Heat Capacity:		
No. of *Cp* ranges = 1		Watanabe (1982), refitted
2.2440600D+02 8.9184D-03 1.5590D+07 0.0000D+00		
-1.9420D+09 0.0000D+00 -5.7576D+04		
Volume (cc/mol) =	43.140	Fei and Saxena (1986)
Thermal expn:		
8.774D-05 -2.609D-09 -3.398D-02 4.435D+00		This study
Isot.compressibility:		
5.540D-07 7.947D-11 7.069D-14 -4.957D-18		This study
P deriv. 4.00000D+00		
T deriv. 5.00000D-04		This study
Melting pt. and enthalpy of melting:		
- -		

NOTE: Beta-fayalite is a fictive component in the system. Its thermochemical and thermophysical properties are either extrapolated from limited solution data or estimated by using analogous data (Fei and Saxena, 1986). The data on H_{1000} and S_{1000} of the olivine-beta transition as estimated by Akaogi et al. (1989) were used as additional constraints in the data evaluation (see Fei, 1990).

Thermochemical Data

T (K)	C_p (J)	H (kJ)	S (J)	α 10^{-4}	K 10^4	γ	V (cc)	δ_T
298	136.1	-1468.00	141.8	0.2288	171.278	1.25	43.1	9.00
300	136.5	-1467.75	142.6	0.2297	171.213	1.25	43.1	8.99
400	151.1	-1453.29	184.1	0.2947	167.566	1.44	43.3	7.60
500	160.5	-1437.68	218.9	0.3622	163.723	1.65	43.4	6.62
600	168.1	-1421.24	248.9	0.4186	159.729	1.81	43.6	6.06
700	174.6	-1404.10	275.3	0.4642	155.626	1.92	43.8	5.74
800	180.1	-1386.36	299.0	0.5011	151.451	2.00	44.0	5.53
900	185.0	-1368.09	320.5	0.5311	147.239	2.05	44.2	5.40
1000	189.4	-1349.37	340.2	0.5559	143.020	2.08	44.4	5.30
1100	193.3	-1330.23	358.4	0.5764	138.821	2.10	44.7	5.23
1200	196.8	-1310.72	375.4	0.5937	134.663	2.10	45.0	5.16
1300	200.1	-1290.87	391.3	0.6083	130.568	2.09	45.2	5.11
1400	203.0	-1270.72	406.2	0.6208	126.549	2.08	45.5	5.06
1500	205.8	-1250.28	420.3	0.6314	122.620	2.06	45.8	5.01
1600	208.3	-1229.57	433.7	0.6406	118.791	2.03	46.1	4.96
1700	210.7	-1208.62	446.4	0.6485	115.070	2.01	46.4	4.91
1800	213.0	-1187.44	458.5	0.6553	111.461	1.97	46.7	4.86
1900	215.1	-1166.03	470.1	0.6613	107.970	1.94	47.0	4.81
2000	217.1	-1144.42	481.2	0.6664	104.598	1.90	47.3	4.75
2100	219.0	-1122.62	491.8	0.6709	101.346	1.87	47.6	4.70
2200	220.9	-1100.62	502.0	0.6747	98.213	1.83	47.9	4.64
2300	222.7	-1078.44	511.9	0.6780	95.200	1.79	48.3	4.58
2400	224.4	-1056.09	521.4	0.6809	92.304	1.75	48.6	4.52
2500	226.0	-1033.56	530.6	0.6834	89.523	1.71	48.9	4.45
2600	227.6	-1010.88	539.5	0.6854	86.854	1.67	49.3	4.39
2700	229.2	-988.04	548.1	0.6872	84.294	1.63	49.6	4.33
2800	230.7	-965.04	556.5	0.6886	81.839	1.60	50.0	4.26
2900	232.2	-941.89	564.6	0.6898	79.487	1.56	50.3	4.20
3000	233.6	-918.60	572.5	0.6908	77.233	1.52	50.6	4.13

Fe_2SiO_4
β-FAYALITE

<INT VdP (J)>

PRESSURE(GPa)

T(K)	1	5	10	20	30	40	50
298	43015	212694	419882	820133	1204479	1575490	1935044
300	43017	212702	419896	820154	1204502	1575512	1935064
400	43126	213194	420763	821515	1206117	1577216	1936736
500	43265	213829	421904	823399	1208501	1579941	1939697
600	43432	214594	423294	825752	1211569	1583570	1943786
700	43621	215466	424887	828486	1215190	1587923	1948779
800	43828	216424	426644	831525	1219251	1592855	1954490
900	44051	217455	428535	834811	1223667	1598253	1960783
1000	44287	218545	430537	838300	1228376	1604034	1967555
1100	44534	219686	432631	841958	1233330	1610139	1974732
1200	44791	220869	434805	845760	1238493	1616521	1982258
1300	45056	222091	437047	849687	1243838	1623147	1990091
1400	45328	223345	439348	853723	1249345	1629991	1998203
1500	45607	224628	441703	857858	1254999	1637034	2006569
1600	45893	225938	444105	862081	1260788	1644262	2015172
1700	46183	227271	446550	866387	1266702	1651663	2024000
1800	46479	228627	449034	870770	1272736	1659230	2033042
1900	46780	230002	451556	875225	1278883	1666956	2042291
2000	47085	231396	454111	879749	1285140	1674836	2051740
2100	47394	232807	456699	884339	1291504	1682867	2061385
2200	47707	234234	459318	888995	1297972	1691045	2071223
2300	48023	235677	461967	893713	1304543	1699368	2081249
2400	48343	237135	464644	898493	1311214	1707834	2091462
2500	48666	238606	467348	903333	1317985	1716442	2101859
2600	48993	240092	470080	908234	1324856	1725190	2112439
2700	49322	241590	472838	913195	1331824	1734077	2123198
2800	49654	243101	475622	918214	1338890	1743102	2134136
2900	49989	244625	478432	923292	1346052	1752264	2145250
3000	50327	246161	481267	928429	1353311	1761561	2156540

Fe_2SiO_4

FAYALITE

γ-FORSTERITE

Mg_2SiO_4
γ-FORSTERITE Formula wt. (g) = 140.70

Enthalpy (J/mol) =	-2133200.000	Fei et al. (1990)
Entropy (J/mol.K) =	89.000	Fei et al. (1990)
Heat Capacity:		
No. of *Cp* ranges = 1		Watanabe (1982), refitted
1.5856000D+02 1.2200D-02 -1.2297D+07 0.0000D+00		
1.4841D+09 0.0000D+00 7.9719D+03		
Volume (cc/mol) =	39.650	Ito et al. (1974)
Thermal expn:		
1.225D-05 1.104D-08 2.496D-03 -5.110D-01		This study
Isot.compressibility:		
5.07778D-7 1.3371D-10 -2.9854D-14 2.7822D-17		This study
P deriv. 4.30000D+00		This study
T deriv. 6.00000D-04		This study
Melting pt. and enthalpy of melting:		
- -		

NOTE: The phase relations for the three Mg_2SiO_4 polymorphs (olivine, beta-phase and spinel) have been experimentally investigated by Ringwood and Major (1970), Suito (1972, 1977), Kawada (1977), Ohtani (1979), Fukizawa (1982), Sawamoto (1986b), and Katsura and Ito (1989). The calorimetric data for the olivine/beta-phase transition were determined by Akaogi et al. (1984). The data on beta-phase and spinel were evaluated from the olivine/beta-phase transition by Katsura and Ito (1989) and FROM THE beta/spinel transition by Katsura and Ito (1989) (1473-1873 K) and by Sawamoto (1986b) (1873-2373 K), combined with the calorimetric data by Akaogi et al. (1984) and Akaogi et al. (1989) (see Fei et al., 1990 for discussion). The thermal expansion is consistent with the data of Suzuki et al. (1979).

Mg_2SiO_4

γ-FORSTERITE

Thermochemical Data

T(K)	C_p(J)	H(kJ)	S(J)	α 10^{-4}	K 10^4	γ	V(cc)	δ_T
298	106.6	-2133.20	89.0	0.1816	183.242	1.25	39.6	12.44
300	107.1	-2133.00	89.7	0.1820	183.165	1.24	39.7	12.41
400	129.7	-2121.07	123.8	0.1971	179.126	1.09	39.7	11.19
500	143.3	-2107.36	154.4	0.2072	175.240	1.02	39.8	10.55
600	151.9	-2092.57	181.3	0.2161	171.448	0.99	39.9	10.15
700	157.7	-2077.08	205.2	0.2250	167.704	0.97	40.0	9.90
800	162.0	-2061.08	226.5	0.2340	163.966	0.97	40.1	9.76
900	165.3	-2044.71	245.8	0.2433	160.201	0.97	40.2	9.72
1000	167.9	-2028.05	263.4	0.2528	156.383	0.97	40.3	9.74
1100	170.2	-2011.14	279.5	0.2624	152.493	0.98	40.4	9.82
1200	172.2	-1994.02	294.4	0.2722	148.519	0.98	40.5	9.94
1300	174.0	-1976.72	308.2	0.2822	144.452	0.99	40.6	10.09
1400	175.6	-1959.24	321.2	0.2923	140.291	0.99	40.7	10.26
1500	177.1	-1941.60	333.3	0.3025	136.041	0.99	40.8	10.43
1600	178.6	-1923.81	344.8	0.3127	131.710	0.99	41.0	10.61
1700	180.0	-1905.88	355.7	0.3231	127.308	0.99	41.1	10.77
1800	181.4	-1887.81	366.0	0.3335	122.852	0.99	41.2	10.93
1900	182.7	-1869.60	375.9	0.3440	118.359	0.98	41.4	11.07
2000	184.1	-1851.26	385.3	0.3545	113.849	0.97	41.5	11.18
2100	185.3	-1832.79	394.3	0.3651	109.341	0.96	41.7	11.27
2200	186.6	-1814.19	402.9	0.3757	104.855	0.95	41.8	11.34
2300	187.9	-1795.46	411.3	0.3863	100.412	0.94	42.0	11.38
2400	189.1	-1776.61	419.3	0.3970	96.031	0.92	42.1	11.40
2500	190.4	-1757.64	427.0	0.4077	91.728	0.91	42.3	11.39
2600	191.6	-1738.54	434.5	0.4184	87.519	0.89	42.5	11.35
2700	192.8	-1719.31	441.8	0.4291	83.418	0.87	42.7	11.29
2800	194.1	-1699.97	448.8	0.4399	79.437	0.85	42.8	11.21
2900	195.3	-1680.50	455.6	0.4507	75.585	0.83	43.0	11.11
3000	196.5	-1660.91	462.3	0.4615	71.869	0.81	43.2	10.99

<INT VdP (J)>

T(K)	PRESSURE(GPa) 1	5	10	20	30	40	50
298	39542	195667	386597	756291	1112265	1456715	1791253
300	39544	195672	386606	756303	1112276	1456723	1791254
400	39617	195990	387140	757053	1113037	1457356	1791660
500	39694	196332	387726	757922	1114013	1458326	1792552
600	39776	196692	388349	758878	1115140	1459534	1793787
700	39861	197070	389005	759903	1116387	1460925	1795283
800	39950	197463	389692	760989	1117732	1462462	1796983
900	40043	197873	390407	762127	1119158	1464114	1798839
1000	40139	198299	391149	763311	1120650	1465855	1800812
1100	40240	198741	391918	764535	1122194	1467664	1802869
1200	40344	199198	392710	765792	1123778	1469519	1804980
1300	40452	199671	393526	767078	1125392	1471404	1807120
1400	40565	200158	394363	768388	1127027	1473304	1809268
1500	40681	200661	395220	769718	1128674	1475206	1811406
1600	40802	201178	396096	771063	1130327	1477100	1813520
1700	40927	201709	396991	772422	1131980	1478979	1815600
1800	41055	202254	397902	773791	1133631	1480837	1817637
1900	41189	202813	398830	775169	1135275	1482671	1819629
2000	41326	203386	399773	776555	1136913	1484479	1821574
2100	41468	203971	400732	777948	1138544	1486262	1823474
2200	41614	204570	401706	779348	1140169	1488024	1825333
2300	41765	205183	402694	780757	1141791	1489768	1827156
2400	41920	205808	403698	782176	1143413	1491500	1828954
2500	42079	206446	404716	783606	1145041	1493228	1830735
2600	42243	207098	405751	785051	1146678	1494959	1832512
2700	42411	207762	406802	786513	1148331	1496704	1834296
2800	42584	208440	407870	787995	1150007	1498471	1836102
2900	42761	209132	408956	789502	1151712	1500271	1837942
3000	42943	209837	410062	791038	1153454	1502115	1839832

γ-FAYALITE

Fe_2SiO_4
γ-FAYALITE Formula wt. (g) = 203.78

Enthalpy (J/mol) = -1471500.000 This study
Entropy (J/mol.K) = 141.032 This study
Heat Capacity:
No. of *Cp* ranges = 1 Watanabe (1982)
1.6786000D+02 2.8124D-02 -5.6547D+06 0.0000D+00
7.8238D+08 0.0000D+00 -3.5644D+03
Volume (cc/mol) = 42.020 Marumo et al. (1977)
Thermal expn:
8.89700D-05 2.80300D-09 -3.42600D-02 3.81300D+00 Fei and Saxena (1986)
Isot.compressibility:
3.86000D-07 3.97800D-10 -6.30400D-14 8.27500D-18
P deriv. 4.00000D+00 Sato (1977)
T deriv. 1.00000D-03 This study
Melting pt. and enthalpy of melting:
\- -

NOTE: The olivine-spinel transition in Fe_2SiO_4 has been experimentally studied using the quench technique (Akimoto et al., 1965, 1967) and using the in-situ X-ray diffraction method (Inoue, 1975; Sung and Burns, 1976; and Furnish and Bassett,1983). Recently, Yagi et al.(1987) determined the equilibrium phase boundary between 1073 and 1473 K using a cubic anvil type of high pressure and high temperature X-ray diffraction apparatus combined with synchrotron radiation. The transition pressure P(GPa) = 3.43 + 0.0025T(K). The enthalpy and entropy for Fe2SiO4 spinel were evaluated from the phase equilibrium data (Yagi et al., 1987), constrained by the calorimetric data on H_{1000} and S_{1000} of the olivine-spinel transition as obtained by Akaogi et al. (1989).

Fe_2SiO_4
γ-FAYALITE

Thermochemical Data

T(K)	C_p(J)	H(kJ)	S(J)	α 10^{-4}	K 10^4	γ	V(cc)	δ_T
298	130.2	-1471.50	141.0	0.1779	200.313	1.16	42.0	40.80
300	130.6	-1471.26	141.8	0.1798	200.044	1.16	42.0	40.30
400	147.1	-1457.32	181.8	0.2827	186.719	1.54	42.1	23.20
500	158.4	-1442.02	215.9	0.3710	175.385	1.79	42.3	16.12
600	166.7	-1425.74	245.6	0.4414	165.625	1.96	42.4	12.42
700	173.2	-1408.73	271.8	0.4977	157.132	2.06	42.6	10.16
800	178.6	-1391.14	295.3	0.5435	149.671	2.13	42.9	8.62
900	183.3	-1373.04	316.6	0.5813	143.064	2.18	43.1	7.49
1000	187.5	-1354.49	336.1	0.6133	137.168	2.21	43.4	6.63
1100	191.5	-1335.54	354.2	0.6406	131.871	2.23	43.6	5.95
1200	195.2	-1316.20	371.0	0.6643	127.084	2.24	43.9	5.40
1300	198.7	-1296.51	386.8	0.6852	122.732	2.25	44.2	4.94
1400	202.1	-1276.47	401.6	0.7037	118.755	2.25	44.5	4.56
1500	205.4	-1256.10	415.7	0.7203	115.103	2.25	44.8	4.23
1600	208.6	-1235.40	429.0	0.7353	111.733	2.25	45.2	3.95
1700	211.8	-1214.38	441.8	0.7490	108.610	2.25	45.5	3.70
1800	214.9	-1193.04	454.0	0.7616	105.703	2.25	45.8	3.49
1900	218.0	-1171.40	465.7	0.7732	102.986	2.24	46.2	3.30
2000	221.0	-1149.45	476.9	0.7840	100.438	2.24	46.6	3.14
2100	224.0	-1127.20	487.8	0.7941	98.038	2.24	46.9	2.99
2200	227.0	-1104.65	498.3	0.8035	95.771	2.24	47.3	2.87
2300	230.0	-1081.79	508.4	0.8124	93.621	2.24	47.7	2.76
2400	232.9	-1058.65	518.3	0.8208	91.575	2.23	48.1	2.66
2500	235.9	-1035.21	527.8	0.8288	89.622	2.23	48.5	2.57
2600	238.8	-1011.47	537.2	0.8364	87.752	2.23	48.9	2.49
2700	241.7	-987.44	546.2	0.8437	85.957	2.23	49.3	2.43
2800	244.6	-963.12	555.1	0.8507	84.228	2.23	49.7	2.37
2900	247.6	-938.51	563.7	0.8574	82.558	2.23	50.1	2.32
3000	250.4	-913.61	572.1	0.8638	80.941	2.23	50.6	2.28

<INT VdP (J)>

T(K)	PRESSURE(GPa) 1	5	10	20	30	40	50
298	41915	207580	410491	804103	1183718	1551399	1908697
300	41917	207584	410492	804087	1183672	1551313	1908564
400	42007	207893	410806	803728	1181980	1547823	1902934
500	42138	208411	411549	804290	1181773	1546452	1900124
600	42303	209102	412651	805624	1182818	1546875	1899703
700	42496	209934	414045	807597	1184908	1548801	1901291
800	42711	210883	415682	810109	1187884	1552007	1904592
900	42946	211931	417526	813082	1191625	1556323	1909379
1000	43197	213063	419551	816461	1196039	1561613	1915468
1100	43463	214270	421734	820199	1201053	1567771	1922714
1200	43741	215544	424060	824263	1206606	1574708	1930995
1300	44031	216879	426516	828622	1212651	1582352	1940211
1400	44333	218270	429091	833253	1219146	1590640	1950275
1500	44644	219713	431778	838138	1226057	1599519	1961114
1600	44964	221204	434570	843258	1233353	1608945	1972665
1700	45294	222742	437459	848599	1241010	1618878	1984873
1800	45632	224324	440443	854150	1249005	1629283	1997689
1900	45978	225948	443516	859899	1257317	1640130	2011072
2000	46332	227612	446675	865837	1265930	1651392	2024985
2100	46694	229316	449916	871955	1274828	1663046	2039396
2200	47063	231058	453238	878246	1283998	1675070	2054275
2300	47440	232837	456637	884704	1293427	1687447	2069598
2400	47824	234652	460111	891322	1303105	1700160	2085342
2500	48214	236503	463660	898096	1313021	1713193	2101486
2600	48612	238389	467281	905020	1323168	1726534	2118012
2700	49017	240310	470972	912091	1333536	1740171	2134906
2800	49429	242264	474733	919304	1344120	1754093	2152151
2900	49848	244253	478562	926657	1354912	1768292	2169737
3000	50273	246275	482459	934146	1365908	1782757	2187650

ORTHOENSTATITE

$MgSiO_3$
ORTHOENSTATITE Formula wt. (g) = 100.39

Enthalpy (J/mol) =	-1546290.000	Brousse et al. (1984)
Entropy (J/mol.K) =	66.270	Krupka et al. (1985a, b)

Heat Capacity:
No. of *Cp* ranges = 1 — Haselton (1979), refitted

1.4445000D+02 1.8820D-03 -1.3500D+06 0.0000D+00
4.6120D+08 0.0000D+00 -1.9380D+04

Volume (cc/mol) = 31.276 — Robie et al. (1978)

Thermal expn:
3.871D-05 4.4633D-09 3.4352D-04 -1.7278D+0 — This study

Isot.compressibility:
8.892D-07 1.3584D-10 3.1613D-14 1.14126D-17 — This study

P deriv. 4.20000D+00
T deriv. 1.50000D-04 — This study

Melting pt. and enthalpy of melting:
\- -

NOTE: Calorimetric data: H: -1546.19±1.76, Brousse et al.(1984);S: 66.27±0.1, Krupka et al.(1985a). The enthalpy of enstatite is from Brousse et al. (1984). Although there are other determinations of the enthalpy, the value of Brousse et al. is close to the average of all the values. So, their value is adopted. Entropy and heat capacity of enstatite are from Haselton (1979). All these values are calorimetric.

Thermochemical Data

T (K)	C_p (J)	H (kJ)	S (J)	$\alpha\ 10^{-4}$	K 10^4	γ	V (cc)	δ_T
298	82.2	-1546.29	66.3	0.2176	107.202	0.89	31.3	7.77
300	82.5	-1546.14	66.8	0.2200	107.169	0.90	31.3	7.69
400	95.5	-1537.21	92.4	0.3056	105.338	1.07	31.4	5.74
500	104.9	-1527.16	114.8	0.3472	103.471	1.10	31.5	5.25
600	111.7	-1516.31	134.5	0.3716	101.569	1.09	31.6	5.09
700	116.7	-1504.88	152.1	0.3880	99.632	1.08	31.7	5.06
800	120.5	-1493.01	168.0	0.4001	97.661	1.07	31.8	5.09
900	123.6	-1480.80	182.4	0.4098	95.658	1.05	32.0	5.15
1000	126.1	-1468.32	195.5	0.4179	93.627	1.04	32.1	5.23
1100	128.1	-1455.60	207.6	0.4250	91.569	1.03	32.2	5.32
1200	129.9	-1442.70	218.8	0.4315	89.489	1.01	32.4	5.41
1300	131.4	-1429.64	229.3	0.4375	87.390	1.00	32.5	5.51
1400	132.7	-1416.43	239.1	0.4432	85.276	0.99	32.6	5.61
1500	133.9	-1403.10	248.3	0.4487	83.152	0.97	32.8	5.70
1600	134.9	-1389.65	257.0	0.4539	81.022	0.96	32.9	5.80
1700	135.9	-1376.11	265.2	0.4590	78.891	0.95	33.1	5.88
1800	136.7	-1362.48	273.0	0.4640	76.763	0.93	33.2	5.97
1900	137.5	-1348.77	280.4	0.4689	74.643	0.92	33.4	6.04
2000	138.2	-1334.98	287.5	0.4738	72.535	0.91	33.6	6.11
2100	138.9	-1321.12	294.2	0.4785	70.443	0.89	33.7	6.18
2200	139.5	-1307.20	300.7	0.4833	68.372	0.88	33.9	6.23
2300	140.1	-1293.21	306.9	0.4880	66.325	0.86	34.0	6.28
2400	140.7	-1279.17	312.9	0.4927	64.305	0.85	34.2	6.33
2500	141.2	-1265.08	318.6	0.4973	62.316	0.83	34.4	6.36
2600	141.7	-1250.93	324.2	0.5019	60.361	0.82	34.6	6.39
2700	142.2	-1236.73	329.5	0.5065	58.443	0.80	34.7	6.42
2800	142.6	-1222.49	334.7	0.5111	56.563	0.79	34.9	6.43
2900	143.1	-1208.21	339.7	0.5157	54.724	0.77	35.1	6.44

$MgSiO_3$
ORTHOENSTATITE

<INT VdP (J)>

PRESSURE (GPa)

T(K)	1	5	10	20	30	40	50
298	31132	152997	300093	580251	845621	1099291	1343317
300	31133	153002	300102	580263	845634	1099303	1343326
400	31214	153359	300717	581206	846744	1100476	1344493
500	31315	153806	301507	582481	848337	1102285	1346448
600	31425	154301	302385	583921	850169	1104405	1348787
700	31542	154825	303317	585455	852132	1106693	1351334
800	31663	155369	304284	587046	854172	1109078	1353997
900	31788	155929	305275	588673	856256	1111515	1356721
1000	31917	156500	306285	590324	858365	1113978	1359474
1100	32048	157082	307309	591989	860486	1116450	1362232
1200	32181	157672	308344	593663	862610	1118918	1364980
1300	32317	158270	309387	595341	864729	1121373	1367706
1400	32456	158875	310438	597019	866840	1123808	1370403
1500	32596	159487	311496	598696	868937	1126219	1373064
1600	32739	160104	312558	600370	871019	1128602	1375685
1700	32883	160727	313625	602038	873084	1130956	1378265
1800	33030	161355	314696	603700	875130	1133278	1380800
1900	33178	161989	315770	605355	877157	1135567	1383292
2000	33329	162627	316847	607002	879164	1137825	1385740
2100	33481	163271	317926	608641	881151	1140052	1388145
2200	33636	163919	319008	610273	883118	1142248	1390510
2300	33792	164571	320092	611897	885067	1144416	1392837
2400	33951	165228	321178	613513	886999	1146557	1395128
2500	34111	165889	322265	615123	888915	1148673	1397388
2600	34274	166554	323355	616726	890816	1150768	1399619
2700	34438	167224	324446	618324	892704	1152844	1401826
2800	34605	167898	325540	619916	894582	1154904	1404013
2900	34773	168576	326636	621506	896451	1156951	1406184
3000	34944	169258	327735	623092	898313	1158990	1408346

ORTHODIOPSIDE

$(CaMg)_{.5}SiO_3$
ORTHODIOPSIDE Formula wt. (g) = 108.28

Enthalpy (J/mol) = -1598040.000 Shi et al. (1993)
Entropy (J/mol.K) = 72.380 Shi et al. (1993)
Heat Capacity:
No. of *Cp* ranges = 1 Shi et al. (1993)
1.3173500D+02 1.0468D-02 -3.0775D+05 4.9079D-09
0.0000D+00 0.0000D+00 -2.4601D+04
Volume (cc/mol) = 33.525 Shi et al. (1993)
Thermal expn:
3.20639D-05 8.10965D-09 1.34676D-03 -1.81674D+00
Shi et al. (1993)
Isot.compressibility:
8.41911D-07 1.56010D-10 1.09190D-14 1.23890D-17
Shi et al. (1993)
P deriv. 4.00000D+00
T deriv. 5.00000D-04 This study
Melting pt. and enthalpy of melting:
- -

NOTE: The properties of these phases are adopted from Shi et al. (1993). Their evaluations are based on the phase equilibrium data compiled by Lindsley (1983) on quadrilateral pyroxenes.

Thermochemical Data

T(K)	C_p(J)	H(kJ)	S(J)	α 10^{-4}	K 10^4	γ	V(cc)	δ_T
298	48.9	-1598.040	72.4	0.1856	112.394	1.44	33.5	10.04
300	49.5	-1597.949	72.7	0.1880	112.356	1.44	33.5	9.92
400	72.5	-1591.743	90.4	0.2732	110.271	1.42	33.6	6.89
500	86.5	-1583.742	108.2	0.3155	108.202	1.36	33.7	6.04
600	96.2	-1574.580	124.9	0.3413	106.143	1.31	33.8	5.68
700	103.3	-1564.592	140.2	0.3596	104.089	1.27	33.9	5.49
800	108.9	-1553.973	154.4	0.3740	102.036	1.24	34.1	5.38
900	113.4	-1542.850	167.5	0.3862	99.980	1.21	34.2	5.33
1000	117.3	-1531.307	179.7	0.3970	97.921	1.19	34.3	5.30
1100	120.6	-1519.407	191.0	0.4071	95.857	1.17	34.5	5.30
1200	123.6	-1507.193	201.6	0.4166	93.786	1.16	34.6	5.31
1300	126.2	-1494.699	211.6	0.4257	91.710	1.14	34.8	5.33
1400	128.7	-1481.952	221.1	0.4345	89.628	1.13	34.9	5.35
1500	130.9	-1468.971	230.0	0.4432	87.542	1.12	35.1	5.38
1600	133.0	-1455.775	238.5	0.4517	85.454	1.10	35.2	5.41
1700	135.0	-1442.375	246.7	0.4601	83.365	1.09	35.4	5.45
1800	136.8	-1428.785	254.4	0.4685	81.277	1.08	35.5	5.48
1900	138.6	-1415.012	261.9	0.4768	79.194	1.07	35.7	5.51
2000	140.3	-1401.066	269.0	0.4850	77.118	1.05	35.9	5.54
2100	142.0	-1386.952	275.9	0.4932	75.051	1.04	36.1	5.57
2200	143.5	-1372.677	282.6	0.5014	72.998	1.03	36.2	5.59
2300	145.1	-1358.245	289.0	0.5096	70.961	1.02	36.4	5.61
2400	146.6	-1343.662	295.2	0.5177	68.942	1.00	36.6	5.63
2500	148.0	-1328.930	301.2	0.5259	66.945	0.99	36.8	5.64
2600	149.5	-1314.054	307.0	0.5340	64.973	0.97	37.0	5.65
2700	150.9	-1299.035	312.7	0.5421	63.028	0.96	37.2	5.65
2800	152.3	-1283.878	318.2	0.5502	61.113	0.95	37.4	5.65
2900	153.6	-1268.585	323.6	0.5583	59.230	0.93	37.6	5.64
3000	154.9	-1253.157	328.8	0.5664	57.382	0.92	37.8	5.63

$(CaMg)_{.5}SiO_3$
ORTHODIOPSIDE

				<INT VdP (J)>			
				PRESSURE (GPa)			
T(K)	1	5	10	20	30	40	50
298	33378	164146	322164	623420	908918	1181826	1444290
300	33379	164150	322172	623428	908925	1181829	1444287
400	33455	164478	322720	624214	909779	1182648	1445004
500	33551	164904	323465	625402	911262	1184349	1446879
600	33659	165385	324317	626813	913101	1186556	1449430
700	33774	165902	325240	628369	915174	1189102	1452436
800	33895	166445	326214	630031	917420	1191896	1455779
900	34021	167011	327230	631777	919801	1194886	1459384
1000	34151	167595	328281	633594	922293	1198035	1463202
1100	34285	168196	329363	635472	924881	1201320	1467198
1200	34423	168813	330474	637404	927553	1204721	1471347
1300	34564	169445	331612	639386	930300	1208225	1475627
1400	34709	170091	332774	641414	933116	1211822	1480024
1500	34857	170752	333962	643485	935996	1215503	1484524
1600	35009	171426	335172	645598	938935	1219261	1489120
1700	35164	172114	336406	647751	941931	1223092	1493802
1800	35323	172815	337662	649943	944981	1226991	1498566
1900	35485	173529	338941	652173	948083	1230955	1503405
2000	35650	174257	340241	654440	951237	1234983	1508319
2100	35819	174998	341564	656745	954442	1239074	1513303
2200	35991	175752	342909	659086	957697	1243226	1518358
2300	36167	176520	344275	661465	961002	1247439	1523483
2400	36347	177300	345665	663882	964359	1251714	1528677
2500	36530	178094	347076	666337	967767	1256052	1533943
2600	36716	178902	348510	668831	971229	1260453	1539280
2700	36906	179723	349968	671364	974744	1264921	1544693
2800	37100	180558	351448	673938	978314	1269456	1550182
2900	37297	181407	352953	676554	981942	1274060	1555751
3000	37498	182269	354482	679213	985628	1278737	1561402

ORTHOFERROSILITE

$FeSiO_3$
ORTHOFERROSILITE Formula wt. (g) = 131.93

Enthalpy (J/mol) = -1193450.000 This study
Entropy (J/mol.K) = 95.900 This study
Heat Capacity:
No. of *Cp* ranges = 1 Watnabe (1982), refitted
1.0960000D+02 1.3880D-02 -9.1160D+06 0.0000D+00
1.0330D+09 0.0000D+00 1.0980D+04
Volume (cc/mol) = 32.950 Robie et al. (1978)
Thermal expn:
4.95565D-05 8.76170D-09 -1.18560D-02 1.56019D+00
This study
Isot.compressibility:
9.27714D-07 3.70379D-10 -1.01000D-13 2.90970D-17
This study
P deriv. 4.20000D+00
T deriv. 0.00000D+00 This study
Melting pt. and enthalpy of melting:
- -

NOTE: Calorimetric data: S: 94.6 ±0.1, Bohlen and Boettcher (1981) (derived indirectly). Ferrosilite breaks down to give fayalite and quartz according to the following reaction

Fer = Fay + Q

This reaction was studied experimentally by Bohlen et al. (1980). However, when orthopyroxene solutions break down there is a three-phase stability field where the orthopyroxene solution is present along with the olivine solution and quartz. This phenomenon was recently investigated by Bohlen and Boettcher (1981). Free energy of ferrosilite is closely constrained by the above reaction. Enthalpy (adjusted by 300 J) and entropy are the recommended values of Bohlen and Boettcher (1981) and the heat capacity is calculated from the above reaction. The compressibility data is consistent with the data of Bass and Weidner (1984) and Weidner (1986).

$FeSiO_3$

ORTHOFERROSILITE

Thermochemical Data

T(K)	C_p(J)	H(kJ)	S(J)	α 10^{-4}	K 10^4	γ	V(cc)	δ_T
298	87.0	-1193.450	95.9	0.2995	97.093	1.11	33.0	10.30
300	87.3	-1193.289	96.4	0.3000	97.038	1.11	33.0	10.27
400	101.8	-1183.770	123.7	0.3317	94.200	1.03	33.1	8.62
500	110.3	-1173.131	147.4	0.3647	91.635	1.02	33.2	7.32
600	115.7	-1161.813	168.0	0.3939	89.296	1.04	33.3	6.36
700	119.4	-1150.048	186.2	0.4194	87.148	1.05	33.4	5.65
800	122.2	-1137.962	202.3	0.4418	85.159	1.07	33.6	5.10
900	124.5	-1125.626	216.8	0.4619	83.302	1.09	33.7	4.68
1000	126.4	-1113.082	230.1	0.4802	81.553	1.11	33.9	4.34
1100	128.1	-1100.357	242.2	0.4971	79.895	1.12	34.1	4.08
1200	129.7	-1087.468	253.4	0.5127	78.308	1.13	34.2	3.87
1300	131.2	-1074.425	263.8	0.5275	76.779	1.15	34.4	3.72
1400	132.6	-1061.237	273.6	0.5415	75.294	1.16	34.6	3.60
1500	134.0	-1047.907	282.8	0.5549	73.842	1.17	34.8	3.51
1600	135.4	-1034.439	291.5	0.5677	72.414	1.18	35.0	3.45
1700	136.7	-1020.835	299.7	0.5802	71.001	1.18	35.2	3.42
1800	138.0	-1007.097	307.6	0.5922	69.597	1.19	35.4	3.40
1900	139.4	-993.226	315.1	0.6040	68.194	1.20	35.6	3.41
2000	140.7	-979.222	322.3	0.6154	66.789	1.20	35.8	3.42
2100	142.0	-965.086	329.2	0.6266	65.378	1.20	36.0	3.45
2200	143.3	-950.818	335.8	0.6377	63.958	1.21	36.3	3.50
2300	144.7	-936.418	342.2	0.6485	62.527	1.21	36.5	3.54
2400	146.0	-921.886	348.4	0.6592	61.084	1.21	36.7	3.60
2500	147.3	-907.222	354.4	0.6697	59.628	1.20	37.0	3.66
2600	148.6	-892.426	360.2	0.6801	58.162	1.20	37.2	3.72
2700	149.9	-877.498	365.8	0.6904	56.684	1.20	37.5	3.79
2800	151.3	-862.437	371.3	0.7005	55.198	1.19	37.8	3.85
2900	152.6	-847.244	376.6	0.7106	53.704	1.18	38.0	3.92
3000	153.9	-831.917	381.8	0.7206	52.207	1.17	38.3	3.98

<INT VdP (J)>

PRESSURE (GPa)

T(K)	1	5	10	20	30	40	50
298	32783	160843	314953	607434	883569	1146919	1399817
300	32784	160850	314963	607445	883574	1146916	1399804
400	32883	161249	315578	608151	884091	1147068	1399476
500	32993	161708	316318	609124	885024	1147794	1399880
600	33114	162223	317174	610343	886340	1149044	1400953
700	33245	162787	318134	611777	887992	1150751	1402606
800	33384	163396	319184	613401	889935	1152855	1404764
900	33531	164045	320314	615191	892134	1155306	1407359
1000	33685	164729	321518	617130	894558	1158061	1410337
1100	33846	165447	322787	619200	897182	1161081	1413648
1200	34013	166195	324116	621389	899981	1164336	1417249
1300	34186	166972	325501	623685	902937	1167795	1421104
1400	34365	167776	326938	626077	906032	1171434	1425178
1500	34550	168606	328422	628555	909248	1175227	1429438
1600	34740	169461	329951	631112	912571	1179154	1433857
1700	34935	170339	331521	633737	915986	1183193	1438406
1800	35136	171239	333129	636425	919481	1187326	1443060
1900	35342	172161	334774	639168	923043	1191534	1447796
2000	35552	173103	336452	641958	926659	1195800	1452591
2100	35768	174066	338162	644791	930319	1200108	1457424
2200	35989	175048	339900	647659	934013	1204443	1462275
2300	36215	176048	341666	650556	937731	1208791	1467126
2400	36446	177067	343456	653478	941463	1213139	1471960
2500	36682	178103	345270	656419	945200	1217475	1476761
2600	36923	179156	347106	659375	948935	1221786	1481513
2700	37169	180226	348961	662340	952660	1226063	1486205
2800	37420	181311	350835	665311	956368	1230297	1490824
2900	37676	182413	352725	668284	960053	1234478	1495359
3000	37937	183529	354630	671254	963709	1238599	1499802

LOW-CLINOENSTATITE

$MgSiO_3$
LOW-CLINOENSTATITE Formula wt. (g) = 100.39

Enthalpy (J/mol) =	-1548715.113	Shi et al. (1993)
Entropy (J/mol.K) =	63.230	Shi et al. (1993)
Heat Capacity:		
No. of *Cp* ranges = 1		Shi et al. (1993)
1.3807968D+02 8.7088D-03 4.0591D+05 -1.3523D-06		
0.0000D+00 0.0000D+00 -1.8888D+04		
Volume (cc/mol) =	31.282	Robie et al. (1978)
Thermal expn:		
3.67653D-05 2.36466D-08 -2.00989D-02 4.88850D+00		
		Shi et al. (1993)
Isot.compressibility:		
9.37603D-07 -3.72785D-15 1.65566D-13 -4.28415D-17		
		Shi et al. (1993)
P deriv. 4.52467D+00		
T deriv. 5.53033D-04		Shi et al. (1993)
Melting pt. and enthalpy of melting:		
1830.00000	61505.00000	Robie et al. (1978)

NOTE: Calorimetric data: H: -1544.15±2.09 S: 67.86+0.42 Kiseleva et al. (1979). Gasparik (1990) discussed the phase relations in the $MgSiO_3$ system with three polymorphs (clino): a low-clino, a high-P form and a high-T clino. This phase is the low-clinoenstatite.

$MgSiO_3$
LOW-CLINOENSTATITE
Thermochemical Data

T(K)	C_p(J)	H(kJ)	S(J)	α 10^{-4}	K 10^4	γ	V(cc)	δ_T
298	81.8	-1548.715	63.2	0.3140	105.132	1.28	31.3	2.92
300	82.1	-1548.564	63.7	0.3118	105.114	1.26	31.3	2.96
400	96.7	-1539.563	89.5	0.2653	104.020	0.90	31.4	4.39
500	105.9	-1529.402	112.2	0.2794	102.708	0.86	31.5	4.90
600	112.5	-1518.465	132.1	0.3103	101.220	0.90	31.5	4.97
700	117.4	-1506.963	149.8	0.3458	99.598	0.95	31.7	4.86
800	121.2	-1495.028	165.7	0.3820	97.883	1.01	31.8	4.68
900	124.3	-1482.746	180.2	0.4175	96.110	1.07	31.9	4.46
1000	127.0	-1470.177	193.4	0.4520	94.311	1.13	32.0	4.23
1100	129.2	-1457.368	205.6	0.4854	92.514	1.19	32.2	3.98
1200	131.1	-1444.350	217.0	0.5179	90.745	1.25	32.3	3.72
1300	132.8	-1431.150	227.5	0.5494	89.025	1.31	32.5	3.46
1400	134.3	-1417.791	237.4	0.5801	87.371	1.37	32.7	3.19
1500	135.7	-1404.288	246.8	0.6101	85.798	1.44	32.9	2.92
1600	136.9	-1390.658	255.6	0.6395	84.319	1.50	33.1	2.65
1700	138.0	-1376.911	263.9	0.6683	82.946	1.58	33.3	2.38
1800	139.0	-1363.060	271.8	0.6967	81.687	1.66	33.6	2.11
1900	139.9	-1349.113	279.3	0.7247	80.552	1.75	33.8	1.84
2000	140.7	-1335.079	286.5	0.7523	79.546	1.85	34.0	1.57
2100	141.5	-1320.966	293.4	0.7796	78.679	1.97	34.3	1.30
2200	142.2	-1306.781	300.0	0.8066	77.957	2.10	34.6	1.03
2300	142.8	-1292.530	306.4	0.8334	77.388	2.25	34.9	0.76
2400	143.4	-1278.219	312.4	0.8599	76.982	2.44	35.2	0.49
2500	143.9	-1263.853	318.3	0.8862	76.747	2.67	35.5	0.21
2600	144.4	-1249.439	324.0	0.9124	76.697	2.95	35.8	-0.07
2700	144.8	-1234.980	329.4	0.9384	76.845	3.30	36.1	-0.35
2800	145.2	-1220.481	334.7	0.9642	77.210	3.78	36.5	-0.64
2900	145.5	-1205.947	339.8	0.9899	77.813	4.43	36.8	-0.95
3000	145.8	-1191.383	344.7	1.0155	78.680	5.38	37.2	-1.27

$MgSiO_3$
LOW-CLINOENSTATITE

<INT VdP (J)>

T(K)	PRESSURE (GPa) 1	5	10	20	30	40	50
298	31135	152980	300032	580220	845895	1100173	1345105
300	31137	152988	300047	580248	845932	1100219	1345158
400	31223	153384	300776	581526	847666	1102353	1347654
500	31305	153760	301458	582710	849275	1104350	1350024
600	31395	154168	302198	583998	851041	1106567	1352684
700	31496	154624	303028	585453	853051	1109112	1355764
800	31608	155133	303957	587095	855338	1112025	1359308
900	31732	155697	304989	588935	857916	1115327	1363339
1000	31867	156315	306126	590977	860796	1119031	1367874
1100	32014	156987	307369	593226	863986	1123149	1372928
1200	32172	157714	308718	595687	867495	1127695	1378518
1300	32341	158494	310175	598365	871333	1132681	1384662
1400	32521	159330	311741	601266	875509	1138123	1391378
1500	32712	160220	313418	604395	880034	1144035	1398685
1600	32914	161165	315209	607759	884919	1150434	1406605
1700	33127	162167	317116	611365	890177	1157336	1415157
1800	33351	163226	319141	615220	895818	1164758	1424364
1900	33586	164344	321288	619333	901856	1172716	1434248
2000	33832	165521	323561	623710	908304	1181230	1444833
2100	34090	166759	325962	628360	915175	1190319	1456142
2200	34360	168060	328495	633293	922484	1200002	1468202
2300	34642	169424	331164	638518	930246	1210301	1481042
2400	34935	170854	333973	644044	938478	1221239	1494689
2500	35241	172352	336927	649883	947197	1232841	1509178
2600	35559	173919	340031	656045	956421	1245133	1524541
2700	35891	175557	343289	662543	966171	1258145	1540819
2800	36235	177268	346707	669391	976470	1271908	1558052
2900	36593	179056	350291	676602	987341	1286457	1576288
3000	36964	180922	354047	684193	998812	1301832	1595579

PROTOENSTATITE

$MgSiO_3$
PROTOENSTATITE Formula wt. (g) = 100.39

Enthalpy (J/mol) = -1545793.900 Shi et al. (1992)
Entropy (J/mol.K) = 66.960 Shi et al. (1992)
Heat Capacity
No. of *Cp* ranges = 1 Shi et al. (1992)
1.4437008D+02 1.8919D-03 -1.3495D+06 0.0000D+00
4.6265D+08 0.0000D+00 -1.9563D+04
Volume (cc/mol) = 32.458 Shi et al. (1992)
Thermal expn.
-3.29500D-06 3.50000D-08 0.00000D+00 0.00000D+00
This study
Isot.bulk mod.
8.00000D-07 1.35840D-10 3.16130D-14 1.14130D-17
This study
P deriv., T deriv.:
4.00000D+00 4.00000D-04 This study
Melting pt. and enthalpy of melting:
6000.00000 -

NOTE: This is a low-pressure phase and the data on compressibility and thermal expansion are not reliable or useful with pressures more than 10 kbar.

Thermochemical Data

T (K)	C_p (J)	H (kJ)	S (J)	α 10^{-4}	K 10^4	γ	V (cc)	δ_T
600	111.3	-1515.958	134.9	0.1771	111.688	0.57	31.8	11.74
700	116.3	-1504.564	152.4	0.2121	109.350	0.64	31.9	10.15
800	120.2	-1492.728	168.2	0.2471	106.981	0.71	32.0	9.02
900	123.3	-1480.546	182.6	0.2821	104.583	0.78	32.0	8.18
1000	125.8	-1468.086	195.7	0.3171	102.159	0.85	32.1	7.52
1100	127.9	-1455.397	207.8	0.3521	99.714	0.92	32.2	6.99
1200	129.7	-1442.517	219.0	0.3871	97.252	0.98	32.4	6.56
1300	131.2	-1429.472	229.4	0.4220	94.778	1.05	32.5	6.20
1400	132.5	-1416.285	239.2	0.4571	92.297	1.11	32.6	5.89
1500	133.7	-1402.972	248.4	0.4921	89.814	1.18	32.8	5.62
1600	134.8	-1389.548	257.1	0.5271	87.335	1.24	33.0	5.38
1700	135.7	-1376.024	265.3	0.5621	84.864	1.31	33.1	5.17
1800	136.6	-1362.410	273.0	0.5971	82.406	1.38	33.3	4.98
1900	137.4	-1348.713	280.4	0.6321	79.968	1.45	33.5	4.80
2000	138.1	-1334.939	287.5	0.6671	77.553	1.52	33.8	4.64
2100	138.8	-1321.096	294.3	0.7021	75.167	1.60	34.0	4.49
2200	139.4	-1307.187	300.7	0.7370	72.813	1.68	34.2	4.35
2300	140.0	-1293.216	306.9	0.7721	70.496	1.76	34.5	4.22
2400	140.6	-1279.188	312.9	0.8071	68.218	1.85	34.8	4.10
2500	141.1	-1265.106	318.7	0.8421	65.984	1.95	35.1	3.98
2600	141.6	-1250.971	324.2	0.8771	63.797	2.05	35.4	3.87
2700	142.1	-1236.788	329.6	0.9121	61.657	2.16	35.7	3.76
2800	142.5	-1222.558	334.7	0.9471	59.569	2.29	36.0	3.66
2900	143.0	-1208.283	339.7	0.9821	57.532	2.43	36.4	3.56
3000	143.4	-1193.965	344.6	1.0171	55.550	2.59	36.7	3.46

HIGH-T CLINOENSTATITE

$MgSiO_3$
HTCLINOENSTATITE Formula wt. (g) = 100.39

Enthalpy (J/mol) = -1543295.2 Shi et al. (1992)
Entropy (J/mol.K) = 68.332 Shi et al. (1992)
Heat Capacity
No. of *Cp* ranges = 1 Shi et al. (1992)
1.4433281D+02 1.8794D-03 -1.3529D+06 0.0000D+00
4.6254D+08 0.0000D+00 -1.9514D+04
Volume (cc/mol) = 31.282 Shi et al. (1992)
Thermal expn.:
5.42920D-05 3.10143D-09 -2.25441D-02 6.87659D+00
Shi et al. (1992)
Isot.bulk mod.
5.00000D-07 1.35840D-10 3.16130D-14 1.14130D-17
Shi et al. (1993)
P deriv., T deriv.:
4.000D+00 1.50000D-04 Shi et al. (1993)
Melting pt. and enthalpy of melting:
- -

NOTE: This is high-temperature clinoenstatite; assumed as clinoenstatite. This is a low-pressure phase and the data on compressibility and thermal expansion are not reliable or useful with pressures more than 10 kbar.

Thermochemical Data

T(K)	C_p(J)	H(kJ)	S(J)	α 10^{-4}	K 10^4	γ	V(cc)	δ_T
600	111.3	-1513.445	136.3	0.3768	167.968	1.88	31.7	8.30
700	116.4	-1502.050	153.8	0.3829	162.736	1.78	31.8	8.37
800	120.2	-1490.213	169.6	0.3934	157.543	1.74	31.9	8.34
900	123.3	-1478.030	184.0	0.4052	152.397	1.71	32.1	8.29
1000	125.8	-1465.571	197.1	0.4172	147.304	1.68	32.2	8.24
1100	127.9	-1452.883	209.2	0.4289	142.275	1.66	32.3	8.19
1200	129.7	-1440.004	220.4	0.4400	137.315	1.64	32.5	8.15
1300	131.2	-1426.961	230.8	0.4505	132.434	1.62	32.6	8.11
1400	132.5	-1413.776	240.6	0.4604	127.640	1.60	32.8	8.08
1500	133.7	-1400.466	249.8	0.4697	122.940	1.58	32.9	8.05
1600	134.7	-1387.045	258.5	0.4785	118.340	1.56	33.1	8.03
1700	135.7	-1373.525	266.7	0.4868	113.848	1.53	33.2	8.00
1800	136.5	-1359.914	274.4	0.4947	109.469	1.50	33.4	7.98
1900	137.3	-1346.221	281.8	0.5022	105.208	1.47	33.6	7.95
2000	138.1	-1332.451	288.9	0.5094	101.068	1.44	33.7	7.92
2100	138.7	-1318.612	295.6	0.5163	97.052	1.41	33.9	7.89
2200	139.4	-1304.707	302.1	0.5229	93.164	1.38	34.1	7.85
2300	140.0	-1290.742	308.3	0.5292	89.403	1.35	34.3	7.81
2400	140.5	-1276.719	314.3	0.5353	85.772	1.32	34.4	7.77
2500	141.0	-1262.641	320.0	0.5413	82.270	1.28	34.6	7.72
2600	141.5	-1248.512	325.6	0.5470	78.897	1.25	34.8	7.67
2700	142.0	-1234.335	330.9	0.5526	75.650	1.22	35.0	7.61
2800	142.5	-1220.110	336.1	0.5580	72.530	1.19	35.2	7.56
2900	142.9	-1205.841	341.1	0.5633	69.534	1.15	35.4	7.49
3000	143.3	-1191.529	346.0	0.5684	66.658	1.12	35.6	7.43

HIGH-P CLINOENSTATITE

$MgSiO_3$
HPCLINOENSTATITE Formula wt. (g) = 100.39

Enthalpy (J/mol) =	-1544826.400	Shi et al. (1992)
Entropy (J/mol.K) =	65.200	Shi et al. (1992)
Heat Capacity		
No. of *Cp* ranges = 1		Shi et al. (1992)
1.4440140D+02 1.8812D-03 -1.3609D+06 0.0000D+00		
4.5980D+08 0.0000D+00 -1.9350D+04		
Volume (cc/mol) =	30.918	Shi et al. (1992)
Thermal expn.		
3.87702D-05 4.50016D-09 3.43917D-04 -1.72362D+00		Shi et al. (1992)
Isot.bulk mod.		
8.89581D-07 1.35473D-10 3.17162D-14 1.15236D-17		Shi et al. (1992)
P deriv., T deriv.:		
4.20000D+00 4.20000D-04		
Melting pt. and enthalpy of melting:		
6000.00000	61505.00000	

NOTE: Evaluated from the phase equilibrium data of Gasparik (1990).

Thermochemical Data

T (K)	C_p(J)	H(kJ)	S(J)	$\alpha\ 10^{-4}$	K 10^4	γ	V(cc)	δ_T
298	82.1	-1544.826	65.2	0.2188	107.170	0.89	30.9	7.71
300	82.4	-1544.674	65.7	0.2212	107.137	0.89	30.9	7.64
400	95.5	-1535.752	91.3	0.3066	105.309	1.06	31.0	5.72
500	104.9	-1525.708	113.7	0.3481	103.446	1.09	31.1	5.22
600	111.6	-1514.865	133.4	0.3726	101.546	1.08	31.2	5.07
700	116.6	-1503.440	151.0	0.3889	99.611	1.07	31.3	5.04
800	120.5	-1491.576	166.9	0.4011	97.641	1.06	31.5	5.07
900	123.5	-1479.368	181.2	0.4107	95.639	1.04	31.6	5.14
1000	126.0	-1466.886	194.4	0.4189	93.607	1.03	31.7	5.22
1100	128.1	-1454.176	206.5	0.4261	91.549	1.02	31.9	5.31
1200	129.9	-1441.276	217.7	0.4326	89.467	1.00	32.0	5.41
1300	131.4	-1428.213	228.2	0.4387	87.366	0.99	32.1	5.50
1400	132.7	-1415.009	238.0	0.4444	85.249	0.98	32.3	5.60
1500	133.9	-1401.681	247.1	0.4498	83.122	0.96	32.4	5.70
1600	134.9	-1388.242	255.8	0.4551	80.989	0.95	32.6	5.79
1700	135.8	-1374.705	264.0	0.4603	78.854	0.94	32.7	5.88
1800	136.7	-1361.077	271.8	0.4653	76.722	0.92	32.9	5.96
1900	137.5	-1347.368	279.2	0.4702	74.598	0.91	33.0	6.04
2000	138.2	-1333.583	286.3	0.4751	72.486	0.90	33.2	6.11
2100	138.9	-1319.728	293.1	0.4799	70.390	0.88	33.3	6.18
2200	139.5	-1305.809	299.5	0.4847	68.314	0.87	33.5	6.23
2300	140.1	-1291.828	305.7	0.4894	66.262	0.85	33.7	6.29
2400	140.7	-1277.791	311.7	0.4941	64.238	0.84	33.8	6.33
2500	141.2	-1263.699	317.5	0.4988	62.245	0.82	34.0	6.37
2600	141.7	-1249.557	323.0	0.5035	60.286	0.81	34.2	6.40
2700	142.2	-1235.365	328.4	0.5081	58.364	0.79	34.3	6.42
2800	142.6	-1221.127	333.6	0.5127	56.480	0.78	34.5	6.43
2900	143.0	-1206.845	338.6	0.5173	54.638	0.76	34.7	6.44
3000	143.5	-1192.519	343.4	0.5219	52.838	0.75	34.9	6.45

$MgSiO_3$
HPCLINOENSTATITE

<INT VdP (J)>

T(K)	1	5	10	20	30	40	50
				PRESSURE(GPa)			
298	30776	151244	296654	573597	835920	1086674	1327893
300	30777	151250	296663	573609	835933	1086685	1327902
400	30857	151605	297277	574563	837077	1087930	1329187
500	30957	152049	298067	575866	838757	1089909	1331420
600	31066	152541	298949	577350	840716	1092279	1334166
700	31182	153063	299885	578942	842843	1094884	1337225
800	31302	153605	300859	580605	845079	1097644	1340489
900	31426	154162	301860	582316	847389	1100509	1343896
1000	31554	154732	302881	584063	849753	1103450	1347404
1100	31684	155312	303919	585837	852156	1106446	1350984
1200	31816	155901	304970	587631	854589	1109481	1354617
1300	31951	156499	306034	589442	857043	1112545	1358287
1400	32088	157104	307107	591266	859514	1115631	1361984
1500	32227	157716	308190	593102	861999	1118733	1365700
1600	32369	158334	309282	594946	864494	1121847	1369429
1700	32512	158959	310381	596800	866997	1124969	1373167
1800	32657	159590	311488	598661	869509	1128100	1376912
1900	32805	160227	312602	600529	872027	1131236	1380662
2000	32954	160870	313722	602405	874553	1134379	1384417
2100	33106	161518	314850	604287	877086	1137529	1388176
2200	33259	162172	315985	606178	879626	1140686	1391942
2300	33414	162832	317126	608076	882176	1143851	1395715
2400	33572	163497	318275	609983	884735	1147028	1399497
2500	33731	164168	319430	611900	887306	1150216	1403292
2600	33892	164844	320594	613827	889890	1153419	1407101
2700	34056	165527	321765	615766	892488	1156638	1410927
2800	34221	166215	322944	617717	895102	1159877	1414774
2900	34389	166908	324132	619681	897735	1163137	1418646
3000	34558	167608	325328	621660	900388	1166422	1422544

CLINODIOPSIDE

$(CaMg)_{0.5}SiO_3$
CLINODIOPSIDE Formula wt. (g) = 108.28

Enthalpy (J/mol) = -1600910.000 Robie et al. (1978)
Entropy (J/mol.K) = 71.550 Robie et al. (1978)
Heat Capacity:
No. of *Cp* ranges = 1 Robie et al. 1978), refitted
1.2166500D+02 8.1400D-03 -6.8095D+06 0.0000D+00
9.3840D+08 0.0000D+00 1.2772D+02
Volume (cc/mol) = 33.050 Robie et al. (1978)
Thermal expn:
2.51000D-05 9.82100D-09 5.19100D-03 -2.36500D+00
This study
Isot.compressibility:
9.08200D-07 -8.38400D-11 1.16500D-13 2.29100D-18
This study
P deriv. 4.00000D+00 As in clinoenstatite
T deriv. 5.00000D-04 This study
Melting pt. and enthalpy of melting:
\- \-

NOTE: Calorimetric data: H: -1600.15 ±0.86,S: 72.15 ±0.42, Robie et al. (1978). The properties listed in Robie et al. (1978) are adopted. These data are from calorimetric sources.

Thermochemical Data

T(K)	C_p(J)	H(kJ)	S(J)	$\alpha\ 10^{-4}$	K 10^4	γ	V(cc)	δ_T
298	83.3	-1600.91	71.5	0.1883	111.904	0.84	33.0	-0.82
300	83.6	-1600.76	72.1	0.1907	111.908	0.85	33.1	-0.78
400	97.3	-1591.66	98.1	0.2722	111.926	1.05	33.1	0.43
500	106.3	-1581.45	120.9	0.3093	111.646	1.10	33.2	1.24
600	112.2	-1570.51	140.8	0.3307	111.070	1.12	33.3	1.96
700	116.4	-1559.07	158.4	0.3456	110.207	1.12	33.4	2.63
800	119.5	-1547.27	174.2	0.3575	109.068	1.13	33.6	3.26
900	122.0	-1535.18	188.4	0.3679	107.668	1.13	33.7	3.85
1000	124.1	-1522.88	201.4	0.3775	106.028	1.14	33.8	4.38
1100	125.8	-1510.38	213.3	0.3867	104.168	1.14	33.9	4.87
1200	127.4	-1497.72	224.3	0.3957	102.113	1.14	34.1	5.31
1300	128.7	-1484.92	234.6	0.4046	99.887	1.14	34.2	5.69
1400	130.0	-1471.98	244.2	0.4135	97.518	1.14	34.4	6.03
1500	131.2	-1458.91	253.2	0.4224	95.030	1.13	34.5	6.32
1600	132.3	-1445.74	261.7	0.4313	92.449	1.12	34.6	6.57
1700	133.4	-1432.45	269.7	0.4403	89.798	1.12	34.8	6.77
1800	134.4	-1419.06	277.4	0.4493	87.100	1.11	34.9	6.94
1900	135.4	-1405.56	284.7	0.4584	84.375	1.10	35.1	7.06
2000	136.4	-1391.97	291.6	0.4675	81.643	1.09	35.3	7.15
2100	137.4	-1378.28	298.3	0.4766	78.919	1.07	35.4	7.22
2200	138.3	-1364.49	304.7	0.4858	76.219	1.06	35.6	7.25
2300	139.2	-1350.61	310.9	0.4950	73.555	1.05	35.8	7.26
2400	140.1	-1336.65	316.9	0.5042	70.937	1.03	36.0	7.24
2500	141.0	-1322.59	322.6	0.5135	68.375	1.02	36.1	7.21
2600	141.9	-1308.44	328.1	0.5228	65.875	1.00	36.3	7.16
2700	142.8	-1294.20	333.5	0.5321	63.443	0.99	36.5	7.10
2800	143.7	-1279.88	338.7	0.5415	61.084	0.97	36.7	7.02
2900	144.5	-1265.47	343.8	0.5509	58.799	0.95	36.9	6.93
3000	145.4	-1250.97	348.7	0.5603	56.592	0.94	37.1	6.84

$(CaMg)_{0.5}SIO_3$
CLINODIOPSIDE

<INT VdP (J)>

T(K)	1	5	10	20	30	40	50
			PRESSURE(GPa)				
298	32904	161806	317550	614423	895727	1164598	1423157
300	32905	161812	317561	614446	895761	1164642	1423211
400	32983	162198	318323	615940	897974	1167566	1426842
500	33080	162667	319239	617719	900602	1171047	1431187
600	33185	163175	320221	619604	903374	1174715	1435770
700	33297	163706	321237	621527	906184	1178420	1440395
800	33413	164252	322271	623459	908982	1182094	1444968
900	33532	164810	323316	625384	911745	1185701	1449444
1000	33655	165378	324370	627296	914465	1189229	1453803
1100	33781	165955	325430	629192	917136	1192671	1458036
1200	33910	166541	326496	631074	919760	1196029	1462146
1300	34043	167136	327569	632941	922340	1199308	1466140
1400	34178	167739	328649	634797	924881	1202517	1470028
1500	34317	168352	329736	636645	927391	1205666	1473825
1600	34459	168975	330833	638488	929875	1208766	1477545
1700	34605	169607	331939	640331	932343	1211829	1481205
1800	34753	170249	333056	642177	934801	1214867	1484823
1900	34905	170902	334185	644031	937260	1217893	1488415
2000	35061	171566	335329	645898	939726	1220921	1491999
2100	35219	172240	336487	647783	942209	1223962	1495592
2200	35382	172927	337661	649689	944717	1227030	1499210
2300	35548	173626	338854	651621	947258	1230136	1502870
2400	35718	174337	340066	653584	949840	1233293	1506587
2500	35891	175061	341299	655582	952472	1236510	1510376
2600	36068	175799	342554	657620	955159	1239800	1514250
2700	36249	176551	343834	659702	957910	1243171	1518223
2800	36434	177318	345139	661831	960732	1246633	1522306
2900	36623	178099	346471	664012	963630	1250196	1526512
3000	36816	178897	347831	666249	966610	1253868	1530851

CLINOHEDENBERGITE

$(CaFe)_{0.5}SiO_3$
CLINOHEDENBERGITE Formula wt. (g) = 124.05

Enthalpy (J/mol) = -1419410.000		Saxena et al. (1986)
Entropy (J/mol.K) = 85.145		Saxena et al. (1986)
Heat Capacity:		
No. of *Cp* ranges = 1		This study
1.5465000D+02 7.0250D-03 1.5395D+07 0.0000D+00		
-2.3930D+09 0.0000D+00 -4.6610D+04		
Volume (cc/mol) = 34.610		Saxena et al. (1986)
Thermal expn:		
8.82146D-05 3.71690D-09 -3.84310D-02 5.55973D+00		This study
Isot.compressibility:		
8.58338D-07 1.13695D-10 8.31098D-14 2.65640D-18		This study
P deriv. 4.20000D+00		
T deriv. 0.00000D+00		This study
Melting pt. and enthalpy of melting:		
- -		

NOTE: The enthalpy and entropy are as recommended by Saxena et al. (1986) adjusted from the data of Helgeson et al. (1978) to fit the equilibrium data on quadrilateral pyroxenes from Lindsley (1983). Recently, Haselton et al. (1987) determined the heat capacity of hedenbergite by low-temperature adiabatic and differential scanning calorimetry in the range 9-647 K. The heat capacity recommended by Saxena et al. (1986) is consistent with this later experimental determination. The difference between the heat capacity of Saxena et al. and Haselton et al. at 647 K is 0.8 J/cation.

$(CaFe)_{0.5}SiO_3$
CLINOHEDENBERGITE

Thermochemical Data

T(K)	C_p(J)	H(kJ)	S(J)	α 10^{-4}	K 10^4	γ	V(cc)	δ_T
298	83.3	-1419.41	85.1	0.2297	111.149	1.07	34.6	7.93
300	83.8	-1419.26	85.7	0.2300	111.111	1.06	34.6	7.94
400	99.8	-1409.94	112.4	0.2837	109.018	1.09	34.7	6.97
500	107.4	-1399.55	135.5	0.3545	106.804	1.25	34.8	5.99
600	112.9	-1388.53	155.6	0.4184	104.488	1.40	34.9	5.40
700	117.4	-1377.01	173.3	0.4726	102.087	1.51	35.1	5.05
800	121.4	-1365.07	189.3	0.5184	99.617	1.60	35.3	4.84
900	124.9	-1352.75	203.8	0.5572	97.095	1.66	35.5	4.70
1000	128.1	-1340.10	217.1	0.5906	94.536	1.71	35.7	4.61
1100	130.9	-1327.15	229.4	0.6196	91.954	1.75	35.9	4.54
1200	133.5	-1313.92	241.0	0.6451	89.362	1.77	36.1	4.50
1300	135.9	-1300.44	251.7	0.6677	86.773	1.79	36.4	4.46
1400	138.2	-1286.74	261.9	0.6880	84.197	1.80	36.6	4.43
1500	140.2	-1272.81	271.5	0.7064	81.643	1.81	36.9	4.40
1600	142.2	-1258.69	280.6	0.7231	79.121	1.81	37.1	4.38
1700	144.0	-1244.38	289.3	0.7385	76.637	1.80	37.4	4.35
1800	145.7	-1229.89	297.6	0.7527	74.197	1.79	37.7	4.32
1900	147.4	-1215.23	305.5	0.7659	71.808	1.78	38.0	4.30
2000	148.9	-1200.42	313.1	0.7782	69.473	1.77	38.3	4.27
2100	150.4	-1185.45	320.4	0.7898	67.195	1.76	38.6	4.24
2200	151.9	-1170.33	327.4	0.8007	64.977	1.74	38.9	4.20
2300	153.3	-1155.07	334.2	0.8111	62.822	1.72	39.2	4.17
2400	154.6	-1139.68	340.8	0.8209	60.730	1.70	39.5	4.13
2500	155.9	-1124.16	347.1	0.8302	58.702	1.68	39.8	4.09
2600	157.1	-1108.51	353.2	0.8392	56.739	1.66	40.2	4.05
2700	158.3	-1092.73	359.2	0.8478	54.840	1.63	40.5	4.01
2800	159.5	-1076.84	365.0	0.8561	53.006	1.61	40.8	3.97
2900	160.7	-1060.83	370.6	0.8640	51.235	1.59	41.2	3.93
3000	161.8	-1044.70	376.1	0.8717	49.526	1.56	41.6	3.89

<INT VdP (J)>

PRESSURE(GPa)

T(K)	1	5	10	20	30	40	50
298	34456	169430	332520	643528	938470	1220642	1492258
300	34458	169436	332530	643543	938485	1220655	1492268
400	34542	169800	333143	644425	939440	1221556	1493026
500	34649	170272	333956	645667	940890	1223071	1494507
600	34780	170854	334977	647283	942855	1225226	1496740
700	34932	171534	336178	649217	945255	1227918	1499603
800	35101	172295	337529	651413	948009	1231043	1502966
900	35286	173127	339006	653825	951049	1234512	1506725
1000	35485	174019	340591	656417	954322	1238257	1510797
1100	35695	174963	342268	659157	957785	1242225	1515117
1200	35916	175953	344025	662025	961408	1246375	1519635
1300	36147	176985	345852	665001	965162	1250673	1524315
1400	36387	178055	347743	668071	969029	1255094	1529124
1500	36635	179159	349689	671224	972992	1259620	1534041
1600	36891	180295	351687	674449	977038	1264233	1539047
1700	37154	181460	353732	677739	981157	1268921	1544129
1800	37424	182654	355820	681089	985340	1273676	1549277
1900	37702	183873	357949	684492	989582	1278490	1554482
2000	37985	185118	360117	687945	993877	1283357	1559739
2100	38275	186387	362320	691445	998222	1288273	1565045
2200	38572	187679	364558	694988	1002612	1293236	1570396
2300	38874	188993	366829	698573	1007047	1298244	1575792
2400	39182	190329	369132	702198	1011525	1303295	1581232
2500	39496	191687	371465	705862	1016044	1308389	1586716
2600	39816	193065	373829	709563	1020604	1313527	1592247
2700	40142	194463	376222	713302	1025205	1318709	1597824
2800	40473	195882	378645	717078	1029848	1323936	1603451
2900	40810	197320	381095	720890	1034532	1329210	1609130
3000	41153	198778	383574	724739	1039259	1334532	1614863

JADEITE

$NaAlSi_2O_6$
JADEITE Formula wt. (g) = 202.14

Enthalpy (J/mol) =	-3029870.000	Hemingway et al. (1981)
Entropy (J/mol.K) =	134.720	Robie et al. (1978)
Heat Capacity:		
No. of *Cp* ranges = 1		Robie et al. (1978), refitted

1.9337000D+02 3.4080D-02 -3.0733D+07 0.0000D+00
4.1913D+09 0.0000D+00 4.2914D+04

Volume (cc/mol) =	60.400	Robie et al. (1978)

Thermal expn:
4.83199D-05 1.06450D-08 -2.43639D-02 4.29321D+00
This study

Isot.compressibility:
9.86639D-07 2.28862D-11 7.20230D-14 -3.37680D-18
This study

P deriv.	4.20000D+00	Birch (1966)
T deriv.	0.00000D+00	This study

Melting pt. and enthalpy of melting:
\- \-

NOTE: Calorimetric data: H: -3029.88±4.2, Hemingway et al. (1981) (derived indirectly); S: 133.48±1.26, Robie et al. (1978). Entropy and *Cp* of jadeite are as listed in Robie et al. (1978).

Thermochemical Data

T(K)	C_p(J)	H(kJ)	S(J)	α 10^{-4}	K 10^4	γ	V(cc)	δ_T
298	159.9	-3029.87	134.7	0.1807	100.022	0.69	60.4	3.59
300	160.4	-3029.57	135.7	0.1800	100.010	0.68	60.4	3.62
400	187.7	-3012.12	185.7	0.1850	99.295	0.59	60.5	4.23
500	206.8	-2992.33	229.8	0.2209	98.458	0.64	60.6	4.12
600	219.4	-2970.97	268.7	0.2603	97.507	0.71	60.8	3.96
700	228.0	-2948.58	303.2	0.2973	96.451	0.78	60.9	3.85
800	234.4	-2925.44	334.1	0.3309	95.300	0.84	61.1	3.79
900	239.5	-2901.73	362.0	0.3613	94.063	0.90	61.4	3.76
1000	243.8	-2877.56	387.5	0.3889	92.750	0.94	61.6	3.74
1100	247.6	-2852.99	410.9	0.4143	91.369	0.99	61.8	3.73
1200	251.1	-2828.05	432.6	0.4377	89.930	1.03	62.1	3.72
1300	254.4	-2802.77	452.8	0.4596	88.441	1.06	62.4	3.71
1400	257.6	-2777.17	471.8	0.4801	86.913	1.09	62.7	3.71
1500	260.7	-2751.26	489.7	0.4995	85.352	1.12	63.0	3.69
1600	263.7	-2725.04	506.6	0.5180	83.766	1.14	63.3	3.68
1700	266.8	-2698.51	522.7	0.5357	82.162	1.16	63.6	3.66
1800	269.8	-2671.68	538.0	0.5527	80.548	1.18	64.0	3.63
1900	272.8	-2644.55	552.7	0.5691	78.929	1.20	64.3	3.61
2000	275.8	-2617.12	566.8	0.5850	77.310	1.21	64.7	3.57
2100	278.9	-2589.39	580.3	0.6005	75.697	1.22	65.1	3.54
2200	281.9	-2561.35	593.3	0.6155	74.095	1.24	65.5	3.50
2300	284.9	-2533.01	605.9	0.6302	72.506	1.25	65.9	3.46
2400	288.0	-2504.36	618.1	0.6446	70.935	1.26	66.3	3.41
2500	291.1	-2475.40	629.9	0.6587	69.385	1.27	66.8	3.37
2600	294.2	-2446.14	641.4	0.6726	67.858	1.28	67.2	3.32
2700	297.3	-2416.57	652.6	0.6863	66.356	1.28	67.7	3.27
2800	300.4	-2386.69	663.4	0.6997	64.882	1.29	68.1	3.22
2900	303.5	-2356.49	674.0	0.7130	63.437	1.30	68.6	3.16
3000	306.7	-2325.98	684.4	0.7261	62.022	1.30	69.1	3.11

$NaAlSi_2O_6$
JADEITE

<INT VdP (J)>

	PRESSURE(GPa)						
T(K)	1	5	10	20	30	40	50
298	60103	295033	578013	1115653	1623748	2108657	2574564
300	60105	295042	578030	1115681	1623786	2108702	2574615
400	60209	295519	578893	1117138	1625682	2110936	2577116
500	60329	296062	579870	1118777	1627802	2113418	2579876
600	60471	296711	581044	1120755	1630374	2116447	2583265
700	60637	297467	582413	1123075	1633404	2120031	2587291
800	60824	298323	583964	1125706	1636849	2124115	2591890
900	61031	299268	585678	1128617	1640662	2128639	2596989
1000	61256	300295	587540	1131777	1644801	2133550	2602527
1100	61498	301398	589536	1135162	1649233	2138807	2608452
1200	61755	302571	591658	1138754	1653932	2144376	2614726
1300	62027	303809	593895	1142537	1658875	2150231	2621318
1400	62313	305110	596242	1146499	1664047	2156353	2628206
1500	62613	306470	598694	1150631	1669436	2162726	2635374
1600	62926	307887	601246	1154926	1675032	2169341	2642810
1700	63251	309360	603895	1159378	1680828	2176189	2650507
1800	63589	310887	606639	1163984	1686821	2183267	2658461
1900	63939	312467	609475	1168740	1693006	2190571	2666668
2000	64300	314099	612402	1173645	1699383	2198101	2675131
2100	64674	315783	615420	1178697	1705950	2205857	2683850
2200	65060	317519	618527	1183896	1712709	2213841	2692830
2300	65457	319305	621723	1189243	1719659	2222056	2702075
2400	65866	321142	625009	1194737	1726805	2230506	2711589
2500	66286	323030	628384	1200380	1734147	2239194	2721381
2600	66719	324969	631848	1206174	1741689	2248125	2731455
2700	67163	326960	635403	1212119	1749434	2257306	2741821
2800	67619	329002	639050	1218219	1757387	2266741	2752485
2900	68087	331096	642788	1224476	1765550	2276436	2763456
3000	68567	333242	646619	1230891	1773929	2286399	2774743

ALMANDINE

$Fe_3Al_2Si_3O_{12}$
ALMANDINE Formula wt. (g) = 497.76

Enthalpy (J/mol) = -5276337.000 This study
Entropy (J/mol.K) = 336.300 This study
Heat Capacity:
No. of *Cp* ranges = 1 Chatterjee (1987), refitted
7.1300000D+02 -8.8500D-02 1.4580D+07 1.5600D-05
0.0000D+00 0.0000D+00 -1.4960D+05
Volume (cc/mol) = 115.284 Skinner (1956)
Thermal expn:
3.08360D-05 6.65900D-09 -6.10600D-03 6.45300D-01
This study
Isot.compressibility:
4.80760D-07 3.04050D-10 1.63500D-15 3.99610D-17
This study
P deriv. 5.60000D+00 Birch (1966), Soga (1967)
T deriv. 0.00000D+00 This study
Melting pt. and enthalpy of melting:
\- -

NOTE: Calorimetric data: H: -5249.82±10.71, Chatillon-Colinet et al.(1983a); S: 339.69±0.51, Metz et al.(1983). *Cp* of almandine is estimated from the Fe-Mg exchange reaction

En + Alm = Fer + Pyr (a)

by assuming that -DCp of the reaction is zero. Although this is normally not true, it is preferred over the *Cp* determined by Metz et al. (as in Bohlen et al., 1986) since when their determined *Cp* values are extrapolated to high T's such as for the range of experiments of Lee and Ganguly (1987) (see Ganguly and Saxena, 1988) they become too high to fit the observed equilibria. The substantial adjustment in H and S is needed to be consistent with the experimental phase equilibria data on the reactions involving rutile and ilmenite and anorthite and forsterite as considered later.

$Fe_3Al_2Si_3O_{12}$

ALMANDINE

Thermochemical Data

T(K)	C_p(J)	H(kJ)	S(J)	α 10^{-4}	K 10^4	γ	V(cc)	δ_T
298	350.3	-5276.34	336.3	0.1960	174.637	1.13	115.3	28.13
300	351.2	-5275.69	338.5	0.1965	174.459	1.13	115.3	28.04
400	397.2	-5238.18	446.1	0.2227	165.235	1.08	115.5	24.08
500	431.8	-5196.64	538.6	0.2453	156.693	1.04	115.8	21.44
600	456.7	-5152.15	619.7	0.2645	148.719	1.02	116.1	19.63
700	474.7	-5105.53	691.5	0.2809	141.222	0.99	116.4	18.35
800	488.0	-5057.36	755.8	0.2954	134.137	0.97	116.8	17.41
900	497.8	-5008.05	813.9	0.3084	127.411	0.95	117.1	16.69
1000	505.1	-4957.89	866.7	0.3203	121.006	0.93	117.5	16.14
1100	510.6	-4907.09	915.1	0.3314	114.892	0.91	117.9	15.69
1200	514.7	-4855.82	959.7	0.3419	109.048	0.89	118.3	15.33
1300	517.9	-4804.18	1001.0	0.3518	103.457	0.87	118.7	15.03
1400	520.3	-4752.27	1039.5	0.3613	98.108	0.85	119.1	14.76
1500	522.1	-4700.15	1075.5	0.3704	92.990	0.82	119.5	14.53
1600	523.5	-4647.87	1109.2	0.3793	88.098	0.80	120.0	14.31
1700	524.7	-4595.45	1141.0	0.3879	83.424	0.78	120.4	14.11
1800	525.6	-4542.94	1171.0	0.3963	78.964	0.76	120.9	13.92
1900	526.5	-4490.33	1199.5	0.4045	74.713	0.74	121.4	13.72
2000	527.2	-4437.65	1226.5	0.4126	70.667	0.71	121.9	13.53
2100	528.0	-4384.88	1252.2	0.4206	66.820	0.69	122.4	13.34
2200	528.8	-4332.04	1276.8	0.4284	63.168	0.67	122.9	13.14
2300	529.7	-4279.12	1300.3	0.4362	59.704	0.65	123.5	12.95
2400	530.7	-4226.10	1322.9	0.4439	56.423	0.62	124.0	12.74
2500	531.7	-4172.98	1344.6	0.4514	53.319	0.60	124.6	12.54
2600	533.0	-4119.75	1365.5	0.4590	50.386	0.58	125.1	12.33
2700	534.4	-4066.38	1385.6	0.4664	47.615	0.56	125.7	12.12
2800	535.9	-4012.87	1405.1	0.4738	45.002	0.54	126.3	11.90
2900	537.7	-3959.19	1423.9	0.4812	42.538	0.52	126.9	11.68
3000	539.7	-3905.32	1442.2	0.4885	40.216	0.50	127.5	11.47

<INT VdP (J)>

	PRESSURE(GPa)						
T(K)	1	5	10	20	30	40	50
298	114958	568643	1123341	2198114	3234939	4240632	5219955
300	114961	568656	1123355	2198103	3234879	4240508	5219754
400	115185	569445	1124266	2197934	3232486	4235158	5210951
500	115437	570372	1125474	2198523	3231534	4232150	5205583
600	115712	571414	1126925	2199730	3231749	4231020	5202946
700	116009	572552	1128577	2201446	3232922	4231417	5202505
800	116323	573770	1130397	2203587	3234888	4233066	5203845
900	116653	575057	1132361	2206090	3237522	4235754	5206641
1000	116997	576406	1134450	2208904	3240722	4239308	5210632
1100	117355	577809	1136648	2211990	3244409	4243592	5215611
1200	117725	579262	1138947	2215318	3248516	4248495	5221412
1300	118107	580760	1141336	2218861	3252992	4253927	5227901
1400	118500	582301	1143809	2222602	3257796	4259818	5234974
1500	118904	583882	1146361	2226523	3262894	4266112	5242548
1600	119319	585501	1148990	2230615	3268261	4272767	5250561
1700	119743	587158	1151691	2234868	3273879	4279749	5258964
1800	120177	588850	1154465	2239276	3279732	4287034	5267722
1900	120622	590578	1157310	2243837	3285811	4294605	5276809
2000	121075	592342	1160226	2248547	3292109	4302450	5286210
2100	121538	594142	1163214	2253405	3298623	4310562	5295913
2200	122011	595977	1166275	2258414	3305351	4318938	5305914
2300	122492	597850	1169411	2263573	3312294	4327577	5316212
2400	122983	599759	1172623	2268885	3319454	4336482	5326810
2500	123484	601707	1175913	2274353	3326834	4345656	5337713
2600	123993	603694	1179283	2279979	3334439	4355105	5348929
2700	124512	605721	1182735	2285768	3342271	4364835	5360466
2800	125040	607789	1186272	2291723	3350338	4374853	5372333
2900	125578	609900	1189896	2297848	3358643	4385166	5384540
3000	126125	612055	1193609	2304146	3367193	4395782	5397099

PYROPE

$Mg_3Al_2Si_3O_{12}$
PYROPE   Formula wt. (g) = 403.14

Enthalpy (J/mol) =	-6291540.000	Charlu et al. (1975)
Entropy (J/mol.K) =	266.270	Haselton and Westrum (1980)
Heat Capacity		
No. of *Cp* ranges = 1		Haselton (1979)
4.7680000D+02 3.1670D-02 -2.1200D+07 0.0000D+00		
2.1680D+09 0.0000D+00 -1.1700D+03		
Volume (cc/mol) =	113.280	Robie et al. (1978)
Thermal expn.		
9.91000D-05 1.16500D-08 1.06240D-02 -2.50000D+00		
		Saxena and Shen (1992)
Isot.bulk mod.		
5.54600D-07 8.06000D-11 -4.60000D-15 1.56700D-17		
		Saxena and Shen (1992)
P deriv., T deriv.:		
4.80000D+00 0.00000D+00		Bonzcar et al. (1977)
Melting pt. and enthalpy of melting:		
1830.18005 81000.00000		Robie et al. (1978)

NOTE:Calorimetric data: H: -6290.76+1.59, Charlu et al. (1975); S: 266.28, Haselton and Westrum (1980). The enthalpy is from Charlu et al. (1975), the entropy is from Haselton and Westrum (1980) and the heat capacity is from Haselton (1979). The values are calorimetric. The thermal expansion is consistent with Skinner (1956). Similarly, the compressibility data are consistent with the data of Bonczar et al. (1977).

$Mg_3Al_2Si_3O_{12}$

PYROPE

Thermochemical Data

T (K)	C_p (J)	H (kJ)	S (J)	α 10^{-4}	K 10^4	γ	V (cc)	δ_T
298	325.6	-6291.540	266.3	1.1008	172.820	8.45	113.3	1.29
300	327.1	-6290.936	268.3	1.1023	172.775	8.44	113.3	1.29
400	387.9	-6254.882	371.6	1.1469	170.327	7.85	114.6	1.25
500	422.8	-6214.198	462.2	1.1617	167.867	7.75	115.9	1.27
600	445.0	-6170.732	541.4	1.1685	165.374	7.92	117.3	1.30
700	460.4	-6125.422	611.2	1.1733	162.829	8.27	118.7	1.35
800	471.8	-6078.790	673.5	1.1779	160.216	8.77	120.1	1.41
900	480.8	-6031.145	729.6	1.1830	157.521	9.44	121.5	1.47
1000	488.3	-5982.681	780.6	1.1887	154.734	10.30	122.9	1.54
1100	494.7	-5933.527	827.5	1.1951	151.849	11.41	124.4	1.62
1200	500.4	-5883.769	870.8	1.2020	148.860	12.84	125.9	1.70
1300	505.5	-5833.472	911.0	1.2094	145.766	14.75	127.4	1.78
1400	510.3	-5782.679	948.7	1.2172	142.567	17.38	129.0	1.87
1500	514.7	-5731.426	984.0	1.2255	139.269	21.21	130.6	1.96
1600	519.0	-5679.738	1017.4	1.2340	135.875	27.26	132.2	2.05
1700	523.1	-5627.634	1049.0	1.2429	132.395	38.18	133.8	2.14
1800	527.0	-5575.131	1079.0	1.2520	128.839	63.70	135.5	2.23
1900	530.8	-5522.241	1107.6	1.2613	125.217	191.26	137.2	2.31
2000	534.5	-5468.974	1134.9	1.2709	121.542	-192.55	139.0	2.39
2100	538.2	-5415.338	1161.1	1.2806	117.828	-64.31	140.8	2.47
2200	541.8	-5361.341	1186.2	1.2904	114.089	-38.72	142.6	2.55
2300	545.3	-5306.987	1210.4	1.3004	110.338	-27.77	144.5	2.61
2400	548.8	-5252.282	1233.6	1.3105	106.591	-21.70	146.4	2.68
2500	552.3	-5197.229	1256.1	1.3207	102.860	-17.84	148.3	2.74
2600	555.7	-5141.832	1277.8	1.3311	99.159	-15.17	150.3	2.79
2700	559.1	-5086.094	1298.9	1.3415	95.500	-13.22	152.3	2.84
2800	562.5	-5030.017	1319.3	1.3520	91.895	-11.73	154.4	2.88
2900	565.8	-4973.604	1339.1	1.3625	88.353	-10.56	156.5	2 91
3000	569.1	-4916.856	1358.3	1.3731	84.883	-9.61	158.6	2.94

<INT VdP (J)>

T (K)	PRESSURE (GPa) 1	5	10	20	30	40	50
298	112955	558630	1103182	2156808	3171086	4152794	5106764
300	112978	558742	1103400	2157223	3171685	4153565	5107700
400	114257	564979	1115538	2180389	3205090	4196609	5159904
500	115580	571433	1128096	2204345	3239625	4241101	5213856
600	116929	578013	1140891	2228730	3274751	4286326	5268668
700	118301	584696	1153873	2253437	3310305	4332062	5324060
800	119694	591476	1167031	2278438	3346234	4378232	5379928
900	121109	598354	1180364	2303725	3382522	4424806	5436228
1000	122547	605336	1193879	2329302	3419164	4471772	5492934
1100	124009	612424	1207580	2355172	3456158	4519117	5550028
1200	125496	619623	1221474	2381339	3493503	4566834	5607488
1300	127009	626937	1235567	2407806	3531195	4614909	5665295
1400	128550	634371	1249862	2434575	3569230	4663331	5723425
1500	130119	641927	1264365	2461648	3607603	4712085	5781856
1600	131717	649610	1279078	2489025	3646307	4761158	5840564
1700	133346	657421	1294007	2516705	3685336	4810533	5899526
1800	135006	665365	1309153	2544687	3724681	4860197	5958719
1900	136699	673444	1324520	2572971	3764336	4910136	6018122
2000	138424	681662	1340110	2601555	3804295	4960337	6077717
2100	140184	690020	1355926	2630439	3844551	5010788	6137487
2200	141978	698521	1371970	2659621	3885099	5061481	6197419
2300	143808	707169	1388246	2689102	3925934	5112406	6257503
2400	145676	715965	1404755	2718880	3967054	5163560	6317731
2500	147581	724913	1421500	2748956	4008457	5214939	6378101
2600	149524	734014	1438484	2779331	4050143	5266543	6438613
2700	151507	743273	1455709	2810008	4092113	5318375	6499271
2800	153531	752691	1473179	2840987	4134370	5370439	6560085
2900	155596	762271	1490896	2872274	4176919	5422743	6621067
3000	157704	772017	1508864	2903871	4219767	5475299	6682233

GROSSULAR

$Ca_3Al_2Si_3O_{12}$
GROSSULAR Formula wt. (g) = 450.46

Enthalpy (J/mol) =	-6634308.000	Charlu et al. (1978)
Entropy (J/mol.K) =	256.118	Haselton and Westrum (1980)
Heat Capacity:		
No. of *Cp* ranges = 1		Haselton (1979), refitted
5.4260000D+02 1.2940D-02 -3.1860D+06 0.0000D+00		
2.7770D+08 0.0000D+00 -5.6020D+04		
Volume (cc/mol) =	125.310	Robie et al. (1978)
Thermal expn:		
2.52900D-05 6.97000D-09 2.72400D-03 -1.47940D+00		This study
Isot.compressibility:		
5.97200D-07 -1.84000D-11 7.49000D-14 6.47000D-18		This study
P deriv. 4.00000D+00		This study
T deriv. 0.00000D+00		
Melting pt. and enthalpy of melting:		
- -		

NOTE: Calorimetric data: H: -6632.76±2.79, Charlu et al. (1978); S: 260.13, Haselton and Westrum (1980). The enthalpy is from Charlu et al. (1975), the entropy is from Haselton and Westrum (1980) and the heat capacity is from Haselton (1979). The values are calorimetric. Thermal expansion was checked against Skinner (1956). Compressibility data of Halleck (1973) are taken into consideration.

$Ca_3Al_2Si_3O_{12}$

GROSSULAR

Thermochemical Data

T(K)	C_p(J)	H(kJ)	S(J)	α 10^{-4}	K 10^4	γ	V(cc)	δ_T
298	333.2	-6634.308	256.1	0.1986	167.072	1.26	125.3	2.35
300	334.6	-6633.690	258.2	0.2002	167.058	1.26	125.3	2.36
400	392.2	-6597.073	363.1	0.2564	166.047	1.38	125.6	2.89
500	426.5	-6556.012	454.6	0.2831	164.600	1.40	125.9	3.57
600	449.4	-6512.146	534.5	0.2990	162.728	1.40	126.3	4.27
700	465.9	-6466.337	605.1	0.3104	160.452	1.40	126.7	4.96
800	478.5	-6419.089	668.2	0.3196	157.796	1.39	127.1	5.62
900	488.4	-6370.725	725.1	0.3276	154.793	1.38	127.5	6.24
1000	496.6	-6321.459	777.0	0.3350	151.476	1.37	127.9	6.82
1100	503.5	-6271.445	824.7	0.3421	147.885	1.36	128.4	7.34
1200	509.4	-6220.795	868.8	0.3490	144.060	1.34	128.8	7.82
1300	514.6	-6169.591	909.7	0.3557	140.041	1.33	129.3	8.23
1400	519.2	-6117.899	948.1	0.3624	135.870	1.31	129.7	8.60
1500	523.3	-6065.771	984.0	0.3690	131.586	1.29	130.2	8.91
1600	527.1	-6013.246	1017.9	0.3757	127.226	1.28	130.7	9.18
1700	530.6	-5960.358	1050.0	0.3823	122.825	1.26	131.2	9.39
1800	533.8	-5907.134	1080.4	0.3889	118.415	1.23	131.7	9.56
1900	536.9	-5853.598	1109.3	0.3956	114.024	1.21	132.2	9.69
2000	539.7	-5799.768	1137.0	0.4022	109.678	1.19	132.7	9.78
2100	542.4	-5745.661	1163.3	0.4089	105.398	1.16	133.3	9.84
2200	545.0	-5691.291	1188.6	0.4156	101.201	1.14	133.8	9.87
2300	547.4	-5636.671	1212.9	0.4223	97.104	1.12	134.4	9.86
2400	549.8	-5581.809	1236.3	0.4290	93.118	1.09	135.0	9.83
2500	552.1	-5526.717	1258.8	0.4357	89.252	1.07	135.6	9.78
2600	554.2	-5471.402	1280.5	0.4424	85.514	1.04	136.2	9.71
2700	556.4	-5415.871	1301.4	0.4491	81.907	1.02	136.8	9.62
2800	558.4	-5360.131	1321.7	0.4559	78.436	0.99	137.4	9.52
2900	560.4	-5304.187	1341.3	0.4627	75.101	0.97	138.0	9.40
3000	562.4	-5248.044	1360.3	0.4694	71.902	0.94	138.7	9.28

<INT VdP (J)>

PRESSURE(GPa)

T(K)	1	5	10	20	30	40	50
298	124938	617610	1218873	2379601	3493406	4567974	5608903
300	124943	617632	1218915	2379679	3493516	4568113	5609068
400	125231	619017	1221560	2384557	3500339	4576668	5619188
500	125568	620622	1224597	2390071	3507945	4586079	5630187
600	125930	622334	1227805	2395786	3515688	4595503	5641025
700	126309	624109	1231096	2401526	3523311	4604600	5651286
800	126702	625928	1234429	2407211	3530693	4613212	5660776
900	127105	627780	1237783	2412797	3537772	4621258	5669396
1000	127518	629660	1241146	2418262	3544513	4628694	5677096
1100	127940	631564	1244512	2423590	3550896	4635496	5683851
1200	128372	633490	1247875	2428773	3556910	4641655	5689653
1300	128812	635438	1251233	2433805	3562550	4647169	5694505
1400	129262	637406	1254584	2438684	3567816	4652041	5698421
1500	129720	639394	1257927	2443409	3572710	4656282	5701418
1600	130187	641402	1261261	2447980	3577236	4659903	5703521
1700	130664	643429	1264585	2452399	3581402	4662922	5704759
1800	131149	645476	1267899	2456666	3585217	4665358	5705164
1900	131644	647542	1271204	2460785	3588690	4667232	5704773
2000	132147	649627	1274498	2464761	3591835	4668567	5703623
2100	132661	651732	1277783	2468596	3594663	4669391	5701757
2200	133183	653857	1281060	2472297	3597189	4669728	5699215
2300	133715	656001	1284328	2475868	3599427	4669609	5696043
2400	134257	658164	1287588	2479316	3601394	4669060	5692285
2500	134808	660348	1290842	2482646	3603105	4668112	5687986
2600	135369	662552	1294090	2485865	3604578	4666796	5683191
2700	135940	664776	1297334	2488981	3605830	4665141	5677946
2800	136521	667021	1300574	2492001	3606878	4663177	5672295
2900	137112	669286	1303812	2494931	3607740	4660936	5666284
3000	137713	671573	1307050	2497781	3608434	4658447	5659955

ANDRADITE

$Ca_3Fe_2Si_3O_{12}$
ANDRADITE Formula wt. (g) = 508.19

Enthalpy (J/mol) = -5769999.0 Zhang and Saxena (1991)
Entropy (J/mol.K) = 316.350 Robie et al. (1978)
Heat Capacity:
No. of *Cp* ranges = 1 Robie et al. (1978)
8.0924000D+02 -7.0251D-02 -6.7890D+05 0.0000D+00
0.0000D+00 -7.4030D+03 0.0000D+00
Volume (cc/mol) = 131.850 Helgeson et al. (1978)
Thermal expn:
2.15300D-05 1.08950D-08 2.12400D-03 -7.27800D-01
This study
Isot.compressibility:
4.92300D-07 2.26900D-10 -4.80300D-14 4.64900D-17
This study
P deriv. 4.00000D+00
T deriv. 0.00000D+00
Melting pt. and enthalpy of melting:
- -

NOTE: Enthalpy optimized by Zhang and Saxena (1991) on the basis of the following reactions:

Ad = Woll + Mag + O2
Ad + Mag + Q = Hed + O2
Ad + Q = Hed + Woll + O2
Ad = Woll + Hem

Volume is from the data base of Helgeson et al. (1978) (optimized). Other values are calorimetric. Thermal expansion and compressibility are as for grossular.

$Ca_3Fe_2Si_3O_{12}$
ANDRADITE

Thermochemical Data

T(K)	C_p(J)	H(kJ)	S(J)	α 10^{-4}	K 10^4	γ	V(cc)	δ_T
298	351.9	-5770.00	316.4	0.2371	179.561	1.61	131.8	15.95
300	353.2	-5769.35	318.5	0.2379	179.436	1.61	131.9	15.89
400	406.7	-5731.12	428.1	0.2665	172.906	1.52	132.2	13.68
500	440.3	-5688.65	522.8	0.2831	166.791	1.45	132.6	12.59
600	463.0	-5643.41	605.2	0.2959	160.981	1.40	132.9	11.94
700	478.9	-5596.28	677.8	0.3071	155.390	1.37	133.3	11.54
800	490.2	-5547.79	742.5	0.3176	149.951	1.34	133.8	11.30
900	498.4	-5498.33	800.8	0.3280	144.614	1.33	134.2	11.17
1000	504.2	-5448.19	853.6	0.3382	139.342	1.31	134.6	11.14
1100	508.2	-5397.55	901.9	0.3484	134.111	1.30	135.1	11.16
1200	510.8	-5346.59	946.2	0.3587	128.907	1.30	135.6	11.23
1300	512.2	-5295.44	987.1	0.3690	123.726	1.29	136.1	11.33
1400	512.7	-5244.19	1025.1	0.3793	118.569	1.28	136.6	11.43
1500	512.4	-5192.93	1060.5	0.3897	113.445	1.27	137.1	11.55
1600	511.5	-5141.73	1093.5	0.4001	108.365	1.26	137.7	11.65
1700	510.0	-5090.65	1124.5	0.4105	103.345	1.25	138.2	11.75
1800	508.1	-5039.74	1153.6	0.4210	98.403	1.24	138.8	11.83
1900	505.7	-4989.04	1181.0	0.4315	93.554	1.22	139.4	11.88
2000	503.0	-4938.60	1206.9	0.4420	88.818	1.21	140.0	11.91
2100	500.0	-4888.45	1231.4	0.4526	84.209	1.19	140.6	11.91
2200	496.7	-4838.61	1254.5	0.4631	79.742	1.18	141.3	11.89
2300	493.2	-4789.11	1276.5	0.4737	75.430	1.16	141.9	11.84
2400	489.4	-4739.98	1297.5	0.4844	71.282	1.14	142.6	11.77
2500	485.4	-4691.24	1317.4	0.4950	67.305	1.12	143.3	11.67
2600	481.3	-4642.90	1336.3	0.5057	63.506	1.10	144.0	11.55
2700	477.0	-4594.98	1354.4	0.5163	59.885	1.08	144.8	11.42
2800	472.5	-4547.50	1371.7	0.5270	56.446	1.06	145.5	11.26
2900	468.0	-4500.48	1388.2	0.5377	53.185	1.04	146.3	11.09
3000	463.3	-4453.91	1404.0	0.5484	50.101	1.02	147.1	10.91

<INT VdP (J)>

PRESSURE(GPa)

T(K)	1	5	10	20	30	40	50
298	131486	650470	1284794	2511780	3691593	4831649	5937419
300	131491	650493	1284828	2511815	3691605	4831622	5937340
400	131812	651831	1286939	2514243	3693205	4831617	5935182
500	132162	653314	1289355	2517400	3696132	4833725	5936105
600	132531	654893	1291970	2521034	3699948	4837273	5939155
700	132917	656548	1294735	2525021	3704420	4841886	5943778
800	133317	658268	1297623	2529284	3709396	4847309	5949588
900	133733	660048	1300618	2533767	3714759	4853344	5956286
1000	134162	661882	1303705	2538424	3720417	4859835	5963634
1100	134605	663769	1306873	2543218	3726294	4866654	5971437
1200	135062	665705	1310112	2548119	3732328	4873696	5979540
1300	135534	667687	1313415	2553101	3738469	4880877	5987816
1400	136019	669714	1316773	2558143	3744675	4888132	5996170
1500	136518	671783	1320181	2563230	3750918	4895411	6004530
1600	137031	673893	1323634	2568350	3757177	4902683	6012847
1700	137558	676042	1327129	2573497	3763439	4909926	6021092
1800	138100	678230	1330664	2578666	3769700	4917133	6029253
1900	138656	680455	1334238	2583860	3775963	4924309	6037334
2000	139225	682717	1337851	2589082	3782236	4931465	6045352
2100	139810	685016	1341504	2594340	3788533	4938623	6053333
2200	140408	687351	1345200	2599643	3794873	4945808	6061312
2300	141021	689725	1348942	2605005	3801278	4953054	6069333
2400	141649	692136	1352734	2610440	3807774	4960396	6077442
2500	142291	694587	1356582	2615964	3814387	4967873	6085689
2600	142948	697079	1360492	2621594	3821146	4975525	6094128
2700	143619	699613	1364470	2627347	3828080	4983393	6102811
2800	144306	702192	1368523	2633242	3835219	4991519	6111791
2900	145008	704818	1372659	2639297	3842592	4999942	6121121
3000	145725	707494	1376885	2645531	3850229	5008702	6130851

ANORTHITE

$CaAl_2Si_2O_8$
ANORTHITE Formula wt. (g) = 278.22

Enthalpy (J/mol) =	-4232500.000	Chatterjee (1987)
Entropy (J/mol.K) =	203.000	Chatterjee (1987), Robie et al. (1978)
Heat Capacity:		
No. of Cp ranges =	1	Chatterjee (1987), Holland (1981)
2.9090000D+02 2.7600D-02 -3.4080D+07 0.0000D+00		
5.2180D+09 0.0000D+00 2.9625D+04		
Volume (cc/mol) =	100.200	Czank and Schultz (1971)
Thermal expn:		
2.34041D-05 1.21057D-08 5.20652D-03 -2.19267D+00		This study
Isot.compressibility:		
9.33561D-07 2.17769D-10 7.87580D-15 2.51490D-17		This study
P deriv.	4.00000D+00	
T deriv.	0.00000D+00	
Melting pt. and enthalpy of melting:		
1830.00000	81000.00000	Robie et al. (1978)

NOTE: Calorimetric data: H: -4227.83±2.01, Charlu et al. (1978); S: 199.3 ±0.3, Robie et al. (1978). A reaction involving anorthite and grossular is as follows:

An = Gr + Ky + Q (a)

This was studied by many workers. Koziol and Newton's (1988a) results constrain the boundary within tight reversal brackets. Gasparik's (1984c) results place the reaction boundary a few kbars above that of Goldsmith's (1980). Gasparik (1984b) also studied the following reaction for the silica unsaturated system:

Gr + Ky = An + Cor (b)

For reactions involving anorthite, wollastonite and gehlenite see note under gehlenite. The following reaction is an important mechanism in the plagioclase-lherzolite to spinel-lherzolite transition

An + Fo = Sp + opx + cpx (c)

This reaction was studied by Herzberg (1976) and Kushiro and Yoder (1966). The heat capacity of anorthite, as listed in Robie et al. (1978) up to 1800 K, shows an anomalous extrapolation above 1200 K. In Chatterjee (1989), the data up to 1200 K are fitted to the Fei-Saxena equation for high-T extrapolation. The resulting equation gives very similar results when compared to the equation of Holland (1981). Entropy of anorthite is from Robie et al. (1978) and enthalpy is calculated from reaction (a). The low pressure equilibria of anorthite in the CaO-MgO-Al_2O_3-SiO_2 system [reaction (c)] are reproduced by the data recommended in this study.

$CaAl_2Si_2O_8$

ANORTHITE

Thermochemical Data

T(K)	C_p(J)	H(kJ)	S(J)	α 10^{-4}	K 10^4	γ	V(cc)	δ_T
298	212.0	-4232.50	203.0	0.1981	100.014	0.94	100.2	11.57
300	212.5	-4232.11	204.3	0.2003	99.972	0.95	100.2	11.45
400	244.5	-4209.25	269.9	0.2756	97.700	1.12	100.4	8.37
500	269.4	-4183.47	327.3	0.3110	95.460	1.13	100.7	7.51
600	286.3	-4155.64	378.0	0.3325	93.241	1.12	101.1	7.13
700	298.2	-4126.38	423.1	0.3484	91.035	1.11	101.4	6.94
800	307.0	-4096.10	463.5	0.3617	88.834	1.10	101.8	6.84
900	313.7	-4065.05	500.0	0.3738	86.635	1.09	102.2	6.79
1000	319.3	-4033.39	533.4	0.3852	84.434	1.09	102.5	6.77
1100	323.9	-4001.23	564.0	0.3964	82.229	1.08	102.9	6.77
1200	328.1	-3968.62	592.4	0.4075	80.020	1.08	103.4	6.78
1300	331.8	-3935.63	618.8	0.4185	77.807	1.08	103.8	6.80
1400	335.2	-3902.28	643.5	0.4295	75.592	1.07	104.2	6.82
1500	338.4	-3868.59	666.8	0.4406	73.378	1.07	104.7	6.85
1600	341.5	-3834.59	688.7	0.4517	71.166	1.07	105.1	6.87
1700	344.5	-3800.29	709.5	0.4629	68.961	1.06	105.6	6.89
1800	347.4	-3765.69	729.3	0.4741	66.768	1.05	106.1	6.91
1900	350.3	-3730.81	748.1	0.4854	64.589	1.05	106.6	6.92
2000	353.0	-3695.64	766.2	0.4967	62.430	1.04	107.2	6.93
2100	355.8	-3660.20	783.5	0.5081	60.295	1.03	107.7	6.93
2200	358.5	-3624.48	800.1	0.5195	58.188	1.02	108.3	6.92
2300	361.2	-3588.49	816.1	0.5310	56.114	1.01	108.8	6.90
2400	363.9	-3552.23	831.5	0.5425	54.077	1.00	109.4	6.88
2500	366.6	-3515.70	846.4	0.5540	52.079	0.98	110.0	6.85
2600	369.3	-3478.91	860.9	0.5656	50.125	0.97	110.6	6.81
2700	372.0	-3441.84	874.8	0.5772	48.217	0.96	111.3	6.77
2800	374.7	-3404.51	888.4	0.5888	46.358	0.94	111.9	6.72
2900	377.3	-3366.91	901.6	0.6005	44.550	0.93	112.6	6.66
3000	380.0	-3329.05	914.4	0.6121	42.794	0.91	113.3	6.60

<INT VdP (J)>

PRESSURE(GPa)

T(K)	1	5	10	20	30	40	50
298	99707	489408	958670	1849544	2690456	3492001	4261202
300	99710	489422	958690	1849560	2690457	3491979	4261152
400	99942	490365	960147	1851234	2691661	3492282	4260262
500	100226	491555	962079	1853794	2694129	3494205	4261333
600	100537	492874	964249	1856778	2697183	3496861	4263272
700	100868	494277	966566	1860004	2700551	3499890	4265635
800	101213	495743	968988	1863386	2704101	3503115	4268202
900	101572	497263	971493	1866874	2707757	3506434	4270842
1000	101943	498830	974069	1870440	2711472	3509780	4273470
1100	102327	500442	976706	1874062	2715211	3513104	4276023
1200	102722	502097	979400	1877724	2718946	3516367	4278450
1300	103130	503793	982145	1881414	2722656	3519537	4280711
1400	103549	505529	984938	1885119	2726320	3522588	4282772
1500	103981	507305	987775	1888831	2729923	3525498	4284604
1600	104424	509119	990653	1892539	2733450	3528245	4286182
1700	104880	510972	993570	1896237	2736890	3530815	4287488
1800	105349	512862	996522	1899918	2740230	3533194	4288504
1900	105830	514789	999507	1903575	2743463	3535371	4289220
2000	106323	516753	1002524	1907204	2746581	3537338	4289628
2100	106830	518752	1005570	1910799	2749580	3539090	4289723
2200	107349	520787	1008643	1914358	2752456	3540625	4289506
2300	107882	522857	1011743	1917879	2755206	3541942	4288978
2400	108428	524962	1014867	1921358	2757830	3543045	4288146
2500	108987	527102	1018014	1924796	2760330	3543938	4287019
2600	109560	529275	1021185	1928191	2762709	3544627	4285609
2700	110147	531483	1024377	1931546	2764970	3545122	4283930
2800	110747	533724	1027591	1934860	2767119	3545434	4282000
2900	111362	536000	1030826	1938138	2769162	3545574	4279837
3000	111991	538309	1034083	1941380	2771109	3545557	4277462

ALBITE

$NaAlSi_3O_8$
ALBITE Formula wt. (g) = 262.23

Enthalpy (J/mol) =	-3929860.000	Hemingway et al. (1981)
Entropy (J/mol.K) =	221.800	Hemingway et al. (1981)

Heat Capacity:
No. of *Cp* ranges = 1 Hemingway et al. (1981), refitted

3.0974000D+02 1.5270D-02 -2.6160D+07 0.0000D+00
4.1092D+09 0.0000D+00 8.8361D+03

Volume (cc/mol) = 100.430 Robie et al. (1978)

Thermal expn:
2.52098D-05 8.32390D-09 3.10442D-03 -1.89468D+00
This study

Isot.compressibility:
1.40528D-06 2.07138D-10 2.46170D-14 1.52582D-17
This study

P deriv. 4.00000D+00
T deriv. 0.00000D+00
Melting pt. and enthalpy of melting:
- -

NOTE: Calorimetric data: H: -3929.56±1.71; S: 220.3 ±0.4. Hemingway et al. (1981) (derived from aluminum avoidance model). The following reaction, studied by Holland (1979) relates jadeite and albite:

$$Ab = Jd + Q \qquad (a)$$

The free energies of albite and jadeite were evaluated by Hemingway et al. (1981). Holland (1979) also attempted to evaluate the free energy of albite on the basis of reaction (a). Hemingway et al.'s values on albite are based on a careful consideration of the low-albite->albite transition. Their values on jadeite and albite are thus consistent with both the transition and the reaction (a). However, they did not consider the effect of thermal expansion and bulk modulus of these two phases. When these two parameters are considered, their second model, which assumes that the aluminum avoidance principle holds for albite, fits the experimental equilibria the best. Thus, the values recommended by Hemingway et al. (1981) are adopted for albite.

$NaAlSi_3O_8$

ALBITE

Thermochemical Data

T(K)	C_p(J)	H(kJ)	S(J)	$\alpha\ 10^{-4}$	K 10^4	γ	V(cc)	δ_T
298	204.7	-3929.86	221.8	0.1679	68.044	0.56	100.4	9.15
300	205.3	-3929.48	223.1	0.1700	68.025	0.57	100.4	9.04
400	238.6	-3907.24	286.8	0.2446	66.977	0.70	100.6	6.41
500	263.3	-3882.07	342.9	0.2800	65.923	0.71	100.9	5.73
600	280.0	-3854.85	392.5	0.3012	64.863	0.72	101.2	5.45
700	291.6	-3826.24	436.5	0.3160	63.793	0.71	101.5	5.33
800	300.2	-3796.63	476.1	0.3279	62.713	0.71	101.8	5.28
900	306.6	-3766.27	511.8	0.3381	61.623	0.71	102.2	5.26
1000	311.8	-3735.34	544.4	0.3474	60.522	0.71	102.5	5.26
1100	316.0	-3703.94	574.3	0.3562	59.410	0.71	102.9	5.28
1200	319.6	-3672.16	602.0	0.3647	58.287	0.71	103.3	5.31
1300	322.8	-3640.03	627.7	0.3730	57.153	0.71	103.7	5.34
1400	325.6	-3607.61	651.7	0.3811	56.010	0.71	104.0	5.38
1500	328.1	-3574.92	674.3	0.3892	54.858	0.71	104.4	5.41
1600	330.5	-3541.99	695.5	0.3973	53.699	0.71	104.9	5.45
1700	332.7	-3508.83	715.6	0.4053	52.534	0.71	105.3	5.48
1800	334.8	-3475.46	734.7	0.4133	51.364	0.71	105.7	5.52
1900	336.8	-3441.88	752.9	0.4213	50.192	0.70	106.2	5.55
2000	338.7	-3408.11	770.2	0.4294	49.017	0.70	106.6	5.58
2100	340.5	-3374.15	786.7	0.4374	47.844	0.70	107.1	5.60
2200	342.3	-3340.01	802.6	0.4454	46.672	0.70	107.5	5.63
2300	344.1	-3305.69	817.9	0.4535	45.505	0.69	108.0	5.64
2400	345.8	-3271.19	832.6	0.4615	44.343	0.69	108.5	5.66
2500	347.5	-3236.52	846.7	0.4696	43.189	0.69	109.0	5.67
2600	349.2	-3201.69	860.4	0.4777	42.045	0.68	109.5	5.67
2700	350.9	-3166.68	873.6	0.4857	40.911	0.68	110.1	5.67
2800	352.5	-3131.51	886.4	0.4938	39.790	0.67	110.6	5.67
2900	354.1	-3096.18	898.8	0.5019	38.683	0.67	111.2	5.66
3000	355.7	-3060.69	910.8	0.5101	37.591	0.66	111.7	5.65

<INT VdP (J)>

T(K)	PRESSURE(GPa) 1	5	10	20	30	40	50
298	99709	485625	944001	1800872	2598474	3351491	4068998
300	99712	485636	944017	1800885	2598476	3351477	4068966
400	99913	486437	945243	1802333	2599666	3352131	4068907
500	100166	487482	946932	1804636	2602065	3354324	4070695
600	100446	488653	948854	1807352	2605033	3357224	4073320
700	100744	489903	950915	1810302	2608306	3360491	4076366
800	101057	491211	953073	1813398	2611759	3363959	4079628
900	101380	492564	955304	1816593	2615320	3367536	4082993
1000	101714	493957	957593	1819860	2618946	3371163	4086389
1100	102057	495385	959932	1823178	2622609	3374803	4089767
1200	102410	496847	962316	1826535	2626286	3378425	4093093
1300	102772	498339	964740	1829920	2629962	3382008	4096337
1400	103142	499861	967199	1833325	2633624	3385535	4099478
1500	103522	501412	969693	1836743	2637260	3388990	4102497
1600	103910	502991	972217	1840168	2640862	3392361	4105380
1700	104308	504597	974771	1843595	2644421	3395639	4108114
1800	104714	506231	977352	1847019	2647932	3398813	4110688
1900	105129	507891	979960	1850438	2651387	3401878	4113094
2000	105554	509577	982591	1853846	2654783	3404826	4115324
2100	105987	511288	985246	1857243	2658115	3407654	4117375
2200	106430	513025	987922	1860624	2661380	3410357	4119243
2300	106881	514787	990619	1863987	2664575	3412935	4120926
2400	107342	516574	993336	1867332	2667700	3415384	4122424
2500	107813	518385	996073	1870657	2670752	3417707	4123738
2600	108292	520221	998827	1873961	2673732	3419902	4124871
2700	108782	522081	1001600	1877243	2676640	3421974	4125827
2800	109281	523964	1004390	1880503	2679478	3423923	4126610
2900	109789	525872	1007196	1883742	2682247	3425755	4127228
3000	110308	527803	1010020	1886959	2684949	3427474	4127687

QUARTZ

SiO_2
QUARTZ Formula wt. (g) = 60.09

Enthalpy (J/mol) = -910700.000+-1000.0 Robie et al. (1978)
Entropy (J/mol.K) = 41.460+-0.2 Robie et al. (1978)
Heat Capacity
No. of *Cp* ranges = 2 Robie et al. (1978)
8.1144700D+01 1.8280D-02 -1.8099D+05 5.4058D-06
0.0000D+00 -6.9846D+02 0.0000D+00
Temp. and enthalpy of transition:
848.00 3000.00
7.5812000D+01 1.2050D-03 1.7310D+06 0.0000D+00
1.2020D+08 0.0000D+00 -1.2130D+04
Volume (cc/mol) = 22.688 Robie et al. (1978)
Thermal expn.
2.75130D-05 2.98680D-08 5.57220D-06 9.11810D-02
This study
Isot.bulk mod.(b0, b1*T, b2*T^2, b3*T^3):
2.55600D-06 1.15570D-11 1.01280D-16 8.89180D-19
This study
P deriv., T deriv.:
6.40000D+00 1.00000D-02 Soga (1968)
Melting pt. and enthalpy of melting:
\- -

NOTE: All the values were determined through calorimetry. The experimental data on the transformation: α-quartz = β-quartz are available from Fenner (1913), Cohen and Klement (1967) and Mirwald and Massonne (1980). These data were used to estimate the data for beta-quartz.

SiO_2
QUARTZ

Thermochemical Data

T(K)	C_p(J)	H(kJ)	S(J)	α 10^{-4}	K 10^4	γ	V(cc)	δ_T
298	44.6	-910.700	41.5	0.3746	39.070	0.75	22.7	0.12
300	44.8	-910.617	41.7	0.3751	39.070	0.75	22.7	0.12
400	53.3	-905.690	55.9	0.4004	39.052	0.68	22.8	0.12
500	59.7	-900.031	68.5	0.4282	39.033	0.65	22.9	0.11
600	65.0	-893.789	79.8	0.4570	39.014	0.64	23.0	0.11
700	69.8	-887.042	90.2	0.4861	38.995	0.64	23.1	0.10
800	74.3	-879.837	99.8	0.5156	38.975	0.64	23.2	0.10
900	65.7	-869.821	111.6	0.5451	38.954	0.78	23.3	0.10
1000	66.7	-863.197	118.6	0.5748	38.933	0.82	23.5	0.10
1100	67.6	-856.477	125.0	0.6045	38.910	0.87	23.6	0.10
1200	68.4	-849.674	130.9	0.6342	38.887	0.92	23.7	0.10
1300	69.1	-842.796	136.4	0.6640	38.863	0.97	23.9	0.10
1400	69.8	-835.851	141.6	0.6938	38.838	1.02	24.1	0.10
1500	70.3	-828.846	146.4	0.7236	38.811	1.08	24.2	0.10
1600	70.9	-821.785	151.0	0.7534	38.784	1.15	24.4	0.10
1700	71.3	-814.674	155.3	0.7832	38.755	1.22	24.6	0.10
1800	71.8	-807.517	159.4	0.8131	38.725	1.29	24.8	0.10
1900	72.2	-800.316	163.3	0.8429	38.693	1.38	25.0	0.10
2000	72.6	-793.075	167.0	0.8727	38.660	1.47	25.2	0.10
2100	73.0	-785.796	170.5	0.9026	38.626	1.58	25.4	0.10
2200	73.3	-778.481	173.9	0.9324	38.589	1.70	25.7	0.10
2300	73.6	-771.132	177.2	0.9623	38.551	1.84	25.9	0.10
2400	74.0	-763.752	180.4	0.9921	38.512	1.99	26.2	0.11
2500	74.3	-756.341	183.4	1.0220	38.470	2.18	26.4	0.11
2600	74.5	-748.901	186.3	1.0519	38.427	2.40	26.7	0.11
2700	74.8	-741.433	189.1	1.0817	38.381	2.66	27.0	0.11
2800	75.1	-733.938	191.8	1.1116	38.334	2.99	27.3	0.11
2900	75.3	-726.417	194.5	1.1414	38.284	3.41	27.6	0.12
3000	75.6	-718.871	197.0	1.1713	38.232	3.95	27.9	0.12

<INT VdP (J)>

T(K)	PRESSURE(GPa) 1	5	10	20	30	40	50
298	22414	107756	207663	393213	566101	730098	887269
300	22415	107764	207678	393240	566140	730149	887331
400	22503	108221	208672	395518	569894	735507	894384
500	22598	108765	209983	398898	575822	744298	906263
600	22700	109374	211489	402850	582791	754657	920277
700	22810	110032	213112	407061	590155	765541	934939
800	22927	110733	214811	411373	597601	776457	949563
900	23051	111469	216564	415715	604998	787211	963884
1000	23184	112239	218362	420062	612309	797751	977841
1100	23324	113042	220201	424414	619537	808090	991459
1200	23472	113877	222083	428777	626704	818269	1004798
1300	23628	114745	224010	433167	633841	828338	1017930
1400	23792	115647	225985	437597	640978	838346	1030928
1500	23964	116584	228014	442082	648145	848342	1043858
1600	24145	117556	230099	446636	655370	858369	1056781
1700	24334	118566	232245	451272	662678	868466	1069751
1800	24532	119614	234455	456004	670092	878669	1082819
1900	24739	120702	236734	460842	677634	889010	1096028
2000	24955	121830	239085	465796	685324	899519	1109418
2100	25180	123002	241512	470878	693178	910223	1123026
2200	25415	124216	244019	476097	701215	921146	1136885
2300	25911	125476	246608	481461	709449	932312	1151027
2400	26165	126783	249285	486981	717896	943743	1165480
2500	26431	128138	252051	492663	726571	955459	1180271
2600	26706	129542	254910	498517	735487	967480	1195428
2700	26993	130998	257867	504550	744657	979826	1210975
2800	27291	132506	260924	510772	754096	992514	1226937
2900	27600	134069	264086	517189	763816	1005565	1243338
3000	27921	135687	267355	523811	773830	1018995	1260200

β-QUARTZ

SiO_2
β-QUARTZ Formula wt. (g) = 60.09

Enthalpy (J/mol) = -910497.000+-1000.0 Robie et al. (1978)
Entropy (J/mol.K) = 41.700+-0.2
Heat Capacity
No. of *Cp* ranges = 2 Robie et al. (1978)
8.1144700D+01 1.8280D-02 -1.8099D+05 5.4058D-06
0.0000D+00 -6.9846D+02 0.0000D+00
Temp. and enthalpy of transition:
848.00 844.00
7.8812000D+01 1.2050D-03 1.7310D+06 0.0000D+00
1.2020D+08 0.0000D+00 -1.2130D+04
Volume (cc/mol) = 22.875 Robie et al. (1978)
Thermal expn. (a0, a1*T, a2*T^-1, a3*T^-2):
2.06036D-05 3.4694D-08 1.30754D-04 -1.6376D-00
This study
Isot.bulk mod.(b0, b1*T, b2*T^2, b3*T^3):
1.3970D-06 2.41191D-09 -2.1518D-12 6.4543D-16
This study
P deriv., T deriv.:
5.34600D+00 0.00000D+00 Soga (1968)
Melting pt. and enthalpy of melting:
\- -

NOTE: All the values were determined through calorimetry. The experimental data on the transformation α-quartz = β-quartz are available from Fenner (1913), Cohen and Klement (1967) and Mirwald and Massonne (1980). These data were used to estimate the data for β-quartz.

SiO_2
β- QUARTZ

Thermochemical Data

T (K)	C_p (J)	H (kJ)	S (J)	α 10^{-4}	K 10^4	γ	V (cc)	δ_T
298	44.6	-910.497	41.7	0.1296	51.495	0.34	22.9	51.67
300	44.8	-910.414	42.0	0.1325	51.431	0.35	22.9	50.26
400	53.3	-905.487	56.1	0.2457	48.572	0.52	22.9	19.77
500	59.7	-899.828	68.7	0.3166	46.605	0.57	23.0	10.95
600	65.0	-893.586	80.1	0.3709	45.271	0.60	23.1	6.43
700	69.8	-886.839	90.4	0.4173	44.398	0.63	23.2	3.70
800	74.3	-879.634	100.1	0.4596	43.863	0.65	23.3	1.99
900	68.7	-871.618	109.5	0.4995	43.568	0.77	23.4	0.93
1000	69.7	-864.694	116.8	0.5379	43.430	0.82	23.5	0.36
1100	70.6	-857.674	123.5	0.5753	43.375	0.88	23.6	0.16
1200	71.4	-850.571	129.7	0.6121	43.327	0.94	23.8	0.25
1300	72.1	-843.393	135.4	0.6484	43.216	1.01	23.9	0.60
1400	72.8	-836.148	140.8	0.6843	42.970	1.07	24.1	1.14
1500	73.3	-828.843	145.8	0.7200	42.523	1.14	24.2	1.85
1600	73.9	-821.482	150.6	0.7556	41.821	1.20	24.4	2.67
1700	74.3	-814.071	155.1	0.7909	40.824	1.25	24.6	3.57
1800	74.8	-806.614	159.3	0.8262	39.514	1.29	24.8	4.49
1900	75.2	-799.113	163.4	0.8614	37.898	1.32	25.0	5.39
2000	75.6	-791.572	167.3	0.8965	36.009	1.34	25.2	6.23
2100	76.0	-783.993	171.0	0.9315	33.899	1.34	25.5	6.96
2200	76.3	-776.378	174.5	0.9665	31.635	1.32	25.7	7.58
2300	76.6	-768.729	177.9	1.0015	29.288	1.29	26.0	8.06
2400	77.0	-761.049	181.2	1.0364	26.928	1.25	26.2	8.41
2500	77.3	-753.338	184.3	1.0713	24.613	1.19	26.5	8.63
2600	77.5	-745.598	187.4	1.1062	22.392	1.14	26.8	8.73
2700	77.8	-737.830	190.3	1.1410	20.298	1.07	27.1	8.73
2800	78.1	-730.035	193.1	1.1758	18.353	1.01	27.4	8.65
2900	78.3	-722.214	195.9	1.2107	16.566	0.95	27.8	8.51
3000	78.6	-714.368	198.5	1.2455	14.941	0.89	28.1	8.31

<INT VdP (J)>

PRESSURE (GPa)

T (K)	1	5	10	20	30	40	50
298	22661	109705	212364	403734	582179	751321	913232
300	22662	109703	212354	403700	582117	751228	913107
400	22694	109684	212036	402425	579658	747476	908004
500	22749	109820	212092	402041	578655	745766	905537
600	22821	110072	212433	402339	578767	745617	905082
700	22907	110421	213005	403188	579776	746721	906238
800	23004	110852	213773	404494	581521	748846	908702
900	23113	111354	214706	406178	583867	751798	912220
1000	23233	111919	215778	408167	586691	755402	916564
1100	23362	112537	216964	410395	589880	759495	921517
1200	23501	113202	218241	412796	593319	763911	926865
1300	23649	113906	219583	415304	596893	768487	932392
1400	23806	114640	220968	417851	600490	773057	937882
1500	23971	115398	222371	420371	603993	777457	943117
1600	24144	116171	223768	422800	607291	781527	947891
1700	24325	116952	225137	425075	610279	785116	952007
1800	24513	117734	226456	427140	612864	788095	955296
1900	24709	118510	227705	428945	614966	790354	957623
2000	24911	119273	228867	430453	616529	791817	958893
2100	25120	120018	229929	431638	617516	792443	959060
2200	25336	120742	230882	432487	617916	792224	958123
2300	25557	121439	231721	432999	617740	791188	956129
2400	25785	122108	232445	433187	617022	789393	953161
2500	26018	122748	233056	433073	615811	786917	949332
2600	26257	123359	233562	432688	614169	783858	944773
2700	26501	123942	233972	432069	612166	780319	939625
2800	26751	124499	234297	431256	609874	776409	934030
2900	27006	125032	234551	430292	607367	772233	928127
3000	27267	125546	234748	429217	604715	767891	922043

COESITE

SiO_2
COESITE Formula wt. (g) = 60.09

Enthalpy (J/mol) =	-906900.000	This study
Entropy (J/mol.K) =	40.500	This study
Heat Capacity		
No. of *Cp* ranges = 1		Robie et al. (1978)
7.8000000D+01 0.000D-04 -3.1000D+06 0.0000D+00		
5.8580D+08 0.0000D+00 -6.6890D+03		
Volume (cc/mol) =	20.640	Robie et al. (1978)
Thermal expn.		
5.4300D-05 5.00D-09 0.000D-00 0.000D-00		Saxena and Shen (1992)
Isot.bulk mod.		
1.0500D-07 2.000D -10 0 0.000D-00 0.00D-00		Saxena and Shen (1992)
P deriv., T deriv.:		
8.40000D+00 0.00000D-03		Levien and Prewitt (1981)
Melting pt. and enthalpy of melting:		
- -		

NOTE: Akaogi and Navrotsky (1984), using new calorimetric data, calculated phase relations in SiO_2 (quartz, coesite and stishovite). Fei and Saxena (1986) started with the thermochemical data on quartz from Robie et al. (1978), and evaluated the thermochemical data on coesite and stishovite; these data were then combined with the phase equilibrium data for the quartz-coesite transformation (Bohlen and Boettcher, 1982),for the coesite-stishovite transformation (Yagi and Akimoto, 1976) and with the calorimetric data from Akaogi and Navrotsky (1984).

SiO_2

COESITE

Thermochemical Data

T (K)	C_p (J)	H (kJ)	S (J)	α 10^{-4}	K 10^4	γ	V (cc)	δ_T
298	42.8	-906.900	40.5	0.0692	90.120	0.30	20.6	26.04
300	43.0	-906.821	40.8	0.0693	90.090	0.30	20.6	26.00
400	51.1	-902.105	54.3	0.0743	88.496	0.27	20.7	23.82
500	56.9	-896.688	66.3	0.0793	86.957	0.25	20.7	21.93
600	61.0	-890.783	77.1	0.0843	85.470	0.24	20.7	20.28
700	63.8	-884.536	86.7	0.0893	84.034	0.24	20.7	18.82
800	65.9	-878.043	95.4	0.0943	82.645	0.25	20.7	17.53
900	67.5	-871.365	103.2	0.0993	81.301	0.25	20.7	16.37
1000	68.8	-864.546	110.4	0.1043	80.000	0.25	20.8	15.34
1100	69.8	-857.614	117.0	0.1093	78.740	0.26	20.8	14.41
1200	70.6	-850.593	123.1	0.1143	77.519	0.26	20.8	13.56
1300	71.3	-843.497	128.8	0.1193	76.336	0.27	20.8	12.80
1400	71.9	-836.339	134.1	0.1243	75.188	0.27	20.9	12.10
1500	72.3	-829.129	139.1	0.1293	74.074	0.28	20.9	11.46
1600	72.8	-821.874	143.8	0.1343	72.993	0.28	20.9	10.87
1700	73.1	-814.580	148.2	0.1393	71.942	0.29	20.9	10.33
1800	73.4	-807.253	152.4	0.1443	70.922	0.29	21.0	9.83
1900	73.7	-799.896	156.4	0.1493	69.930	0.30	21.0	9.37
2000	74.0	-792.513	160.2	0.1543	68.966	0.31	21.0	8.94
2100	74.2	-785.106	163.8	0.1593	68.027	0.31	21.1	8.54
2200	74.4	-777.678	167.2	0.1643	67.114	0.32	21.1	8.17
2300	74.6	-770.232	170.5	0.1693	66.225	0.32	21.1	7.82
2400	74.7	-762.768	173.7	0.1743	65.359	0.33	21.2	7.50
2500	74.9	-755.289	176.8	0.1793	64.516	0.33	21.2	7.20
2600	75.0	-747.795	179.7	0.1843	63.694	0.34	21.3	6.91
2700	75.1	-740.289	182.5	0.1893	62.893	0.34	21.3	6.64
2800	75.2	-732.770	185.3	0.1943	62.112	0.35	21.3	6.39
2900	75.3	-725.241	187.9	0.1993	61.350	0.35	21.4	6.16
3000	75.4	-717.701	190.5	0.2043	60.606	0.36	21.4	5.93

<INT VdP (J)>

PRESSURE (GPa)

T (K)	1	5	10	20	30	40	50
298	20529	100728	197567	383042	560489	731868	898358
300	20529	100729	197568	383040	560483	731858	898344
400	20542	100762	197583	382938	560205	731366	897615
500	20556	100801	197609	382859	559960	730921	896945
600	20571	100845	197645	382802	559750	730522	896334
700	20586	100894	197692	382766	559573	730170	895782
800	20603	100948	197750	382753	559431	729863	895288
900	20621	101008	197818	382761	559321	729602	894852
1000	20641	101073	197897	382792	559246	729387	894474
1100	20661	101143	197986	382844	559204	729217	894153
1200	20682	101218	198086	382918	559195	729092	893889
1300	20704	101298	198197	383013	559219	729011	893681
1400	20727	101384	198318	383130	559277	728975	893530
1500	20752	101475	198450	383269	559367	728984	893434
1600	20777	101572	198592	383430	559490	729037	893395
1700	20804	101673	198746	383612	559646	729134	893411
1800	20831	101780	198910	383816	559835	729275	893482
1900	20860	101893	199084	384041	560057	729460	893609
2000	20890	102010	199269	384288	560311	729689	893790
2100	20921	102133	199465	384557	560598	729962	894026
2200	20953	102262	199672	384847	560917	730278	894317
2300	20986	102396	199890	385159	561270	730638	894662
2400	21020	102535	200118	385493	561654	731041	895061
2500	21055	102680	200357	385849	562072	731488	895515
2600	21091	102830	200608	386226	562522	731978	896023
2700	21129	102985	200869	386625	563004	732512	896585
2800	21167	103147	201141	387045	563519	733088	897201
2900	21207	103313	201423	387488	564067	733709	897872
3000	21248	103486	201717	387952	564647	734372	898596

SiO$_2$

COESITE

STISHOVITE

SiO_2
STISHOVITE Formula wt. (g) = 60.09

Enthalpy (J/mol) = -864000.000 This study
Entropy (J/mol.K) = 29.500 This study
Heat Capacity: Robie et al. (1978), refitted
5.8120000D+01 7.0020D-03 -1.2689D+07 0.0000D+00
1.7928D+09 0.0000D+00 1.7010D+04
Volume (cc/mol) = 14.010 Robie et al. (1978)
Thermal expn:
2.30000D-06 1.20000D-08 6.20000D-03 -1.13000D+00
This study
Isot.compressibility:
2.95400D-07 8.96100D-11 -3.29000D-14 2.33100D-17
This study
P deriv. 6.00000D+00
T deriv. 1.00000D-03 This study
Melting pt. and enthalpy of melting:
\- -

NOTE: Akaogi and Navrotsky (1984), using new calorimetric data, calculated phase relations in SiO_2 (quartz, coesite and stishovite). Fei and Saxena (1986) started with the thermochemical data on quartz from Robie et al. (1978), and evaluated the thermochemical data on coesite and stishovite; these data were then combined with the phase equilibrium data for the quartz-coesite transformation (Bohlen and Boettcher, 1982), for the coesite-stishovite transformation (Yagi and Akimoto, 1976) and with the calorimetric data from Akaogi and Navrotsky (1984). Data on thermal expansion and compressibility are consistent with those of Ito et al. (1974) and Weidner et al. (1982), respectively.

SiO_2

STISHOVITE

Thermochemical Data

T (K)	C_p (J)	H (kJ)	S (J)	$\alpha\ 10^{-4}$	K 10^4	γ	V (cc)	δ_T
298	42.2	-864.000	29.5	0.1396	312.685	1.46	14.0	17.07
300	42.3	-863.922	29.8	0.1401	312.547	1.46	14.0	16.99
400	52.2	-859.188	43.3	0.1554	305.370	1.29	14.0	14.64
500	59.2	-853.593	55.8	0.1618	298.602	1.16	14.1	13.69
600	63.7	-847.428	67.0	0.1669	292.093	1.09	14.1	13.18
700	66.7	-840.900	77.1	0.1725	285.713	1.06	14.1	12.89
800	68.7	-834.128	86.1	0.1788	279.356	1.04	14.1	12.77
900	70.1	-827.186	94.3	0.1859	272.931	1.04	14.2	12.78
1000	71.2	-820.116	101.7	0.1937	266.368	1.05	14.2	12.89
1100	72.1	-812.946	108.5	0.2020	259.614	1.06	14.2	13.09
1200	72.9	-805.692	114.9	0.2108	252.630	1.07	14.2	13.34
1300	73.6	-798.364	120.7	0.2200	245.396	1.08	14.3	13.64
1400	74.3	-790.970	126.2	0.2295	237.907	1.09	14.3	13.95
1500	74.9	-783.515	131.3	0.2393	230.170	1.10	14.3	14.26
1600	75.4	-776.000	136.2	0.2493	222.208	1.10	14.4	14.56
1700	76.0	-768.428	140.8	0.2596	214.051	1.10	14.4	14.83
1800	76.6	-760.800	145.1	0.2700	205.742	1.10	14.4	15.07
1900	77.1	-753.115	149.3	0.2805	197.327	1.10	14.5	15.27
2000	77.7	-745.375	153.3	0.2912	188.857	1.09	14.5	15.42
2100	78.2	-737.579	157.1	0.3020	180.386	1.08	14.6	15.52
2200	78.8	-729.727	160.7	0.3128	171.966	1.07	14.6	15.57
2300	79.4	-721.818	164.2	0.3238	163.646	1.06	14.7	15.58
2400	79.9	-713.853	167.6	0.3349	155.473	1.04	14.7	15.53
2500	80.5	-705.830	170.9	0.3460	147.489	1.02	14.8	15.44
2600	81.1	-697.750	174.1	0.3572	139.728	1.00	14.8	15.31
2700	81.7	-689.612	177.1	0.3684	132.220	0.97	14.9	15.14
2800	82.3	-681.415	180.1	0.3797	124.989	0.95	14.9	14.93
2900	82.9	-673.159	183.0	0.3910	118.052	0.92	15.0	14.70
3000	83.5	-664.844	185.8	0.4024	111.421	0.90	15.0	14.44

<INT VdP (J)>

T (K)	PRESSURE (GPa) 1	5	10	20	30	40	50
298	13987	69509	138010	272334	403544	532053	658181
300	13988	69511	138012	272338	403548	532057	658183
400	14008	69602	138173	272585	403829	532335	658429
500	14030	69701	138349	272868	404174	532710	658812
600	14052	69804	138533	273171	404555	533144	659281
700	14076	69910	138725	273490	404964	533621	659813
800	14100	70021	138924	273824	405397	534133	660393
900	14125	70136	139131	274171	405851	534674	661011
1000	14151	70255	139346	274532	406322	535238	661658
1100	14179	70379	139569	274905	406809	535820	662327
1200	14207	70508	139800	275290	407309	536415	663008
1300	14237	70642	140038	275684	407819	537020	663697
1400	14268	70781	140285	276088	408337	537630	664386
1500	14301	70926	140540	276501	408862	538241	665072
1600	14334	71076	140802	276922	409391	538852	665751
1700	14370	71231	141072	277349	409923	539460	666419
1800	14407	71392	141349	277783	410457	540065	667075
1900	14445	71557	141633	278224	410993	540664	667719
2000	14485	71729	141924	278670	411532	541259	668350
2100	14526	71905	142222	279122	412072	541850	668970
2200	14569	72087	142528	279581	412614	542439	669581
2300	14613	72274	142841	280047	413161	543027	670186
2400	14659	72466	143160	280520	413713	543617	670788
2500	14707	72664	143488	281001	414272	544212	671391
2600	14756	72868	143823	281491	414841	544814	672000
2700	14807	73077	144166	281992	415420	545428	672619
2800	14859	73291	144517	282504	416013	546057	673255
2900	14913	73511	144877	283029	416623	546705	673911
3000	14969	73737	145246	283568	417251	547376	674594

TRIDYMITE

SiO_2
TRIDYMITE Formula wt. (g) = 60.09

Enthalpy (J/mol) = -907771.100 This study
Entropy (J/mol.K) = 43.648 This study
Heat Capacity: This study
No. of Cp ranges = 6
2.5157400D+01 2.9743D-02 -1.3266D+05 1.3675D-04
0.0000D+00 0.0000D+00 0.0000D+00
Temp. and enthalpy of transition:
388.00 270.00
3.7870100D+01 4.7371D-02 0.0000D+00 1.0101D-06
0.0000D+00 0.0000D+00 0.0000D+00
433.00 150.00
3.5605000D+01 6.1000D-02 3.2405D+05 -2.7753D-05
0.0000D+00 0.0000D+00 0.0000D+00
900.00 0.00
1.2754400D+02 -6.7741D-02 -1.4142D+07 2.2721D-05
0.0000D+00 0.0000D+00 0.0000D+00
1668.00 0.00
7.7587500D+01 -6.0805D-03 -4.4542D+06 2.7787D-06
0.0000D+00 0.0000D+00 0.0000D+00
3300.00 0.00
8.7373000D+01 0.0000D+00 0.0000D+00 0.0000D+00
0.0000D+00 0.0000D+00 0.0000D+00
Volume (cc/mol) = 26.530 Robie et al. (1978)
Thermal expn:
2.34923D-04 -2.80434D-07 -1.07500D-02 1.27930D-01
Swamy et al. (1993)
Isot.compressibility:
3.80000D-06 0.00000D+00 0.00000D+00 0.00000D+00
Swamy et al. (1993)
P deriv. 6.00000D+00
T deriv. 1.00000D-03 This study
Melting pt. and enthalpy of melting:
\- -

NOTE: The thermal expansion here requires two temperature ranges. The data above 693.15 K are: molar volume 27.4936 cm^3; the coefficients for thermal expansion: 0.21797D-04 -0.172494D-07 0.000D-02 0.000D-01. This is a low-pressure phase and the data on compressibility and thermal expansion are not reliable or useful with pressures more than 10 kbar.

SiO_2

TRIDYMITE

Thermochemical Data

T(K)	C_p(J)	H(kJ)	S(J)	α 10^{-4}	K 10^4	γ	V(cc)	δ_T
298	44.7	-907.771	43.6	1.1670	26.316	1.95	26.5	0.00
300	44.9	-907.688	43.9	1.1638	26.316	1.93	26.5	0.00
400	57.0	-902.292	59.3	0.9667	26.316	1.26	26.8	0.00
500	60.5	-896.249	72.8	0.7372	26.316	0.90	27.1	0.00
600	63.1	-890.067	84.0	0.4910	26.316	0.57	27.2	0.00
700	65.4	-883.639	93.9	0.2352	26.316	0.26	27.3	0.00
800	67.1	-877.009	102.8	-0.0266	26.316	-0.03	27.3	0.00
900	67.5	-870.226	110.8	-0.2925	26.316	-0.31	27.3	0.00
1000	68.4	-863.426	117.9	-0.5613	26.316	-0.61	27.2	0.00
1100	68.8	-856.563	124.5	-0.8322	26.316	-0.93	27.0	0.00
1200	69.2	-849.663	130.5	-1.1047	26.316	-1.32	26.7	0.00
1300	69.5	-842.731	136.0	-1.3783	26.316	-1.83	26.4	0.00
1400	70.0	-835.756	141.2	-1.6530	26.316	-2.58	26.0	0.00
1500	70.8	-828.718	146.0	-1.9284	26.316	-3.90	25.5	0.00
1600	71.8	-821.592	150.6	-2.2044	26.316	-7.05	25.0	0.00
1700	73.7	-814.324	155.1	-2.4809	26.316	-24.84	24.4	0.00
1800	74.3	-806.923	159.3	-2.7579	26.316	14.99	23.8	0.00
1900	74.8	-799.469	163.3	-3.0352	26.316	5.82	23.1	0.00
2000	75.4	-791.956	167.2	-3.3129	26.316	3.62	22.4	0.00
2100	76.1	-784.382	170.9	-3.5908	26.316	2.62	21.7	0.00
2200	76.7	-776.742	174.4	-3.8689	26.316	2.04	20.9	0.00
2300	77.5	-769.032	177.8	-4.1472	26.316	1.67	20.0	0.00
2400	78.2	-761.248	181.2	-4.4258	26.316	1.40	19.2	0.00
2500	79.0	-753.386	184.4	-4.7044	26.316	1.21	18.3	0.00
2600	79.9	-745.439	187.5	-4.9832	26.316	1.06	17.5	0.00
2700	80.8	-737.403	190.5	-5.2621	26.316	0.94	16.6	0.00
2800	81.8	-729.274	193.5	-5.5412	26.316	0.84	15.7	0.00
2900	82.8	-721.046	196.4	-5.8203	26.316	0.76	14.9	0.00
3000	83.9	-712.713	199.2	-6.0995	26.316	0.69	14.0	0.00

<INT VdP (J)>

PRESSURE(GPa)

T(K)	1	5	10	20	30	40	50
298	26065	123499	235436	440050	628369	805618	974539
300	26070	123525	235487	440145	628504	805792	974749
400	26351	124853	238018	444876	635260	814453	985226
500	26577	125924	240059	448691	640707	821436	993674
600	26741	126701	241540	451459	644660	826504	999805
700	26838	127162	242420	453104	647009	829516	1003448
800	26866	127296	242674	453579	647687	830385	1004499
900	26823	127093	242288	452857	646656	829064	1002901
1000	26709	126552	241256	450929	643903	825533	998630
1100	26524	125673	239581	447798	639433	819802	991697
1200	26268	124462	237273	443483	633271	811902	982141
1300	25944	122927	234345	438012	625458	801886	970024
1400	25554	121078	230820	431423	616050	789824	955433
1500	25100	118929	226724	423767	605117	775807	938477
1600	24587	116497	222087	415100	592741	759941	919284
1700	24018	113799	216945	405489	579017	742345	897999
1800	23397	110857	211336	395006	564047	723152	874782
1900	22729	107692	205302	383728	547944	702506	849807
2000	22019	104328	198888	371740	530825	680559	823257
2100	21272	100788	192140	359127	512814	657468	795325
2200	20493	97098	185106	345979	494040	633397	766207
2300	19688	93283	177833	332386	474630	608512	736104
2400	18862	89369	170371	318439	454715	582979	705218
2500	18020	85381	162769	304229	434423	556964	673748
2600	17168	81344	155072	289844	413882	530629	641891
2700	16310	77282	147329	275370	393214	504131	609837
2800	15453	73218	139582	260890	372538	477622	577769
2900	14599	69175	131873	246483	351965	451246	545863
3000	13755	65172	124243	232222	331601	425138	514281

CRISTOBALITE

SiO_2
CRISTOBALITE Formula wt. (g) = 60.09

Enthalpy (J/mol) = -907771.100 This study
Enthalpy (J/mol) = -908137.000 This study
Entropy (J/mol.K) = 43.065 This study
Heat Capacity
No. of Cp ranges = 5 This study
-1.3998908D+03 5.7158D+00 2.6288D+07 -6.2449D-03
0.0000D+00 0.0000D+00 0.0000D+00
Temp. and enthalpy of transition:
373.00 0.00
2.1783561D+03 -6.9872D+00 -5.8201D+07 6.4573D-03
0.0000D+00 0.0000D+00 0.0000D+00
453.00 0.00
7.8602125D+03 -2.3634D+01 -2.5501D+08 2.0191D-02
0.0000D+00 0.0000D+00 0.0000D+00
543.00 0.00
7.7587500D+01 -6.0805D-03 -4.4542D+06 2.7787D-06
0.0000D+00 0.0000D+00 0.0000D+00
3300.00 0.00
8.7373000D+01 0.0000D+00 0.0000D+00 0.0000D+00
0.0000D+00 0.0000D+00 0.0000D+00
Volume (cc/mol) = 25.739 Robie et al. (1978)
Thermal expn. (a0, a1*T, a2*T^-1, a3*T^-2):
4.54450D-05 8.15080D-08 4.46200D-03 0.00000D+00
Swamy et al. (1993)
Isot.bulk mod.(b0, b1*T, b2*T^2, b3*T^3):
5.13000D-06 0.00000D+00 0.00000D+00 0.00000D+00
Swamy et al. (1993)
P deriv., T deriv.:
6.00000D+00 0.00000D+00 4000.0 0.00000 Robie et al. (1978)
Melting pt. and enthalpy of melting:

NOTE: The thermal expansion here requires two temperature ranges. The data above 693.15 K are: molar volume 27.2707 cm^3; the coefficients for thermal expansion: 0.32131D-04 -0.161323D-07 -7.202D-03 0.000D-00.

SiO_2

CRISTOBALITE

Thermochemical Data

T(K)	C_p(J)	H(kJ)	S(J)	α 10^{-4}	K 10^4	γ	V(cc)	δ_T
298	44.9	-908.137	43.1	0.8471	19.493	0.97	25.7	0.00
300	44.9	-908.054	43.3	0.8477	19.493	0.97	25.7	0.00
400	52.9	-903.090	57.6	0.8920	19.493	0.88	26.0	0.00
500	70.8	-897.413	70.2	0.9512	19.493	0.71	26.2	0.00
600	62.6	-890.046	83.7	1.0179	19.493	0.88	26.5	0.00
700	65.6	-883.625	93.6	1.0887	19.493	0.93	26.7	0.00
800	67.5	-876.961	102.5	1.1623	19.493	0.99	27.0	0.00
900	68.9	-870.137	110.5	1.2376	19.493	1.07	27.4	0.00
1000	69.8	-863.200	117.8	1.3142	19.493	1.17	27.7	0.00
1100	70.6	-856.178	124.5	1.3916	19.493	1.29	28.1	0.00
1200	71.2	-849.088	130.7	1.4697	19.493	1.44	28.5	0.00
1300	71.7	-841.941	136.4	1.5484	19.493	1.61	28.9	0.00
1400	72.2	-834.741	141.7	1.6274	19.493	1.83	29.4	0.00
1500	72.7	-827.491	146.7	1.7068	19.493	2.10	29.9	0.00
1600	73.2	-820.193	151.4	1.7865	19.493	2.47	30.4	0.00
1700	73.7	-812.844	155.9	1.8663	19.493	2.97	31.0	0.00
1800	74.3	-805.444	160.1	1.9464	19.493	3.71	31.6	0.00
1900	74.8	-797.989	164.2	2.0266	19.493	4.93	32.2	0.00
2000	75.4	-790.477	168.0	2.1069	19.493	7.30	32.9	0.00
2100	76.1	-782.902	171.7	2.1874	19.493	13.99	33.6	0.00
2200	76.7	-775.263	175.3	2.2679	19.493	161.12	34.4	0.00
2300	77.5	-767.553	178.7	2.3485	19.493	-16.95	35.2	0.00
2400	78.2	-759.769	182.0	2.4292	19.493	-8.04	36.0	0.00
2500	79.0	-751.906	185.2	2.5100	19.493	-5.27	36.9	0.00
2600	79.9	-743.959	188.3	2.5908	19.493	-3.91	37.9	0.00
2700	80.8	-735.924	191.4	2.6717	19.493	-3.10	38.9	0.00
2800	81.8	-727.795	194.3	2.7526	19.493	-2.57	39.9	0.00
2900	82.8	-719.566	197.2	2.8336	19.493	-2.19	41.1	0.00
3000	83.9	-711.234	200.0	2.9146	19.493	-1.90	42.3	0.00

<INT VdP (J)>

PRESSURE(GPa)

T(K)	1	5	10	20	30	40	50
298	25145	117623	222380	411936	585161	747527	901823
300	25149	117641	222415	412001	585253	747644	901964
400	25369	118667	224354	415593	590355	754163	909828
500	25603	119765	226430	419437	595817	761139	918244
600	25856	120949	228669	423585	601709	768667	927325
700	26130	122230	231090	428069	608079	776804	937142
800	26426	123613	233705	432914	614960	785595	947748
900	26745	125105	236526	438139	622383	795077	959187
1000	27088	126711	239563	443765	630374	805286	971503
1100	27457	128437	242826	449809	638960	816254	984735
1200	27853	130288	246324	456290	648167	828015	998924
1300	28276	132269	250070	463228	658022	840605	1014112
1400	28729	134386	254072	470642	668554	854059	1030344
1500	29212	136645	258343	478554	679793	868416	1047664
1600	29727	139053	262895	486986	691771	883718	1066124
1700	30275	141616	267741	495962	704521	900006	1085774
1800	30857	144341	272894	505507	718080	917328	1106672
1900	31476	147237	278369	515649	732488	935733	1128875
2000	32134	150312	284182	526417	747784	955273	1152449
2100	32831	153574	290350	537843	764013	976006	1177461
2200	33571	157034	296890	549958	781224	997991	1203985
2300	34355	160700	303823	562800	799466	1021295	1232099
2400	35185	164586	311168	576406	818794	1045986	1261886
2500	36065	168701	318949	590819	839267	1072140	1293438
2600	36997	173059	327188	606081	860947	1099835	1326850
2700	37983	177673	335911	622240	883901	1129159	1362226
2800	39027	182557	345146	639347	908202	1160203	1399677
2900	40133	187728	354922	657456	933926	1193065	1439323
3000	41303	193202	365271	676626	961157	1227852	1481290

SiO_2

CRISTOBALITE

WOLLASTONITE

$CaSiO_3$
WOLLASTONITE Formula wt. (g) = 116.17

Enthalpy (J/mol) = -1632000.000 Haas et al. (1981)
Entropy (J/mol.K) = 82.028 Haas et al. (1981)
Heat Capacity:
No. of *Cp* ranges = 1 Haas et al. (1981), refitted

1.3958000D+02 2.3600D-03 1.4012D+06 0.0000D+00
-2.1722D+08 0.0000D+00 -1.8547D+04

Volume (cc/mol) = 39.930 Robie et al. (1978)
Thermal expn:
3.39404D-05 4.66398D-09 1.45882D-03 -1.86336D+00
This study

Isot.compressibility:
8.63114D-07 1.47260D-10 2.43030D-14 1.45000D-17
This study

P deriv. 4.00000D+00
T deriv. 0.00000D+00
Melting pt. and enthalpy of melting:
- -

NOTE: Calorimetric data: H: -1634.77±0.7; S: 81.03±0.68, Haas et al. (1981). The properties of wollastonite are as listed in Haas et al. (1981) and are closely similar to those of Charlu et al. (1981).

Thermochemical Data

T (K)	C_p (J)	H (kJ)	S (J)	$\alpha\ 10^{-4}$	K 10^4	γ	V (cc)	δ_T
298	85.6	-1632.00	82.0	0.1926	109.943	0.99	39.9	9.45
300	86.0	-1631.84	82.6	0.1950	109.906	1.00	39.9	9.34
400	99.5	-1622.50	109.3	0.2781	107.894	1.22	40.0	6.74
500	107.5	-1612.11	132.5	0.3174	105.861	1.28	40.1	6.09
600	113.0	-1601.07	152.6	0.3399	103.804	1.29	40.3	5.87
700	117.0	-1589.57	170.3	0.3549	101.721	1.29	40.4	5.81
800	120.0	-1577.71	186.2	0.3658	99.612	1.28	40.6	5.83
900	122.5	-1565.58	200.4	0.3746	97.475	1.27	40.7	5.89
1000	124.6	-1553.22	213.5	0.3820	95.313	1.25	40.9	5.97
1100	126.3	-1540.67	225.4	0.3886	93.127	1.24	41.0	6.07
1200	127.8	-1527.97	236.5	0.3946	90.919	1.22	41.2	6.18
1300	129.1	-1515.12	246.8	0.4002	88.693	1.21	41.4	6.29
1400	130.3	-1502.15	256.4	0.4056	86.453	1.19	41.5	6.41
1500	131.3	-1489.07	265.4	0.4108	84.202	1.18	41.7	6.52
1600	132.3	-1475.89	273.9	0.4159	81.945	1.16	41.9	6.63
1700	133.1	-1462.62	281.9	0.4208	79.686	1.15	42.0	6.73
1800	133.9	-1449.27	289.6	0.4257	77.430	1.13	42.2	6.83
1900	134.7	-1435.84	296.8	0.4305	75.182	1.11	42.4	6.93
2000	135.3	-1422.34	303.8	0.4353	72.948	1.09	42.6	7.01
2100	136.0	-1408.77	310.4	0.4401	70.730	1.08	42.8	7.09
2200	136.6	-1395.14	316.7	0.4448	68.535	1.06	43.0	7.16
2300	137.2	-1381.45	322.8	0.4495	66.366	1.04	43.2	7.22
2400	137.7	-1367.70	328.7	0.4542	64.227	1.02	43.3	7.28
2500	138.3	-1353.90	334.3	0.4589	62.123	1.00	43.5	7.32
2600	138.8	-1340.05	339.7	0.4635	60.055	0.98	43.7	7.36
2700	139.3	-1326.15	345.0	0.4682	58.029	0.96	44.0	7.38
2800	139.7	-1312.20	350.0	0.4728	56.045	0.94	44.2	7.40
2900	140.2	-1298.20	355.0	0.4775	54.107	0.92	44.4	7.41
3000	140.6	-1284.16	359.7	0.4821	52.216	0.90	44.6	7.41

$CaSiO_3$
WOLLASTONITE

<INT VdP (J)>

T(K)	1	5	10	20	30	40	50
	PRESSURE (GPa)						
298	39751	195420	383408	741521	1080650	1404651	1716125
300	39752	195426	383417	741532	1080660	1404656	1716124
400	39845	195826	384086	742478	1081654	1405551	1716817
500	39961	196337	384966	743814	1083202	1407152	1718358
600	40090	196903	385949	745334	1085000	1409061	1720260
700	40225	197504	386991	746949	1086919	1411108	1722314
800	40367	198126	388069	748615	1088892	1413207	1724412
900	40512	198765	389172	750306	1090882	1415307	1726493
1000	40661	199417	390292	752007	1092863	1417376	1728516
1100	40813	200080	391423	753707	1094819	1419390	1730452
1200	40968	200751	392562	755397	1096738	1421333	1732280
1300	41126	201430	393706	757072	1098609	1423192	1733984
1400	41286	202115	394853	758726	1100425	1424957	1735551
1500	41449	202807	396001	760354	1102180	1426618	1736971
1600	41614	203504	397149	761954	1103867	1428169	1738236
1700	41781	204206	398295	763523	1105483	1429604	1739338
1800	41951	204912	399437	765057	1107024	1430919	1740272
1900	42123	205623	400576	766555	1108486	1432109	1741036
2000	42298	206338	401710	768014	1109867	1433173	1741624
2100	42474	207055	402838	769433	1111165	1434107	1742038
2200	42653	207777	403960	770811	1112379	1434911	1742274
2300	42835	208501	405074	772145	1113506	1435584	1742335
2400	43018	209228	406181	773436	1114547	1436127	1742221
2500	43204	209957	407279	774682	1115502	1436540	1741935
2600	43393	210688	408368	775884	1116371	1436825	1741478
2700	43583	211422	409447	777040	1117153	1436984	1740856
2800	43776	212157	410517	778151	1117852	1437019	1740073
2900	43971	212894	411576	779216	1118467	1436933	1739132
3000	44168	213632	412625	780237	1119000	1436729	1738040

$CaSiO_3$
WOLLASTONITE

CORUNDUM

Al_2O_3
CORUNDUM Formula wt. (g) = 101.96

Enthalpy (J/mol) = -1675711.000 Robie et al. (1978)
Entropy (J/mol.K) = 50.917 Robie et al. (1978)
Heat Capacity:
No. of *Cp* ranges = 1 Robie et al. (1978)
1.2260000D+02 7.0310D-03 -5.5830D+06 0.0000D+00
5.7070D+08 0.0000D+00 -1.4140D+03
Volume (cc/mol) = 25.575 Robie et al. (1978)
Thermal expn:
4.19600D-05 -8.00000D-10 -1.14060D-02 1.06230D+00
Saxena and Shen (1992)

Isot.compressibility:
3.84690D-07 3.68725D-11 1.28980D-14 2.35000D-18
Saxena and Shen (1992)

P deriv. 4.20000D+00
T deriv. 0.00000D+00
Melting pt. and enthalpy of melting:
- -

NOTE: Hankey and Schuele's (1970) data on compressibility are consistent with the above.

Thermochemical Data

T (K)	C_p (J)	H (kJ)	S (J)	$\alpha\ 10^{-4}$	K 10^4	γ	V (cc)	δ_T
298	78.7	-1675.71	50.9	0.1542	251.957	1.28	25.6	7.34
300	79.1	-1675.57	51.4	0.1550	251.904	1.28	25.6	7.32
400	95.9	-1666.73	76.7	0.1976	248.971	1.23	25.6	6.57
500	105.5	-1656.62	99.2	0.2300	245.915	1.23	25.7	6.24
600	111.6	-1645.74	119.0	0.2542	242.739	1.25	25.7	6.08
700	115.8	-1634.36	136.6	0.2727	239.448	1.26	25.8	6.02
800	118.8	-1622.62	152.2	0.2872	236.046	1.28	25.8	6.01
900	121.2	-1610.62	166.4	0.2988	232.539	1.30	25.9	6.03
1000	123.2	-1598.39	179.3	0.3082	228.932	1.31	26.0	6.07
1100	124.9	-1585.98	191.1	0.3159	225.233	1.33	26.0	6.12
1200	126.3	-1573.42	202.0	0.3223	221.449	1.34	26.1	6.18
1300	127.6	-1560.73	212.2	0.3277	217.588	1.35	26.2	6.24
1400	128.8	-1547.91	221.7	0.3323	213.657	1.35	26.3	6.31
1500	129.9	-1534.97	230.6	0.3363	209.665	1.36	26.3	6.37
1600	130.9	-1521.93	239.0	0.3397	205.621	1.36	26.4	6.43
1700	131.9	-1508.79	247.0	0.3426	201.534	1.36	26.5	6.49
1800	132.8	-1495.55	254.6	0.3451	197.412	1.37	26.6	6.54
1900	133.8	-1482.22	261.8	0.3473	193.264	1.37	26.7	6.59
2000	134.6	-1468.80	268.6	0.3492	189.098	1.36	26.8	6.63
2100	135.5	-1455.30	275.2	0.3509	184.923	1.36	26.9	6.67
2200	136.3	-1441.70	281.6	0.3523	180.747	1.36	26.9	6.70
2300	137.1	-1428.03	287.6	0.3536	176.579	1.35	27.0	6.73
2400	138.0	-1414.28	293.5	0.3547	172.425	1.35	27.1	6.75
2500	138.8	-1400.44	299.1	0.3557	168.293	1.34	27.2	6.76
2600	139.5	-1386.52	304.6	0.3565	164.189	1.33	27.3	6.77
2700	140.3	-1372.53	309.9	0.3572	160.121	1.32	27.4	6.77
2800	141.1	-1358.46	315.0	0.3578	156.094	1.32	27.5	6.77
2900	141.9	-1344.31	320.0	0.3583	152.113	1.31	27.6	6.77
3000	142.6	-1330.09	324.8	0.3588	148.183	1.29	27.8	6.75

Al_2O_3
CORUNDUM

<INT VdP (J)>

T(K)	1	5	10	20	30	40	50
				PRESSURE(GPa)			
298	25524	126647	250992	493536	728889	957992	1181574
300	25525	126651	250998	493547	728903	958009	1181592
400	25569	126861	251391	494233	729812	959083	1182787
500	25624	127119	251874	495094	730970	960477	1184370
600	25685	127412	252426	496082	732309	962103	1186228
700	25753	127731	253029	497165	733781	963892	1188281
800	25824	128072	253671	498318	735348	965799	1190469
900	25899	128428	254344	499524	736985	967790	1192751
1000	25977	128798	255040	500770	738674	969839	1195096
1100	26057	129179	255756	502047	740399	971927	1197479
1200	26140	129569	256488	503347	742150	974039	1199881
1300	26224	129967	257232	504664	743917	976162	1202288
1400	26309	130371	257986	505993	745692	978287	1204687
1500	26396	130780	258749	507330	747470	980406	1207068
1600	26484	131194	259518	508672	749246	982511	1209423
1700	26574	131612	260292	510016	751015	984598	1211743
1800	26664	132034	261071	511359	752773	986660	1214024
1900	26755	132458	261852	512699	754517	988693	1216260
2000	26847	132885	262636	514034	756244	990694	1218445
2100	26939	133314	263420	515363	757951	992660	1220577
2200	27032	133745	264205	516684	759636	994587	1222651
2300	27126	134177	264991	517995	761298	996472	1224665
2400	27221	134611	265775	519295	762934	998314	1226615
2500	27316	135046	266558	520584	764543	1000111	1228501
2600	27411	135481	267340	521860	766123	1001860	1230318
2700	27507	135917	268119	523122	767673	1003561	1232067
2800	27603	136354	268896	524369	769192	1005211	1233745
2900	27700	136791	269669	525601	770679	1006810	1235352
3000	27797	137227	270440	526817	772132	1008357	1236887

SILLIMANITE

Al_2SiO_5
SILLIMANITE Formula wt. (g) = 162.05

Enthalpy (J/mol) =	-2587770.000	Robie and Hemingway (1984b)
Entropy (J/mol.K) =	95.790	Robie and Hemingway (1984b)
Heat Capacity:		
No. of *Cp* ranges = 1		Robie and Hemingway (1984), refitted
1.8387000D+02 1.8150D-02 -1.2360D+07 0.0000D+00		
1.6024D+09 0.0000D+00 3.2052D+03		
Volume (cc/mol) =	49.860	Robie and Hemingway (1984b)
Thermal expn:		
3.48900D-05 7.12700D-09 -1.12130D-02 1.38154D+00		This study
Isot.compressibility:		
6.22896D-07 8.83830D-11 9.68950D-15 3.82780D-18		This study
P deriv. 4.00000D+00		
T deriv. 4.00000D-04		This study
Melting pt. and enthalpy of melting:		
- -		

NOTE: Calorimetric data: H: -2587.77±1.73; S: 95.79±0.14. Robie and Hemingway (1984b). Several experiments on the following equilibria

andalusite = kyanite
kyanite = sillimanite
sillimanite = andalusite

relating the three polymorphs have been performed. However, the results obtained by different workers are very different. Experiments were done by Evans (1965), Newton (1966, 1969), Althaus (1967), Richardson et al. (1968, 1969) and Holdaway (1971). The reason behind this discrepancy is the small difference in the free energies of formation of the three polymorphs which leads to metastability.

Robie and Hemingway (1984b) determined the heat capacities of the three aluminosilicates in the range 10-380 K. The slopes of phase boundaries among these phases match the experimental data of Evans (1965), Newton (1966,1969) and Holdaway (1971). Robie and Hemingway concluded that since Evans's, Newton's and Holdaway's results are consistent with their independently derived calorimetric data, the equilibrium results of these three workers must be correct. The thermodynamic properties given by Robie and Hemingway are adopted here. Robie and Hemingway also concluded that Al and Si are ordered in sillimanite to at least 1100 K. The compressibility data are consistent with the data of Brace et al. (1969).

Al_2SiO_5

SILLIMANITE

Thermochemical Data

T(K)	CP(J)	H(KJ)	S(J)	A E-4	K E4	GR	V(cc)	DT
298	121.4	-2587.77	95.8	0.1495	153.796	0.95	49.9	9.79
300	122.0	-2587.54	96.5	0.1500	153.755	0.95	49.9	9.76
400	146.9	-2574.01	135.3	0.1834	151.505	0.95	49.9	8.09
500	162.7	-2558.46	169.9	0.2155	149.256	1.00	50.0	6.99
600	173.2	-2541.64	200.6	0.2432	147.007	1.05	50.2	6.30
700	180.6	-2523.93	227.9	0.2668	144.754	1.10	50.3	5.84
800	186.2	-2505.57	252.4	0.2873	142.498	1.14	50.4	5.52
900	190.7	-2486.72	274.6	0.3055	140.237	1.17	50.6	5.28
1000	194.5	-2467.46	294.9	0.3219	137.970	1.20	50.7	5.11
1100	197.7	-2447.84	313.6	0.3368	135.697	1.23	50.9	4.98
1200	200.7	-2427.92	330.9	0.3506	133.418	1.25	51.1	4.88
1300	203.3	-2407.72	347.1	0.3635	131.134	1.27	51.3	4.80
1400	205.8	-2387.26	362.2	0.3756	128.845	1.29	51.5	4.73
1500	208.2	-2366.55	376.5	0.3872	126.552	1.31	51.6	4.68
1600	210.5	-2345.62	390.0	0.3982	124.256	1.32	51.9	4.64
1700	212.7	-2324.46	402.8	0.4089	121.958	1.33	52.1	4.61
1800	214.8	-2303.09	415.1	0.4192	119.660	1.34	52.3	4.58
1900	216.9	-2281.51	426.7	0.4291	117.363	1.35	52.5	4.56
2000	218.9	-2259.72	437.9	0.4388	115.069	1.36	52.7	4.54
2100	220.9	-2237.73	448.6	0.4483	112.780	1.37	53.0	4.52
2200	222.9	-2215.54	458.9	0.4576	110.498	1.37	53.2	4.51
2300	224.8	-2193.16	468.9	0.4667	108.224	1.38	53.4	4.49
2400	226.7	-2170.58	478.5	0.4756	105.961	1.38	53.7	4.48
2500	228.7	-2147.81	487.8	0.4844	103.711	1.38	54.0	4.47
2600	230.6	-2124.85	496.8	0.4931	101.474	1.39	54.2	4.45
2700	232.4	-2101.70	505.5	0.5017	99.255	1.39	54.5	4.44
2800	234.3	-2078.37	514.0	0.5102	97.053	1.39	54.8	4.43
2900	236.2	-2054.84	522.3	0.5186	94.872	1.39	55.1	4.41
3000	238.1	-2031.12	530.3	0.5269	92.713	1.38	55.3	4.40

<INT VdP (J)>

PRESSURE(GPa)

T(K)	1	5	10	20	30	40	50
298	49699	245451	483912	943158	1382761	1806089	2215560
300	49700	245457	483921	943171	1382775	1806101	2215567
400	49781	245811	484532	944085	1383802	1807116	2216483
500	49878	246248	485305	945330	1385347	1808858	2218358
600	49990	246758	486223	946865	1387341	1811222	2221043
700	50115	247332	487265	948647	1389712	1814105	2224401
800	50252	247960	488412	950638	1392403	1817426	2228324
900	50398	248635	489653	952813	1395371	1821125	2232732
1000	50554	249354	490977	955149	1398583	1825156	2237563
1100	50718	250112	492376	957632	1402013	1829481	2242769
1200	50889	250907	493845	960247	1405643	1834072	2248313
1300	51068	251735	495378	962985	1409454	1838908	2254164
1400	51254	252596	496972	965839	1413436	1843970	2260298
1500	51446	253487	498623	968801	1417577	1849243	2266696
1600	51645	254408	500330	971867	1421869	1854715	2273340
1700	51850	255357	502089	975031	1426305	1860376	2280217
1800	52062	256334	503901	978291	1430879	1866217	2287317
1900	52279	257338	505762	981644	1435587	1872232	2294629
2000	52502	258369	507673	985087	1440425	1878415	2302147
2100	52731	259426	509631	988618	1445390	1884762	2309864
2200	52966	260509	511637	992235	1450478	1891269	2317775
2300	53206	261618	513690	995939	1455689	1897933	2325876
2400	53452	262751	515789	999726	1461020	1904752	2334164
2500	53704	263910	517934	1003598	1466470	1911723	2342636
2600	53962	265095	520125	1007553	1472040	1918847	2351292
2700	54226	266304	522362	1011591	1477728	1926122	2360129
2800	54495	267538	524644	1015713	1483534	1933548	2369148
2900	54770	268798	526973	1019918	1489458	1941126	2378348
3000	55050	270083	529346	1024206	1495501	1948855	2387731

Al_2SiO_5

SILLIMANITE

KYANITE

Al_2SiO_5
KYANITE Formula wt. (g) = 162.05

Enthalpy (J/mol) = -2596200.000 Robie and Hemingway (1984b)
Entropy (J/mol.K) = 82.300 Robie and Hemingway (1984b)
Heat Capacity:
No. of *Cp* ranges = 1 Robie and Hemingway (1984b), refitted
2.2354000D+02 2.6450D-03 -4.1490D+06 0.0000D+00
7.6010D+08 0.0000D+00 -2.5240D+04
Volume (cc/mol) = 44.150 Robie and Hemingway (1984b)
Thermal expn:
5.39802D-05 -2.05750D-10 -1.50528D-02 1.55300D+00
This study
Isot.compressibility:
6.37481D-07 9.77990D-11 3.40907D-14 4.43050D-18
This study
P deriv. 4.00000D+00
T deriv. 6.00000D-04 This study
Melting pt. and enthalpy of melting:
- -

NOTE: Calorimetric data: H: -2596.01±1.7; S: 82.3 ±0.13, Robie and Hemingway (1984b). See note on sillimanite.

Al_2SiO_5
KYANITE

Thermochemical Data

T(K)	C_p(J)	H(kJ)	S(J)	α 10^{-4}	K 10^4	γ	V(cc)	δ_T
298	121.7	-2596.01	82.3	0.2090	149.301	1.14	44.1	8.52
300	122.3	-2595.78	83.1	0.2100	149.252	1.14	44.2	8.49
400	147.4	-2582.21	121.9	0.2597	146.555	1.16	44.3	7.18
500	163.9	-2566.59	156.7	0.2998	143.790	1.19	44.4	6.48
600	175.1	-2549.61	187.7	0.3308	140.966	1.21	44.5	6.11
700	183.1	-2531.68	215.3	0.3550	138.090	1.23	44.7	5.91
800	189.1	-2513.06	240.1	0.3743	135.170	1.24	44.8	5.81
900	193.8	-2493.90	262.7	0.3899	132.215	1.25	45.0	5.76
1000	197.6	-2474.33	283.3	0.4027	129.232	1.25	45.2	5.75
1100	200.6	-2454.41	302.3	0.4135	126.230	1.25	45.4	5.77
1200	203.2	-2434.22	319.9	0.4227	123.215	1.24	45.6	5.79
1300	205.5	-2413.78	336.2	0.4305	120.197	1.23	45.8	5.83
1400	207.4	-2393.14	351.5	0.4373	117.182	1.22	46.0	5.88
1500	209.1	-2372.31	365.9	0.4433	114.177	1.21	46.2	5.92
1600	210.6	-2351.33	379.4	0.4485	111.188	1.19	46.4	5.97
1700	211.9	-2330.20	392.2	0.4531	108.222	1.18	46.6	6.02
1800	213.1	-2308.95	404.4	0.4573	105.284	1.16	46.8	6.07
1900	214.2	-2287.58	415.9	0.4610	102.380	1.14	47.0	6.11
2000	215.3	-2266.11	426.9	0.4643	99.514	1.12	47.2	6.16
2100	216.2	-2244.53	437.5	0.4673	96.690	1.10	47.4	6.20
2200	217.1	-2222.87	447.5	0.4701	93.913	1.08	47.7	6.24
2300	217.9	-2201.11	457.2	0.4726	91.186	1.06	47.9	6.27
2400	218.7	-2179.28	466.5	0.4748	88.511	1.03	48.1	6.30
2500	219.4	-2157.37	475.4	0.4769	85.891	1.01	48.3	6.33
2600	220.1	-2135.39	484.1	0.4789	83.328	0.99	48.6	6.35
2700	220.8	-2113.35	492.4	0.4806	80.823	0.97	48.8	6.37
2800	221.4	-2091.23	500.4	0.4823	78.379	0.94	49.0	6.39
2900	222.0	-2069.06	508.2	0.4838	75.996	0.92	49.3	6.40
3000	222.6	-2046.83	515.8	0.4852	73.675	0.90	49.5	6.41

Al_2SiO_5
KYANITE

<INT VdP (J)>

T(K)	PRESSURE(GPa) 1	5	10	20	30	40	50
298	44003	217244	428138	833935	1222026	1595497	1956549
300	44005	217251	428151	833954	1222047	1595517	1956566
400	44106	217703	428941	835189	1223517	1597086	1958140
500	44227	218251	429921	836804	1225580	1599487	1960816
600	44364	218875	431047	838709	1228090	1602513	1964314
700	44514	219557	432286	840834	1230938	1606008	1968427
800	44673	220286	433612	843128	1234043	1609860	1973008
900	44840	221052	435008	845554	1237349	1613990	1977953
1000	45015	221849	436460	848085	1240814	1618339	1983183
1100	45195	222671	437957	850701	1244406	1622863	1988640
1200	45380	223514	439492	853386	1248102	1627529	1994281
1300	45570	224376	441060	856128	1251883	1632312	2000071
1400	45763	225253	442654	858918	1255734	1637190	2005984
1500	45960	226145	444271	861748	1259645	1642148	2011998
1600	46161	227048	445909	864612	1263606	1647174	2018098
1700	46364	227963	447564	867506	1267611	1652258	2024269
1800	46570	228887	449234	870426	1271653	1657392	2030503
1900	46778	229820	450917	873368	1275729	1662570	2036790
2000	46989	230761	452613	876330	1279834	1667787	2043124
2100	47201	231709	454319	879311	1283966	1673040	2049500
2200	47416	232664	456035	882309	1288123	1678324	2055914
2300	47633	233625	457761	885322	1292303	1683639	2062363
2400	47852	234592	459495	888350	1296505	1688982	2068845
2500	48073	235564	461236	891393	1300729	1694352	2075359
2600	48295	236541	462986	894449	1304973	1699749	2081903
2700	48519	237523	464742	897518	1309237	1705171	2088476
2800	48744	238509	466506	900601	1313522	1710620	2095079
2900	48972	239500	468276	903697	1317827	1716094	2101712
3000	49200	240494	470053	906806	1322152	1721595	2108374

ANDALUSITE

Al_2SiO_5
ANDALUSITE Formula wt. (g) = 162.05

Enthalpy (J/mol) =	-2591700.000	Robie and Hemingway (1984b)
Entropy (J/mol.K) =	91.420	Robie and Hemingway (1984b)
Heat Capacity:		
No. of *Cp* ranges = 1		Robie and Hemingway (1984b), refitted
2.2170000D+02 2.5810D-03 -2.5590D+06 0.0000D+00		
4.8990D+08 0.0000D+00 -2.6690D+04		
Volume (cc/mol) =	51.520	Robie and Hemingway (1984b)
Thermal expn:		
4.14736D-05 3.39116D-09 -1.33905D-02 1.81290D+00		This study
Isot.compressibility:		
6.27150D-07 1.41400D-10 -2.62790D-15 2.36607D-17		This study
P deriv. 4.00000D+00		
T deriv. 4.00000D-04		This study
Melting pt. and enthalpy of melting:		
- -		

NOTE: Calorimetric data: H: -2591.9 ±1.72; S: 91.39±0.14, Robie and Hemingway (1984b). See note on sillimanite.

Al_2SiO_5
ANDALUSITE

Thermochemical Data

T(K)	C_p(J)	H(kJ)	S(J)	α 10^{-4}	K 10^4	γ	V(cc)	δ_T
298	122.6	-2591.70	91.4	0.1797	149.320	1.13	51.5	12.15
300	123.2	-2591.47	92.2	0.1800	149.260	1.13	51.5	12.12
400	147.7	-2577.83	131.3	0.2068	146.027	1.07	51.6	10.64
500	163.3	-2562.23	166.0	0.2364	142.826	1.08	51.7	9.46
600	173.9	-2545.34	196.8	0.2623	139.635	1.11	51.9	8.72
700	181.6	-2527.54	224.2	0.2842	136.433	1.14	52.0	8.28
800	187.4	-2509.08	248.8	0.3028	133.209	1.15	52.2	8.03
900	191.9	-2490.11	271.2	0.3189	129.949	1.17	52.3	7.91
1000	195.5	-2470.74	291.6	0.3329	126.649	1.18	52.5	7.88
1100	198.5	-2451.03	310.4	0.3453	123.304	1.18	52.7	7.91
1200	201.1	-2431.05	327.7	0.3564	119.914	1.18	52.9	7.98
1300	203.2	-2410.83	343.9	0.3665	116.481	1.18	53.0	8.09
1400	205.1	-2390.41	359.1	0.3758	113.009	1.17	53.2	8.21
1500	206.8	-2369.81	373.3	0.3844	109.506	1.16	53.4	8.35
1600	208.3	-2349.06	386.7	0.3924	105.980	1.15	53.7	8.50
1700	209.6	-2328.16	399.3	0.3999	102.440	1.13	53.9	8.65
1800	210.8	-2307.14	411.3	0.4070	98.898	1.12	54.1	8.79
1900	211.9	-2286.00	422.8	0.4137	95.364	1.10	54.3	8.94
2000	212.9	-2264.76	433.7	0.4201	91.851	1.08	54.5	9.07
2100	213.9	-2243.42	444.1	0.4263	88.369	1.06	54.8	9.19
2200	214.8	-2221.99	454.1	0.4322	84.929	1.03	55.0	9.30
2300	215.6	-2200.47	463.6	0.4379	81.543	1.01	55.2	9.40
2400	216.4	-2178.87	472.8	0.4435	78.219	0.98	55.5	9.48
2500	217.1	-2157.20	481.7	0.4489	74.967	0.96	55.7	9.55
2600	217.8	-2135.45	490.2	0.4541	71.793	0.93	56.0	9.61
2700	218.5	-2113.64	498.4	0.4592	68.706	0.90	56.2	9.65
2800	219.1	-2091.76	506.4	0.4642	65.709	0.88	56.5	9.67
2900	219.7	-2069.82	514.1	0.4691	62.807	0.85	56.8	9.68
3000	220.3	-2047.82	521.5	0.4739	60.005	0.82	57.0	9.68

Al_2SiO_5

ANDALUSITE

<INT VdP (J)>

	PRESSURE (GPa)						
T(K)	1	5	10	20	30	40	50
298	51349	253509	499610	973151	1426032	1861854	2283185
300	51351	253516	499621	973164	1426042	1861856	2283176
400	51446	253919	500273	973996	1426763	1862274	2283156
500	51556	254395	501070	975120	1427944	1863344	2284007
600	51681	254940	502000	976506	1429527	1864972	2285589
700	51818	255544	503040	978104	1431430	1867038	2287739
800	51966	256196	504171	979868	1433578	1869437	2290317
900	52123	256889	505374	981762	1435912	1872083	2293210
1000	52288	257616	506637	983756	1438384	1874907	2296328
1100	52461	258372	507948	985824	1440953	1877854	2299595
1200	52639	259153	509298	987946	1443588	1880878	2302952
1300	52824	259956	510680	990106	1446261	1883940	2306349
1400	53013	260778	512086	992290	1448950	1887011	2309747
1500	53208	261616	513513	994485	1451638	1890065	2313113
1600	53407	262469	514954	996682	1454308	1893081	2316422
1700	53611	263333	516406	998873	1456949	1896045	2319653
1800	53819	264209	517865	1001051	1459552	1898943	2322791
1900	54030	265094	519329	1003211	1462110	1901768	2325827
2000	54246	265987	520795	1005348	1464617	1904514	2328755
2100	54465	266887	522261	1007460	1467071	1907177	2331570
2200	54687	267794	523726	1009545	1469469	1909758	2334275
2300	54913	268706	525187	1011601	1471813	1912257	2336872
2400	55142	269622	526644	1013629	1474103	1914678	2339367
2500	55374	270543	528097	1015629	1476343	1917027	2341767
2600	55609	271467	529544	1017602	1478535	1919309	2344081
2700	55848	272395	530986	1019551	1480685	1921532	2346319
2800	56089	273325	532423	1021476	1482798	1923705	2348493
2900	56333	274258	533855	1023382	1484879	1925836	2350616
3000	56580	275194	535282	1025271	1486934	1927934	2352699

CORDIERITE

$Mg_2Al_4Si_5O_{18}$
CORDIERITE Formula wt. (g) = 584.97

Enthalpy (J/mol) = -9172020.000 Herzberg (1983), Robie et al. (1978)
Entropy (J/mol.K) = 410.880 Herzberg (1983), Robie et al. (1978)
Heat Capacity:
No. of *Cp* ranges = 1
7.6548000D+02 -1.9181D-02 5.7814D+06 0.0000D+00
-5.4662D+08 0.0000D+00 -8.8644D+04
Volume (cc/mol) = 233.220 Robie et al. (1978)
Thermal expn:
\- - - -
Isot.compressibility:
1.00000D-06 0.00000D+00 0.00000D+00 0.00000D+00
P deriv. 4.00000D+00
T deriv. 0.00000D+00
Melting pt. and enthalpy of melting:
\- -

NOTE: Calorimetric data: H: -9161.52±5.85; S: 407.2 ±3.8, Robie et al. (1978). The upper stability limit of Mg-cordierite is given by the following reaction studied by Seifert (1974) and more recently by Herzberg (1983):

$$\text{Cord} + \text{Fo} = \text{En} + \text{Sp} \qquad \text{(a)}$$

This reaction is influenced by Al solubility in enstatite. The heat capacity of cordierite is from Robie et al. (1978). The enthalpy and entropy are adjusted within the error limits of Robie et al. to fit the experimental data on reaction (a). Enstatite is considered an ideal solution between $MgSiO_3$ and Al_2O_3.

$Mg_2Al_4Si_5O_{18}$

CORDIERITE

Thermochemical Data

T(K)	C_p(J)	H(kJ)	S(J)
298	506.9	-9172.02	410.9
300	508.2	-9171.08	414.0
400	563.8	-9117.22	568.6
500	597.4	-9059.04	698.3
600	619.8	-8998.11	809.3
700	635.6	-8935.30	906.1
800	647.3	-8871.13	991.8
900	656.1	-8805.94	1068.5
1000	662.9	-8739.97	1138.0
1100	668.2	-8673.41	1201.5
1200	672.3	-8606.38	1259.8
1300	675.5	-8538.98	1313.7
1400	678.1	-8471.29	1363.9
1500	680.0	-8403.39	1410.8
1600	681.5	-8335.31	1454.7
1700	682.6	-8267.10	1496.0
1800	683.4	-8198.79	1535.1
1900	683.9	-8130.43	1572.0
2000	684.2	-8062.02	1607.1
2100	684.2	-7993.60	1640.5
2200	684.1	-7925.18	1672.3
2300	683.9	-7856.78	1702.7
2400	683.5	-7788.41	1731.8
2500	683.0	-7720.09	1759.7
2600	682.3	-7651.82	1786.5
2700	681.6	-7583.62	1812.3
2800	680.8	-7515.50	1837.0
2900	680.0	-7447.46	1860.9
3000	679.0	-7379.51	1883.9

FERROCORDIERITE

$Fe_2Al_4Si_5O_{18}$
FERROCORDIERITE Formula wt. (g) = 648.05

Enthalpy (J/mol) = -8470600.000 Chatterjee (1989)
Entropy (J/mol.K) = 410.880 Chatterjee (1989)
Heat Capacity:
No. of *Cp* ranges = 1 Chatterjee (1989)
1.0856400D+03 -3.7262D-01 1.3161D+07 0.0000D+00
-1.2848D+10 0.0000D+00 -7.0756D+04
Volume (cc/mol) = 249.840
Thermal expn:
- - - -
Isot.compressibility:
- - - -
Melting pt. and enthalpy of melting:
- -

NOTE: Experimental data on the distribution of Fe and Mg between garnet and cordierite were obtained by Hensen and Green (1973), Holdaway and Lee (1977) and Perchuk and Lavrent'eva (1983). The reaction is as follows:

Fe-cord + Pyr = Cord + Alm (a)

The equation given below represents the average of all the experimental data:

$$\ln K_d = 3039.06/T(K) - 1.242 \quad (1)$$

where K_d is the distribution coefficient. The free energy and volume of Fe-cordierite and the interaction parameters of the cordierite-ferrocordierite solution are calculated assuming $-\Delta Cp$ of reaction (a) as zero and from the following equilibrium relation:

$$-\Delta G^o(P,T) + RT\ln K_d + RT\ln K_\gamma = 0 \quad (2)$$

the only unknown quantity in the term $-\Delta G^o(P,T)$ being the free energy of ferrocordierite at P and T; and in the K_γ term, the activities of cordierite and ferrocordierite; and where $\ln K_d$ is given by equation (1).

$Fe_2Al_4Si_5O_{18}$
FERROCORDIERITE

Thermochemical Data

T(K)	C_p(J)	H(kJ)	S(J)
298	400.5	-8470.60	410.9
300	408.4	-8469.85	413.4
400	641.2	-8414.94	569.8
500	707.7	-8346.81	721.5
600	721.2	-8275.11	852.2
700	713.1	-8203.27	963.0
800	694.6	-8132.82	1057.1
900	670.3	-8064.54	1137.5
1000	642.6	-7998.88	1206.7
1100	612.7	-7936.10	1266.6
1200	581.2	-7876.40	1318.6
1300	548.7	-7819.89	1363.8
1400	515.5	-7766.67	1403.3
1500	481.6	-7716.82	1437.7
1600	447.2	-7670.37	1467.7
1700	412.5	-7627.38	1493.7
1800	377.5	-7587.88	1516.3
1900	342.2	-7551.90	1535.8
2000	306.7	-7519.45	1552.4
2100	271.0	-7490.56	1566.5
2200	235.2	-7465.25	1578.3
2300	199.3	-7443.52	1588.0
2400	163.2	-7425.39	1595.7
2500	127.1	-7410.88	1601.6
2600	90.8	-7399.98	1605.9
2700	54.5	-7392.72	1608.7
2800	18.1	-7389.08	1610.0
2900	-18.3	-7389.09	1610.0
3000	-54.8	-7392.75	1608.7

Fe-PEROVSKITE

$FeSiO_3$
Fe-PEROVSKITE Formula wt. (g) = 131.93

Enthalpy (J/mol) = -1091400.000 This study
Entropy (J/mol.K) = 101.460 This study
Heat Capacity:
No. of *Cp* ranges = 1 This study
1.4222450D+02 1.5367D-03 -1.1788D+06 0.0000D+00
2.7333D+08 0.0000D+00 -1.4556D+04
Volume (cc/mol) = 25.590 Fei (1990)
Thermal expn:
2.62700D-05 1.51980D-08 0.00000D+00 -4.29000D-02
Fei and Saxena (1986)
Isot.compressibility:
3.36700D-07 1.01000D-10 0.00000D+00 0.00000D+00
P deriv. 4.00000D+00
T deriv. 0.00000D+00
Melting pt. and enthalpy of melting:
- -

NOTE: The data on Fe-perovskite are estimated from limited solution data and from analogous data (Fei and Saxena, 1986). The Mg-Fe distribution data between coexisting magnesiowustite and perovskite and the maximum solubility of $FeSiO_3$ in perovskite, determined by Yagi et al. (1979), Ito (1984) and Fei (1990), are taken into consideration in the evaluation.

Thermochemical Data

T(K)	C_p(J)	H(kJ)	S(J)	α 10^{-4}	K 10^4	γ	V(cc)	δ_T
298	90.9	-1091.40	101.5	0.3032	272.618	2.38	25.6	9.08
300	91.2	-1091.23	102.0	0.3035	272.480	2.37	25.6	9.07
400	103.4	-1081.46	130.0	0.3208	265.182	2.17	25.7	8.35
500	111.4	-1070.70	154.0	0.3370	258.264	2.08	25.8	7.74
600	116.9	-1059.27	174.8	0.3527	251.699	2.05	25.8	7.21
700	120.9	-1047.37	193.2	0.3682	245.459	2.04	25.9	6.73
800	124.0	-1035.12	209.5	0.3836	239.521	2.05	26.0	6.31
900	126.4	-1022.60	224.3	0.3990	233.863	2.07	26.1	5.92
1000	128.3	-1009.87	237.7	0.4143	228.467	2.10	26.2	5.57
1100	129.9	-996.96	250.0	0.4295	223.314	2.14	26.4	5.25
1200	131.3	-983.89	261.4	0.4448	218.388	2.19	26.5	4.96
1300	132.5	-970.71	271.9	0.4600	213.675	2.24	26.6	4.69
1400	133.5	-957.41	281.8	0.4753	209.161	2.29	26.7	4.45
1500	134.4	-944.02	291.0	0.4905	204.834	2.35	26.8	4.22
1600	135.2	-930.54	299.7	0.5057	200.682	2.42	27.0	4.01
1700	135.9	-916.98	307.9	0.5209	196.696	2.50	27.1	3.81
1800	136.6	-903.35	315.7	0.5361	192.864	2.58	27.3	3.63
1900	137.2	-889.66	323.1	0.5513	189.179	2.67	27.4	3.47
2000	137.8	-875.92	330.2	0.5666	185.632	2.76	27.6	3.31
2100	138.3	-862.11	336.9	0.5818	182.216	2.87	27.7	3.16
2200	138.8	-848.26	343.4	0.5970	178.923	2.99	27.9	3.03
2300	139.2	-834.36	349.5	0.6122	175.747	3.12	28.1	2.90
2400	139.7	-820.42	355.5	0.6274	172.682	3.27	28.2	2.78
2500	140.1	-806.43	361.2	0.6426	169.722	3.43	28.4	2.67
2600	140.5	-792.40	366.7	0.6578	166.861	3.62	28.6	2.56
2700	140.8	-778.34	372.0	0.6730	164.096	3.83	28.8	2.46
2800	141.2	-764.24	377.1	0.6882	161.421	4.07	29.0	2.37
2900	141.5	-750.10	382.1	0.7034	158.831	4.35	29.2	2.28
3000	141.9	-735.93	386.9	0.7186	156.323	4.67	29.4	2.20

$FeSiO_3$
Fe-PEROVSKITE

	<INT VdP (J)> PRESSURE (GPa)						
T(K)	1	5	10	20	30	40	50
298	25543	126810	251470	494982	731623	962225	1187447
300	25544	126817	251482	495002	731648	962253	1187476
400	25623	127183	252153	496133	733080	963862	1189161
500	25706	127572	252868	497352	734642	965643	1191063
600	25794	127982	253626	498654	736330	967592	1193174
700	25885	128413	254426	500040	738141	969706	1195492
800	25982	128866	255269	501509	740076	971983	1198014
900	26082	129340	256154	503061	742135	974426	1200742
1000	26187	129837	257082	504697	744319	977033	1203676
1100	26297	130354	258052	506418	746628	979807	1206816
1200	26410	130894	259066	508224	749063	982747	1210164
1300	26529	131457	260125	510115	751625	985855	1213720
1400	26652	132042	261227	512094	754315	989133	1217487
1500	26780	132649	262375	514160	757135	992582	1221466
1600	26912	133280	263567	516315	760086	996202	1225658
1700	27049	133935	264806	518559	763168	999997	1230066
1800	27191	134613	266092	520894	766385	1003967	1234691
1900	27338	135315	267424	523321	769736	1008115	1239535
2000	27490	136041	268805	525841	773224	1012442	1244601
2100	27647	136792	270234	528455	776850	1016951	1249891
2200	27809	137569	271712	531164	780616	1021643	1255407
2300	27976	138371	273240	533970	784524	1026520	1261151
2400	28149	139198	274819	536874	788575	1031586	1267128
2500	28326	140052	276449	539878	792772	1036843	1273339
2600	28510	140933	278132	542982	797117	1042292	1279787
2700	28699	141841	279867	546190	801612	1047937	1286475
2800	28893	142776	281657	549501	806258	1053781	1293408
2900	29094	143740	283501	552918	811059	1059825	1300587
3000	29300	144732	285401	556442	816017	1066075	1308017

PERICLASE

MgO
PERICLASE Formula wt. (g) = 40.30

Enthalpy (J/mol) =	-601490.000	Robie et al. (1978)
Entropy (J/mol.K) =	26.940	Robie et al. (1978)
Heat Capacity:		
No. of *Cp* ranges = 1		Robie et al. (1978), refitted
4.5490000D+01 4.7730D-03 -2.1600D+06 0.0000D+00		
1.7440D+08 0.0000D+00 2.4110D+03		
Volume (cc/mol) =	11.250	Robie et al. (1978)
Thermal expn:		
3.64000D-05 8.35000D-09 8.50000D-04 -9.50000D-01		Saxena and Shen (1992)
Isot.compressibility:		
5.87500D-07 1.10100D-10 1.37000D-14 4.80000D-18		Saxena and Shen (1992)
P deriv. 4.17000D+00		
T deriv. 4.00000D-05		This study
Melting pt. and enthalpy of melting:		
3125.00000	-	Robie et al. (1978)

NOTE: For the magnesiowustite solid solution, MgO (periclase) was chosen to mix as a third component with FeO and $FeO_{1.5}$ (see wüstite FeO-$FeO_{1.5}$). T data on thermal expansion and compressibility are consistent with those of Suzuki (1975), Jackson and Niesler (1982) and Anderson and Zou (1989).

MgO
PERICLASE

Thermochemical Data

T(K)	C_p(J)	H(kJ)	S(J)	α 10^{-4}	K 10^4	γ	V(cc)	δ_T
298	37.3	-601.490	26.9	0.3105	160.857	1.53	11.2	6.19
300	37.4	-601.421	27.2	0.3118	160.799	1.53	11.3	6.17
400	42.7	-597.387	38.7	0.3593	157.719	1.53	11.3	5.42
500	45.5	-592.969	48.6	0.3848	154.661	1.53	11.3	5.12
600	47.2	-588.332	57.0	0.4019	151.623	1.52	11.4	4.97
700	48.4	-583.551	64.4	0.4152	148.604	1.52	11.4	4.88
800	49.3	-578.666	70.9	0.4266	145.602	1.52	11.5	4.82
900	50.0	-573.699	76.8	0.4369	142.615	1.52	11.5	4.78
1000	50.7	-568.662	82.1	0.4465	139.645	1.52	11.6	4.75
1100	51.3	-563.563	86.9	0.4557	136.691	1.52	11.6	4.73
1200	51.8	-558.407	91.4	0.4647	133.754	1.52	11.7	4.71
1300	52.4	-553.198	95.6	0.4735	130.834	1.52	11.7	4.70
1400	52.9	-547.938	99.5	0.4821	127.932	1.52	11.8	4.69
1500	53.3	-542.628	103.2	0.4907	125.051	1.51	11.8	4.68
1600	53.8	-537.268	106.6	0.4992	122.191	1.51	11.9	4.67
1700	54.3	-531.861	109.9	0.5077	119.354	1.51	12.0	4.66
1800	54.8	-526.407	113.0	0.5161	116.542	1.50	12.0	4.65
1900	55.3	-520.905	116.0	0.5245	113.757	1.50	12.1	4.64
2000	55.7	-515.356	118.8	0.5329	111.000	1.50	12.2	4.63
2100	56.2	-509.760	121.6	0.5412	108.274	1.49	12.2	4.62
2200	56.7	-504.118	124.2	0.5496	105.581	1.48	12.3	4.61
2300	57.1	-498.429	126.7	0.5579	102.923	1.48	12.4	4.60
2400	57.6	-492.693	129.2	0.5663	100.300	1.47	12.4	4.58
2500	58.1	-486.911	131.5	0.5746	97.716	1.46	12.5	4.57
2600	58.5	-481.083	133.8	0.5830	95.171	1.45	12.6	4.55
2700	59.0	-475.208	136.0	0.5913	92.668	1.45	12.6	4.53
2800	59.4	-469.286	138.2	0.5996	90.207	1.44	12.7	4.51
2900	59.9	-463.318	140.3	0.6080	87.791	1.43	12.8	4.49
3000	60.4	-457.303	142.3	0.6163	85.419	1.42	12.9	4.46

<INT VdP (J)>

	PRESSURE(GPa)						
T(K)	1	5	10	20	30	40	50
298	11215	55419	109328	213320	313053	409252	502438
300	11216	55422	109333	213328	313064	409265	502452
400	11253	55594	109647	213859	313747	410054	503313
500	11294	55786	109999	214460	314531	410975	504337
600	11338	55989	110373	215103	315375	411971	505450
700	11384	56202	110763	215774	316257	413017	506623
800	11431	56421	111166	216468	317171	414101	507840
900	11480	56647	111580	217181	318110	415215	509093
1000	11530	56878	112005	217912	319072	416356	510377
1100	11581	57116	112440	218659	320053	417521	511687
1200	11633	57358	112884	219420	321054	418708	513021
1300	11687	57607	113338	220197	322073	419915	514378
1400	11742	57860	113801	220987	323109	421141	515756
1500	11798	58119	114272	221791	324161	422386	517153
1600	11856	58383	114753	222608	325230	423649	518569
1700	11915	58652	115242	223439	326314	424928	520003
1800	11975	58927	115741	224283	327413	426225	521455
1900	12036	59207	116248	225139	328527	427537	522925
2000	12098	59492	116764	226008	329657	428866	524411
2100	12162	59783	117288	226890	330801	430211	525914
2200	12227	60079	117822	227785	331960	431572	527434
2300	12294	60380	118364	228692	333133	432949	528970
2400	12361	60687	118915	229612	334321	434342	530524
2500	12431	61000	119475	230545	335524	435750	532095
2600	12501	61318	120044	231490	336742	437175	533683
2700	12573	61641	120621	232448	337975	438617	535288
2800	12646	61971	121208	233419	339223	440075	536913
2900	12721	62306	121804	234403	340486	441551	538555
3000	12797	62646	122409	235400	341764	443044	540218

WUSTITE($FeO_{1.5}$)

$FeO_{1.5}$

WUSTITE($FeO_{1.5}$) Formula wt. (g) = 79.85

Enthalpy (J/mol) = -380900.000 Fei and Saxena (1986)

Entropy (J/mol.K) = 54.900 Fei and Saxena (1986)

Heat Capacity:

No. of *Cp* ranges = 3 Haas (1988)

3.9874000D+01 4.6440D-02 -1.6420D+06 0.0000D+00

1.4020D+08 0.0000D+00 3.5750D+03

Temp. and enthalpy of transition:

848.00 499.00

-6.9915000D+01 7.2450D-02 7.8240D+07 0.0000D+00

-7.2840D+06 0.0000D+00 -1.7580D+02

1000.00 0.00

1.9740000D+02 4.3230D-02 6.1970D+07 0.0000D+00

2.4040D+11 0.0000D+00 -4.3270D+05

Volume (cc/mol) = 15.970 Fei and Saxena (1986)

Thermal expn:

2.45300D-05 1.67020D-08 0.00000D+00 2.00000D-03

Fei and Saxena (1986), Haas (1988)

Isot.compressibility:

5.55500D-07 0.00000D+00 0.00000D+00 0.00000D+00

P deriv. 4.00000D+00 Fei and Saxena (1986)

T deriv. 0.00000D+00 This study

Melting pt. and enthalpy of melting:

- -

NOTE: Wüstite has a temperature-composition field within which its composition varies mainly between $Fe_{0.947}O$ and $Fe_{0.85}O$. The system may be modeled by considering wüstite as a solid solution of two fictive, isostructural components FeO and $FeO_{1.5}$. A precise model for the wüstite solid solution particularly for modeling the crystal chemical behavior must include the effect of additional components (other than FeO and $FeO_{1.5}$) and other parameters discussed by Hazen and Jeanloz (1984). The reaction within the field of wüstite solution is

$$\{FeO\} + 0.25O2 = \{FeO_{1.5}\}$$

and in the field towards magnetite, the reaction may be conceived as

$$\{5FeO\} + O_2 = \{2FeO_{1.5}\} + Fe_3O_4$$

where the brackets denote the components in the wüstite solution. The computed W_{ij} values for wüstite are (FeO = 1, $FeO_{1.5}$ = 2):

$$W_{12} = -7229,\ W_{21} = -2142 \text{ (J/mol)}$$

which are the same as in Bjorkman (1984).

$FeO_{1.5}$

WUSTITE ($FeO_{1.5}$)

Thermochemical Data

T(K)	C_p(J)	H(kJ)	S(J)	α 10^{-4}	K 10^4	γ	V(cc)	δ_T
298	52.5	-380.90	54.9	0.2953	180.018	1.64	16.0	0.00
300	52.7	-380.80	55.2	0.2956	180.018	1.64	16.0	0.00
400	59.3	-375.19	71.3	0.3122	180.018	1.55	16.0	0.00
500	64.8	-368.98	85.2	0.3289	180.018	1.50	16.1	0.00
600	69.8	-362.25	97.4	0.3456	180.018	1.48	16.1	0.00
700	74.5	-355.03	108.5	0.3623	180.018	1.47	16.2	0.00
800	79.2	-347.34	118.8	0.3789	180.018	1.46	16.2	0.00
900	91.7	-338.01	129.8	0.3956	180.018	1.33	16.3	0.00
1000	110.3	-329.44	138.8	0.4123	180.018	1.15	16.4	0.00
1100	83.4	-319.91	147.9	0.4290	180.018	1.64	16.4	0.00
1200	70.8	-312.29	154.6	0.4457	180.018	2.08	16.5	0.00
1300	66.8	-305.46	160.0	0.4624	180.018	2.36	16.6	0.00
1400	68.1	-298.75	165.0	0.4791	180.018	2.46	16.7	0.00
1500	72.6	-291.74	169.8	0.4958	180.018	2.43	16.7	0.00
1600	79.0	-284.17	174.7	0.5125	180.018	2.34	16.8	0.00
1700	86.7	-275.89	179.7	0.5292	180.018	2.23	16.9	0.00
1800	95.2	-266.80	184.9	0.5459	180.018	2.12	17.0	0.00
1900	104.0	-256.84	190.3	0.5626	180.018	2.03	17.1	0.00
2000	113.1	-245.99	195.9	0.5793	180.018	1.94	17.2	0.00
2100	122.1	-234.23	201.6	0.5960	180.018	1.88	17.3	0.00
2200	131.2	-221.56	207.5	0.6127	180.018	1.82	17.4	0.00
2300	140.2	-207.99	213.5	0.6294	180.018	1.78	17.5	0.00
2400	149.0	-193.53	219.7	0.6462	180.018	1.75	17.6	0.00
2500	157.7	-178.19	225.9	0.6629	180.018	1.73	17.7	0.00
2600	166.2	-162.00	232.3	0.6796	180.018	1.71	17.9	0.00
2700	174.6	-144.96	238.7	0.6963	180.018	1.71	18.0	0.00
2800	182.8	-127.09	245.2	0.7130	180.018	1.71	18.1	0.00
2900	190.8	-108.41	251.8	0.7297	180.018	1.71	18.2	0.00
3000	198.6	-88.94	258.4	0.7464	180.018	1.72	18.4	0.00

<INT VdP (J)>

PRESSURE (GPa)

T(K)	1	5	10	20	30	40	50
298	15926	78789	155627	304266	447201	585326	719304
300	15926	78793	155635	304282	447225	585358	719344
400	15975	79033	156109	305209	448586	587140	721533
500	16026	79287	156610	306189	450027	589025	723850
600	16080	79554	157139	307223	451547	591015	726295
700	16137	79836	157696	308312	453148	593110	728870
800	16197	80133	158282	309457	454830	595313	731576
900	16260	80444	158896	310658	456595	597623	734415
1000	16326	80769	159539	311915	458443	600042	737388
1100	16395	81110	160212	313230	460376	602572	740497
1200	16467	81466	160914	314603	462394	605213	743743
1300	16542	81836	161646	316035	464499	607967	747127
1400	16620	82223	162409	317526	466691	610836	750653
1500	16701	82624	163203	319078	468971	613821	754321
1600	16785	83042	164028	320691	471342	616924	758134
1700	16873	83476	164884	322366	473803	620146	762094
1800	16964	83926	165773	324103	476357	623489	766202
1900	17058	84392	166695	325905	479005	626954	770460
2000	17156	84875	167649	327771	481748	630544	774872
2100	17257	85376	168637	329703	484588	634261	779440
2200	17362	85893	169660	331702	487525	638106	784165
2300	17470	86428	170717	333768	490563	642082	789050
2400	17582	86981	171809	335904	493701	646190	794099
2500	17697	87553	172937	338110	496943	650433	799313
2600	17816	88142	174102	340387	500290	654814	804697
2700	17939	88751	175304	342736	503743	659334	810251
2800	18066	89378	176543	345160	507305	663996	815980
2900	18197	90025	177821	347659	510978	668802	821887
3000	18332	90692	179138	350234	514763	673757	827975

WUSTITE(FeO)

FeO
WUSTITE(FeO) Formula wt. (g) = 71.85

Enthalpy (J/mol) = -267270.000 Fei and Saxena (1986)
Entropy (J/mol.K) = 57.590 Fei and Saxena (1986)
Heat Capacity:
No. of *Cp* ranges = 1 Haas (1988)
-1.6670000D+01 5.6696D-02 -5.3269D+06 -1.0801D-05
0.0000D+00 0.0000D+00 3.2403D+04
Volume (cc/mol) = 12.250 Fei and Saxena (1986)
Thermal expn:
2.60186D-05 1.46789D-08 2.78637D-03 -4.27635D-01
Fei and Saxena (1986), Haas (1988)
Isot.compressibility:
5.33040D-07 9.32000D-11 0.00000D+00 0.00000D+00
P deriv. 4.00000D+00 Fei and Saxena (1986)
T deriv. 0.00000D+00
Melting pt. and enthalpy of melting:
- -

NOTE: Wüstite has a temperature-composition field within which its composition varies mainly between $Fe_{0.947}O$ and $Fe_{0.85}O$. The system may be modeled by considering wüstite as a solid solution of two fictive, isostructural components FeO and $FeO_{1.5}$. A precise model for the wüstite solid solution particularly for modeling the crystal chemical behavior must include the effect of additional components (other than FeO and $FeO_{1.5}$) and other parameters discussed by Hazen and Jeanloz (1984). The reaction within the field of wüstite solution is

$$\{FeO\} + 0.25O2 = \{FeO_{1.5}\}$$

and in the field towards magnetite, the reaction may be conceived as

$$\{5FeO\} + O_2 = \{2FeO_{1.5}\} + Fe_3O_4$$

where the brackets denote the components in the wüstite solution. The computed W_{ij} values for wüstite are (FeO = 1, $FeO_{1.5}$ = 2):

$$W_{12} = -7229,\ W_{21} = -2142 \text{ (J/mol)}$$

which are the same as in Bjorkman (1984).

FeO

WUSTITE(FeO)

Thermochemical Data

T(K)	C_p(J)	H(kJ)	S(J)	$\alpha\ 10^{-4}$	$K\ 10^4$	γ	V(cc)	δ_T
298	48.0	-267.27	57.6	0.3493	178.308	1.62	12.2	4.76
300	48.2	-267.18	57.9	0.3496	178.253	1.61	12.3	4.75
400	52.0	-262.11	72.4	0.3618	175.340	1.53	12.3	4.52
500	52.5	-256.88	84.1	0.3722	172.521	1.55	12.3	4.32
600	52.7	-251.63	93.7	0.3828	169.791	1.58	12.4	4.13
700	53.1	-246.34	101.8	0.3940	167.146	1.61	12.4	3.95
800	54.0	-240.99	109.0	0.4058	164.582	1.63	12.5	3.78
900	55.0	-235.54	115.4	0.4180	162.096	1.64	12.5	3.61
1000	56.3	-229.97	121.3	0.4306	159.683	1.65	12.6	3.46
1100	57.7	-224.28	126.7	0.4435	157.342	1.65	12.6	3.31
1200	59.1	-218.44	131.8	0.4566	155.068	1.66	12.7	3.17
1300	60.6	-212.45	136.6	0.4699	152.858	1.67	12.8	3.03
1400	62.0	-206.33	141.1	0.4834	150.711	1.68	12.8	2.91
1500	63.3	-200.06	145.4	0.4970	148.624	1.69	12.9	2.79
1600	64.6	-193.67	149.6	0.5108	146.593	1.71	12.9	2.67
1700	65.7	-187.15	153.5	0.5246	144.617	1.74	13.0	2.57
1800	66.7	-180.53	157.3	0.5386	142.694	1.76	13.1	2.47
1900	67.6	-173.81	160.9	0.5526	140.821	1.80	13.2	2.38
2000	68.4	-167.01	164.4	0.5666	138.997	1.84	13.2	2.29
2100	69.0	-160.14	167.8	0.5807	137.219	1.89	13.3	2.20
2200	69.4	-153.21	171.0	0.5949	135.487	1.95	13.4	2.12
2300	69.7	-146.26	174.1	0.6091	133.797	2.02	13.5	2.05
2400	69.8	-139.29	177.0	0.6233	132.149	2.10	13.6	1.98
2500	69.7	-132.31	179.9	0.6376	130.541	2.20	13.6	1.91
2600	69.4	-125.36	182.6	0.6519	128.972	2.32	13.7	1.84
2700	68.9	-118.44	185.2	0.6662	127.440	2.45	13.8	1.78
2800	68.3	-111.58	187.7	0.6806	125.945	2.62	13.9	1.72
2900	67.5	-104.79	190.1	0.6950	124.483	2.82	14.0	1.67
3000	66.4	-98.09	192.4	0.7094	123.056	3.06	14.1	1.62

<INT VdP (J)>

PRESSURE(GPa)

T(K)	1	5	10	20	30	40	50
298	12215	60428	119348	233295	342840	448678	551319
300	12216	60432	119354	233307	342856	448697	551341
400	12259	60634	119732	233972	343746	449765	552554
500	12304	60844	120124	234665	344677	450889	553834
600	12350	61061	120530	235385	345648	452065	555180
700	12397	61286	120950	236133	346661	453296	556594
800	12446	61518	121386	236912	347719	454586	558081
900	12497	61759	121838	237722	348823	455938	559643
1000	12550	62008	122308	238566	349976	457352	561284
1100	12604	62266	122795	239443	351178	458833	563005
1200	12660	62534	123299	240355	352433	460381	564810
1300	12719	62810	123823	241304	353740	461997	566700
1400	12779	63097	124365	242289	355100	463685	568677
1500	12841	63394	124927	243312	356516	465444	570742
1600	12905	63700	125509	244374	357988	467277	572897
1700	12972	64017	126111	245474	359518	469185	575144
1800	13040	64345	126733	246614	361105	471168	577484
1900	13111	64683	127377	247795	362751	473228	579918
2000	13184	65032	128042	249016	364458	475367	582449
2100	13259	65393	128728	250280	366225	477585	585077
2200	13337	65764	129437	251586	368055	479884	587804
2300	13417	66147	130168	252936	369947	482265	590631
2400	13499	66542	130922	254329	371904	484729	593561
2500	13584	66949	131699	255767	373925	487278	596593
2600	13671	67368	132500	257251	376013	489912	599731
2700	13761	67799	133325	258780	378168	492634	602975
2800	13853	68243	134174	260357	380391	495445	606328
2900	13948	68699	135048	261982	382683	498345	609791
3000	14046	69169	135948	263655	385046	501337	613365

FeO

MAGNETITE

Fe_3O_4
MAGNETITE Formula wt. (g) = 231.54

Enthalpy (J/mol) = -1115548.000 Haas (1988)
Entropy (J/mol.K) = 150.640 Haas (1988)
Heat Capacity:
No. of *Cp* ranges = 3 Haas (1988)
7.3372000D+01 2.2668D-01 9.5360D+05 0.0000D+00
6.9880D+06 0.0000D+00 1.5476D+02
Temp. and enthalpy of transition:
848.00 1564.00
-1.1312000D+01 9.8040D-02 1.2456D+08 0.0000D+00
7.4440D+07 0.0000D+00 1.5336D+03
1300.00 0.00
3.4724800D+02 6.3280D-02 7.8960D+07 0.0000D+00
3.5584D+11 0.0000D+00 -5.7920D+05
Volume (cc/mol) = 44.520 Mao et al., (1974)
Thermal expn:
2.61100D-05 3.26020D-08 0.00000D+00 4.00000D-03
Haas (1988)
Isot.compressibility:
5.46400D-07 0.00000D+00 0.00000D+00 0.00000D+00
P deriv. 4.00000D+00 Mao et al., (1974)
T deriv. 0.00000D+00
Melting pt. and enthalpy of melting:
1870.00000 38072.00000 Robie et al. (1978)

NOTE: Magnetite transforms to its high-pressure polymorph at pressure of about 25 GPa at room temperature (Mao et al., 1974). The magnetite and high-pressure magnetite boundary was determined to have a negative slope (dT/dP = -45 K/GPa, Huang and Bassett, 1986). Shock-wave data indicate that hematite transforms to its high-pressure polymorph below pressure of 60 GPa. The transition also was not observed in the diamond-anvil cell below 30 GPa (Mao and Bell, 1977). Some physical properties of these phases are available (e.g. Ahrens et al.; Mao et al., 1974; 1969; Wilburn et al., 1978; Sato and Akimoto, 1979; Finger and Hazen, 1980; Finger et al., 1986). Thermochemical data for magnetite and hematite are from Fei and Saxena (1986).

Fe_3O_4
MAGNETITE

Thermochemical Data

T(K)	C_p(J)	H(kJ)	S(J)	α 10^{-4}	K 10^4	γ	V(cc)	δ_T
298	152.5	-1115.548	150.6	0.3588	174.819	1.87	44.5	4.25
300	152.7	-1115.266	151.6	0.3594	174.770	1.87	44.5	4.24
400	170.5	-1099.139	197.9	0.3918	172.124	1.82	44.7	3.90
500	190.9	-1081.082	238.1	0.4243	169.500	1.75	44.9	3.64
600	212.3	-1060.927	274.8	0.4568	166.886	1.69	45.1	3.43
700	234.2	-1038.602	309.1	0.4894	164.268	1.64	45.3	3.26
800	256.4	-1014.071	341.8	0.5220	161.636	1.60	45.5	3.13
900	232.5	-987.491	373.2	0.5546	158.981	1.90	45.8	3.03
1000	212.9	-965.298	396.6	0.5872	156.296	2.25	46.0	2.94
1100	200.9	-944.659	416.3	0.6198	153.575	2.58	46.3	2.88
1200	194.2	-924.941	433.4	0.6524	150.813	2.90	46.6	2.83
1300	192.7	-905.706	448.8	0.6849	148.008	3.16	46.9	2.79
1400	192.1	-886.520	463.1	0.7175	145.159	3.45	47.2	2.76
1500	196.6	-867.121	476.4	0.7501	142.264	3.64	47.6	2.73
1600	204.2	-847.103	489.4	0.7827	139.325	3.77	48.0	2.71
1700	213.9	-826.211	502.0	0.8153	136.344	3.86	48.3	2.70
1800	224.8	-804.288	514.5	0.8479	133.324	3.92	48.7	2.69
1900	236.4	-781.235	527.0	0.8805	130.269	3.97	49.2	2.68
2000	248.4	-756.997	539.4	0.9132	127.183	4.02	49.6	2.67
2100	260.7	-731.543	551.9	0.9458	124.072	4.08	50.1	2.66
2200	272.9	-704.864	564.3	0.9784	120.942	4.15	50.6	2.65
2300	285.1	-676.960	576.7	1.0110	117.799	4.23	51.1	2.64
2400	297.2	-647.840	589.1	1.0436	114.649	4.33	51.6	2.63
2500	309.2	-617.518	601.4	1.0762	111.500	4.44	52.1	2.62
2600	320.9	-586.011	613.8	1.1088	108.359	4.58	52.7	2.61
2700	332.5	-553.338	626.1	1.1414	105.232	4.74	53.3	2.60
2800	343.9	-519.519	638.4	1.1740	102.125	4.92	53.9	2.58
2900	355.0	-484.574	650.7	1.2066	99.046	5.14	54.6	2.56
3000	366.0	-448.523	662.9	1.2392	96.000	5.40	55.2	2.54

<INT VdP (J)>

T(K)	PRESSURE (GPa) 1	5	10	20	30	40	50
298	44393	219558	433535	847134	1244530	1628311	2000376
300	44396	219572	433561	847180	1244592	1628386	2000462
400	44561	220353	435027	849795	1248138	1632702	2005423
500	44742	221208	436636	852684	1252080	1637530	2011008
600	44937	222137	438389	855847	1256417	1642865	2017208
700	45148	223142	440288	859284	1261144	1648702	2024014
800	45375	224222	442332	862995	1266262	1655035	2031417
900	45617	225378	444523	866979	1271767	1661861	2039410
1000	45876	226612	446862	871238	1277658	1669175	2047986
1100	46151	227925	449350	875771	1283934	1676974	2057138
1200	46443	229316	451989	880579	1290595	1685255	2066861
1300	46752	230788	454780	885664	1297639	1694015	2077150
1400	47078	232342	457724	891027	1305067	1703253	2088001
1500	47421	233979	460824	896669	1312881	1712967	2099412
1600	47782	235699	464081	902593	1321079	1723157	2111378
1700	48162	237506	467497	908801	1329665	1733824	2123900
1800	48560	239399	471075	915295	1338639	1744967	2136975
1900	48978	241381	474818	922078	1348005	1756588	2150606
2000	49414	243453	478727	929154	1357765	1768690	2164792
2100	49871	245617	482805	936525	1367922	1781276	2179536
2200	50348	247875	487056	944195	1378481	1794348	2194842
2300	50846	250229	491482	952170	1389446	1807912	2210714
2400	51365	252681	496088	960453	1400822	1821974	2227157
2500	51906	255234	500875	969048	1412614	1836538	2244177
2600	52470	257888	505848	977963	1424830	1851612	2261783
2700	53056	260648	511011	987201	1437475	1867205	2279984
2800	53666	263514	516368	996770	1450558	1883324	2298789
2900	54301	266491	521923	1006675	1464087	1899980	2318209
3000	54960	269581	527681	1016925	1478070	1917184	2338258

HIGH-MAGNETITE

Fe_3O_4
HIGH-MAGNETITE Formula wt. (g) = 231.54

Enthalpy (J/mol) = -946000.000 Fei and Saxena (1986)
Entropy (J/mol.K) = 180.000 Fei and Saxena (1986)
Heat Capacity:
No. of *Cp* ranges = 3 Haas (1984)
7.3372000D+01 2.2668D-01 9.5360D+05 0.0000D+00
6.9880D+06 0.0000D+00 1.5476D+02
Temp. and enthalpy of transition:
848.00 1996.00
-1.1312000D+01 9.8040D-02 1.2456D+08 0.0000D+00
7.4440D+07 0.0000D+00 1.5336D+03
1000.00 0.00
3.4724800D+02 6.3280D-02 7.8960D+07 0.0000D+00
3.5584D+11 0.0000D+00 -5.7920D+05
Volume (cc/mol) = 36.440 Mao et al., (1974)
Thermal expn:
2.61100D-05 3.26020D-08 0.00000D+00 4.00000D-03
Fei and Saxena (1986), Haas (1988)
Isot.compressibility:
2.23200D-07 0.00000D+00 0.00000D+00 0.00000D+00
P deriv. 4.00000D+00 Ahrens et al., (1969)
T deriv. 0.00000D+00
Melting pt. and enthalpy of melting:
\- -

NOTE: Magnetite transforms to its high-pressure polymorph at pressure of about 25 GPa at room temperature (Mao et al., 1974). The magnetite and high-pressure magnetite boundary was determined to have a negative slope (dT/dP = -45 K/GPa, Huang and Bassett, 1986). Shock-wave data indicate that hematite transforms to its high-pressure polymorph below pressure of 60 GPa. The transition also was not observed in the diamond-anvil cell below 30 GPa (Mao and Bell, 1977). Some physical properties of these phases are available (e.g. Ahrens et al.; Mao et al., 1974; 1969; Wilburn et al., 1978; Sato and Akimoto, 1979; Finger and Hazen, 1980; Finger et al., 1986). Thermochemical data for magnetite and hematite are from Fei and Saxena (1986).

Fe_3O_4

H-MAGNETITE

Thermochemical Data

T(K)	C_p(J)	H(kJ)	S(J)	α 10^{-4}	K 10^4	γ	V(cc)	δ_T
298	152.5	-946.000	180.0	0.3588	401.899	3.58	36.4	9.76
300	152.7	-945.718	180.9	0.3594	401.638	3.58	36.4	9.74
400	170.5	-929.591	227.2	0.3918	387.932	3.44	36.6	8.80
500	190.9	-911.534	267.4	0.4243	374.855	3.27	36.7	8.05
600	212.3	-891.379	304.1	0.4568	362.302	3.12	36.9	7.44
700	234.2	-869.054	338.5	0.4894	350.186	2.99	37.1	6.96
800	256.4	-844.523	371.2	0.5220	338.437	2.87	37.3	6.56
900	232.5	-817.511	403.1	0.5546	327.004	3.42	37.5	6.23
1000	266.1	-795.318	426.5	0.5872	315.843	3.10	37.7	5.95
1100	222.9	-771.096	449.6	0.6198	304.925	4.11	37.9	5.72
1200	201.3	-750.023	467.9	0.6524	294.228	5.08	38.1	5.52
1300	192.7	-730.410	483.6	0.6849	283.739	5.91	38.4	5.35
1400	192.1	-711.225	497.9	0.7175	273.448	6.55	38.7	5.19
1500	196.6	-691.825	511.2	0.7501	263.354	7.00	39.0	5.06
1600	204.2	-671.807	524.2	0.7827	253.457	7.30	39.2	4.94
1700	213.9	-650.916	536.8	0.8153	243.761	7.50	39.6	4.83
1800	224.8	-628.992	549.3	0.8479	234.273	7.64	39.9	4.72
1900	236.4	-605.939	561.8	0.8805	225.000	7.74	40.2	4.62
2000	248.4	-581.701	574.2	0.9132	215.950	7.83	40.6	4.53
2100	260.7	-556.248	586.7	0.9458	207.132	7.93	41.0	4.44
2200	272.9	-529.569	599.1	0.9784	198.552	8.05	41.4	4.35
2300	285.1	-501.665	611.5	1.0110	190.220	8.18	41.8	4.27
2400	297.2	-472.545	623.9	1.0436	182.141	8.34	42.2	4.18
2500	309.2	-442.223	636.2	1.0762	174.320	8.54	42.7	4.10
2600	320.9	-410.716	648.6	1.1088	166.762	8.77	43.1	4.02
2700	332.5	-378.043	660.9	1.1414	159.468	9.05	43.6	3.93
2800	343.9	-344.223	673.2	1.1740	152.441	9.38	44.1	3.85
2900	355.0	-309.278	685.5	1.2066	145.681	9.78	44.7	3.77
3000	366.0	-273.227	697.7	1.2392	139.185	10.25	45.2	3.69

<INT VdP (J)>

PRESSURE(GPa)

T(K)	1	5	10	20	30	40	50
298	36394	181089	360043	712010	1056713	1394821	1726894
300	36397	181100	360065	712047	1056762	1394878	1726956
400	36532	181743	361271	714174	1059578	1398190	1730601
500	36680	182447	362596	716531	1062727	1401935	1734777
600	36840	183211	364040	719115	1066207	1406108	1739473
700	37013	184038	365605	721928	1070014	1410701	1744678
800	37199	184927	367290	724967	1074144	1415706	1750379
900	37398	185879	369096	728234	1078595	1421119	1756569
1000	37610	186895	371025	731727	1083366	1426934	1763236
1100	37835	187975	373076	735446	1088452	1433146	1770372
1200	38074	189120	375252	739393	1093854	1439749	1777971
1300	38328	190332	377553	743566	1099569	1446742	1786024
1400	38595	191611	379981	747968	1105597	1454119	1794526
1500	38876	192958	382536	752600	1111937	1461880	1803473
1600	39173	194374	385221	757461	1118590	1470022	1812860
1700	39484	195861	388038	762554	1125555	1478543	1822684
1800	39810	197419	390987	767881	1132834	1487444	1832943
1900	40152	199050	394071	773443	1140428	1496724	1843636
2000	40510	200756	397292	779244	1148338	1506385	1854764
2100	40885	202536	400653	785285	1156567	1516429	1866327
2200	41276	204395	404154	791569	1165118	1526857	1878328
2300	41684	206332	407800	798100	1173994	1537674	1890769
2400	42109	208349	411592	804880	1183199	1548883	1903657
2500	42553	210449	415534	811915	1192738	1560489	1916996
2600	43015	212633	419628	819208	1202615	1572499	1930793
2700	43495	214902	423877	826764	1212836	1584919	1945056
2800	43996	217261	428286	834588	1223407	1597757	1959794
2900	44516	219709	432857	842684	1234336	1611021	1975019
3000	45056	222250	437593	851059	1245629	1624721	1990741

Fe_3O_4

MAGNETITE

HEMATITE

Fe_2O_3
HEMATITE Formula wt. (g) = 159.69

Enthalpy (J/mol) = -824782.000 Haas (1988)
Entropy (J/mol.K) = 87.400 Haas (1988)
Heat Capacity:
No. of *Cp* ranges = 3 Haas (1988)
7.9740000D+01 9.2880D-02 -3.2840D+06 0.0000D+00
2.8040D+08 0.0000D+00 7.1500D+03
Temp. and enthalpy of transition:
956.00 670.00
-1.3983000D+02 1.4490D-01 1.5648D+08 0.0000D+00
-1.4568D+07 0.0000D+00 -3.5160D+02
1250.00 0.00
3.9480000D+02 8.6460D-02 1.2394D+08 0.0000D+00
4.8080D+11 0.0000D+00 -8.6540D+05
Volume (cc/mol) = 30.260 Liebermann and Schreiber (1968)
Thermal expn:
2.45300D-05 1.67020D-08 0.00000D+00 2.00000D-03
Haas (1988)
Isot.compressibility:
4.83100D-07 0.00000D+00 0.00000D+00 0.00000D+00
P deriv. 4.53000D+00 Liebermann and Schreiber (1968)
T deriv. 0.00000D+00
Melting pt. and enthalpy of melting:
1895.00000 - Robie et al. (1978)

NOTE: Magnetite transforms to its high-pressure polymorph at pressure of about 25 GPa at room temperature (Mao et al., 1974). The magnetite and high-pressure magnetite boundary was determined to have a negative slope (dT/dP = -45 K/GPa, Huang and Bassett, 1986). Shock-wave data indicate that hematite transforms to its high-pressure polymorph below pressure of 60 GPa. The transition also was not observed in the diamond-anvil cell below 30 GPa (Mao and Bell, 1977). Some physical properties of these phases are available (e.g. Ahrens et al.; Mao et al., 1974; 1969; Wilburn et al., 1978; Sato and Akimoto, 1979; Finger and Hazen, 1980; Finger et al., 1986). Thermochemical data for magnetite and hematite are from Fei and Saxena (1986).

Fe_2O_3

HEMATITE

Thermochemical Data

T(K)	C_p(J)	H(kJ)	S(J)	α 10^{-4}	K 10^4	γ	V(cc)	δ_T
298	105.0	-824.782	87.4	0.2953	196.572	1.70	30.3	5.80
300	105.3	-824.587	88.1	0.2956	196.510	1.69	30.3	5.80
400	118.6	-813.361	120.3	0.3122	193.170	1.57	30.4	5.50
500	129.6	-800.938	147.9	0.3289	189.872	1.50	30.5	5.26
600	139.6	-787.476	172.4	0.3456	186.597	1.45	30.6	5.07
700	149.1	-773.041	194.7	0.3623	183.331	1.41	30.7	4.92
800	158.4	-757.665	215.2	0.3789	180.058	1.38	30.8	4.81
900	167.6	-741.365	234.4	0.3956	176.770	1.35	30.9	4.72
1000	161.2	-723.897	252.8	0.4123	173.457	1.46	31.0	4.65
1100	148.6	-708.474	267.5	0.4290	170.112	1.65	31.2	4.61
1200	142.4	-693.971	280.1	0.4457	166.730	1.79	31.3	4.58
1300	133.7	-680.155	291.2	0.4624	163.309	1.99	31.4	4.56
1400	136.2	-666.730	301.1	0.4791	159.847	2.02	31.6	4.55
1500	145.1	-652.710	310.8	0.4958	156.344	1.94	31.7	4.54
1600	158.1	-637.578	320.6	0.5125	152.801	1.82	31.9	4.55
1700	173.5	-621.017	330.6	0.5292	149.223	1.68	32.1	4.55
1800	190.3	-602.835	341.0	0.5459	145.613	1.55	32.2	4.56
1900	208.0	-582.921	351.8	0.5626	141.976	1.44	32.4	4.57
2000	226.1	-561.216	362.9	0.5793	138.318	1.33	32.6	4.57
2100	244.3	-537.696	374.4	0.5960	134.647	1.25	32.8	4.58
2200	262.4	-512.360	386.1	0.6127	130.968	1.17	33.0	4.59
2300	280.3	-485.220	398.2	0.6294	127.290	1.10	33.2	4.59
2400	298.0	-456.300	410.5	0.6462	123.621	1.04	33.4	4.59
2500	315.4	-425.627	423.0	0.6629	119.968	0.99	33.6	4.58
2600	332.4	-393.233	435.7	0.6796	116.339	0.94	33.9	4.57
2700	349.2	-359.150	448.6	0.6963	112.742	0.90	34.1	4.56
2800	365.5	-323.413	461.6	0.7130	109.183	0.86	34.3	4.54
2900	381.6	-286.056	474.7	0.7297	105.671	0.82	34.6	4.52
3000	397.3	-247.110	487.9	0.7464	102.211	0.79	34.8	4.50

<INT VdP (J)>

PRESSURE(GPa)

T(K)	1	5	10	20	30	40	50
298	30183	149460	295538	578978	852550	1117803	1375880
300	30185	149467	295553	579003	852582	1117842	1375923
400	30275	149892	296340	580374	854400	1120003	1378351
500	30371	150342	297176	581839	856350	1122336	1380985
600	30473	150817	298061	583395	858430	1124833	1383818
700	30579	151319	298996	585041	860636	1127489	1386840
800	30691	151846	299979	586775	862965	1130298	1390043
900	30809	152399	301010	588596	865412	1133254	1393418
1000	30932	152978	302090	590503	867976	1136352	1396958
1100	31060	153583	303217	592493	870652	1139586	1400654
1200	31195	154214	304393	594566	873437	1142952	1404500
1300	31335	154872	305616	596720	876330	1146444	1408488
1400	31481	155555	306887	598955	879326	1150058	1412612
1500	31632	156266	308206	601268	882423	1153789	1416865
1600	31790	157003	309572	603660	885619	1157634	1421242
1700	31953	157767	310986	606128	888911	1161588	1425737
1800	32123	158557	312447	608673	892298	1165648	1430346
1900	32299	159375	313956	611293	895777	1169811	1435063
2000	32481	160221	315512	613987	899346	1174073	1439885
2100	32669	161094	317116	616755	903005	1178433	1444808
2200	32864	161995	318768	619597	906750	1182888	1449830
2300	33065	162924	320467	622511	910583	1187437	1454947
2400	33273	163881	322215	625498	914501	1192077	1460158
2500	33487	164867	324011	628557	918503	1196807	1465461
2600	33709	165882	325856	631689	922590	1201628	1470856
2700	33937	166927	327749	634892	926761	1206538	1476342
2800	34172	168000	329692	638168	931016	1211538	1481920
2900	34414	169104	331684	641517	935355	1216628	1487590
3000	34664	170238	333726	644939	939779	1221808	1493353

Fe_2O_3

HEMATITE

Thermodynamic Data

[illegible]

HIGH-HEMATITE

Fe_2O_3

H-HEMATITE Formula wt. (g) = 159.69

Enthalpy (J/mol) = -749200.000 Fei and Saxena (1986)

Entropy (J/mol.K) = 87.400 Fei and Saxena (1986)

Heat Capacity:

No. of *Cp* ranges = 3 Haas (1988)

7.9740000D+01 9.2880D-02 -3.2840D+06 0.0000D+00
2.8040D+08 0.0000D+00 7.1500D+03

Temp. and enthalpy of transition:

956.00 998.00

-1.3983000D+02 1.4490D-01 1.5648D+08 0.0000D+00
-1.4568D+07 0.0000D+00 -3.5160D+02

1000.00 0.00

3.9480000D+02 8.6460D-02 1.2394D+08 0.0000D+00
4.8080D+11 0.0000D+00 -8.6540D+05

Volume (cc/mol) = 26.860 Mao et al., (1974)

Thermal expn:

2.45300D-05 1.67020D-08 0.00000D+00 2.00000D-03

Fei and Saxena (1986), Haas (1988)

Isot.compressibility:

2.51000D-07 0.00000D+00 0.00000D+00 0.00000D+00

P deriv. 4.00000D+00 Ahrens et al. (1969)

T deriv. 0.00000D+00

Melting pt. and enthalpy of melting:

\- -

NOTE: Magnetite transforms to its high-pressure polymorph at pressure of about 25 GPa at room temperature (Mao et al., 1974). The magnetite and high-pressure magnetite boundary was determined to have a negative slope (dT/dP = -45 K/GPa, Huang and Bassett, 1986). Shock-wave data indicate that hematite transforms to its high-pressure polymorph below pressure of 60 GPa. The transition also was not observed in the diamond-anvil cell below 30 GPa (Mao and Bell, 1977). Some physical properties of these phases are available (e.g. Ahrens et al.; Mao et al., 1974; 1969; Wilburn et al., 1978; Sato and Akimoto, 1979; Finger and Hazen, 1980; Finger et al., 1986). Thermochemical data for magnetite and hematite are from Fei and Saxena (1986).

Fe_2O_3

H-HEMATITE

Thermochemical Data

T(K)	C_p(J)	H(kJ)	S(J)	α 10^{-4}	K 10^4	γ	V(cc)	δ_T
298	105.0	-749.200	87.4	0.2953	361.508	2.80	26.9	10.67
300	105.3	-749.005	88.1	0.2956	361.297	2.79	26.9	10.65
400	118.6	-737.779	120.3	0.3122	350.168	2.56	26.9	9.96
500	129.6	-725.356	147.9	0.3289	339.478	2.42	27.0	9.40
600	139.6	-711.894	172.4	0.3456	329.150	2.32	27.1	8.94
700	149.1	-697.459	194.7	0.3623	319.119	2.23	27.2	8.56
800	158.4	-682.083	215.2	0.3789	309.334	2.15	27.3	8.26
900	167.6	-665.783	234.4	0.3956	299.754	2.08	27.4	8.00
1000	220.6	-647.987	253.1	0.4123	290.349	1.59	27.5	7.79
1100	166.8	-628.920	271.3	0.4290	281.097	2.21	27.7	7.61
1200	141.7	-613.674	284.6	0.4457	271.981	2.72	27.8	7.47
1300	133.7	-600.015	295.6	0.4624	262.994	3.00	27.9	7.34
1400	136.2	-586.590	305.5	0.4791	254.130	3.01	28.0	7.23
1500	145.1	-572.570	315.2	0.4958	245.388	2.87	28.2	7.13
1600	158.1	-557.438	324.9	0.5125	236.774	2.65	28.3	7.05
1700	173.5	-540.877	335.0	0.5292	228.291	2.41	28.5	6.96
1800	190.3	-522.695	345.4	0.5459	219.949	2.19	28.6	6.89
1900	208.0	-502.781	356.1	0.5626	211.755	2.00	28.8	6.81
2000	226.1	-481.076	367.3	0.5793	203.720	1.83	28.9	6.74
2100	244.3	-457.556	378.7	0.5960	195.854	1.68	29.1	6.66
2200	262.4	-432.220	390.5	0.6127	188.166	1.56	29.3	6.59
2300	280.3	-405.081	402.6	0.6294	180.666	1.45	29.5	6.51
2400	298.0	-376.160	414.9	0.6462	173.363	1.35	29.7	6.43
2500	315.4	-345.487	427.4	0.6629	166.263	1.26	29.8	6.35
2600	332.4	-313.093	440.1	0.6796	159.373	1.18	30.0	6.26
2700	349.2	-279.010	453.0	0.6963	152.699	1.11	30.3	6.18
2800	365.5	-243.274	466.0	0.7130	146.244	1.05	30.5	6.09
2900	381.6	-205.916	479.1	0.7297	140.010	0.99	30.7	5.99
3000	397.3	-166.970	492.3	0.7464	134.000	0.94	30.9	5.90

<INT VdP (J)>

T(K)	PRESSURE(GPa) 1	5	10	20	30	40	50
298	26823	133391	265045	523550	776232	1023671	1266345
300	26824	133398	265058	523571	776260	1023702	1266377
400	26904	133776	265756	524765	777786	1025429	1268194
500	26990	134177	266498	526041	779431	1027308	1270196
600	27080	134600	267284	527399	781191	1029333	1272373
700	27174	135046	268114	528837	783062	1031497	1274714
800	27274	135515	268987	530352	785040	1033792	1277208
900	27378	136008	269903	531944	787120	1036213	1279847
1000	27488	136523	270862	533610	789300	1038753	1282621
1100	27602	137062	271863	535349	791576	1041405	1285521
1200	27721	137623	272907	537160	793944	1044164	1288537
1300	27846	138208	273993	539042	796401	1047024	1291663
1400	27975	138817	275121	540992	798943	1049980	1294890
1500	28110	139449	276291	543010	801569	1053028	1298211
1600	28250	140105	277503	545094	804274	1056161	1301620
1700	28395	140784	278757	547243	807056	1059377	1305110
1800	28546	141488	280052	549456	809914	1062670	1308676
1900	28702	142216	281389	551733	812844	1066038	1312314
2000	28864	142968	282768	554071	815844	1069477	1316018
2100	29031	143744	284188	556471	818913	1072984	1319785
2200	29204	144545	285650	558932	822050	1076558	1323613
2300	29383	145371	287154	561453	825252	1080195	1327497
2400	29568	146222	288700	564034	828518	1083894	1331437
2500	29758	147099	290287	566673	831848	1087654	1335431
2600	29955	148001	291917	569372	835242	1091475	1339478
2700	30158	148929	293589	572129	838697	1095355	1343579
2800	30367	149882	295304	574945	842216	1099295	1347734
2900	30582	150863	297062	577820	845797	1103296	1351943
3000	30804	151870	298862	580753	849440	1107358	1356210

GARNET(Mg)

$MgSiO_3$
GARNET(Mg) Formula wt. (g) = 100.39

Enthalpy (J/mol) = -1513000.000 This study
Entropy (J/mol.K) = 59.000 This study
Heat Capacity:
No. of *Cp* ranges = 1 Yusa et al. (1993)
1.2757530D+02 7.4603D-03 -1.106D+06 0.0000D+00
1.9762D+09 0.0000D+00 0.0000D+00
Volume (cc/mol) = 28.580 Akaogi et al. (1987)

Thermal expn:
1.10190D-05 7.50750D-09 6.04030D-03 -1.02670D+00
Saxena and Shen (1992)

Isot.compressibility:
5.5000D-07 2.52100D-10 -1.10000D-13 5.58636D-17
Saxena and Shen (1992)

P deriv. 4.0D00
T deriv. 4.0D-04 This study
Melting pt. and enthalpy of melting:
\- -

NOTE: Kato and Kumazawa (1985, 1986) reported that the garnet phase of $MgSiO_3$ is stable at very high temperature. It is tetragonal, but a cubic form may also exist and the cubic (majorite) garnet solid solution between $Mg_4Si_4O_{12}$ ($4MgSiO_3$) and $Mg_3Al_2Si_3O_{12}$ (pyrope) is reasonably well documented (Ringwood, 1967; Akaogi and Akimoto, 1977; Liu, 1977; Akaogi et al, 1987). Akaogi et al. (1989) estimated the enthalpy of the pyroxene-cubic garnet transition as 36.5 KJ (per mole $MgSiO_3$) and the bulk modulus of cubic garnet as 154 GPa (assuming K' = 4.0) from enthalpies of solution and volume compression along the $Mg_4Si_4O_{12}$-$Mg_3Al_2Si_3O_{12}$ join. Their extrapolated molar volume for cubic garnet is 28.50 cm^3/mol ($MgSiO_3$), slightly larger than volumes near 28.1 to 28.2 cm^3/mol for tetragonal garnet (Kato and Kumazawa, 1985, 1986). Because the enthalpy of the cubic-tetragonal transition is unknown, we shall use data for the cubic phase in subsequent calculations. Lacking thermal expansion data, we assume α the same as for pyrope (Leitner et al., 1980). Heat capacities for $MgSiO_3$ garnet was recently determined by Yusa et al. (1993).

$MgSiO_3$
GARNET(Mg)

Thermochemical Data

T(K)	C_p(J)	H(kJ)	S(J)	α 10^{-4}	K 10^4	γ	V(cc)	δ_T
298	79.9	-1513.000	59.0	0.2197	162.110	1.28	28.6	14.86
300	80.1	-1512.852	59.5	0.2200	162.012	1.28	28.6	14.82
400	92.3	-1504.248	84.2	0.2271	157.031	1.12	28.6	13.20
500	102.9	-1494.462	106.0	0.2275	152.548	0.98	28.7	12.34
600	110.5	-1483.773	125.4	0.2274	148.428	0.89	28.8	11.78
700	116.0	-1472.436	142.9	0.2281	144.565	0.83	28.8	11.42
800	120.1	-1460.622	158.7	0.2297	140.868	0.79	28.9	11.24
900	123.3	-1448.443	173.0	0.2322	137.266	0.76	29.0	11.22
1000	126.0	-1435.974	186.1	0.2354	133.696	0.74	29.0	11.34
1100	128.1	-1423.267	198.2	0.2392	130.113	0.72	29.1	11.58
1200	130.0	-1410.359	209.5	0.2435	126.478	0.71	29.2	11.92
1300	131.6	-1397.276	219.9	0.2482	122.765	0.69	29.3	12.33
1400	133.1	-1384.039	229.8	0.2532	118.958	0.68	29.3	12.81
1500	134.4	-1370.661	239.0	0.2585	115.050	0.67	29.4	13.32
1600	135.7	-1357.155	247.7	0.2641	111.040	0.65	29.5	13.84
1700	136.8	-1343.529	256.0	0.2698	106.937	0.64	29.6	14.37
1800	137.9	-1329.791	263.8	0.2757	102.756	0.63	29.6	14.87
1900	139.0	-1315.945	271.3	0.2818	98.516	0.61	29.7	15.35
2000	140.0	-1301.997	278.4	0.2880	94.241	0.60	29.8	15.79
2100	140.9	-1287.951	285.3	0.2943	89.955	0.58	29.9	16.18
2200	141.9	-1273.809	291.9	0.3007	85.686	0.56	30.0	16.51
2300	142.8	-1259.574	298.2	0.3072	81.458	0.55	30.1	16.78
2400	143.7	-1245.249	304.3	0.3138	77.298	0.53	30.2	16.98
2500	144.6	-1230.834	310.2	0.3204	73.227	0.51	30.3	17.13
2600	145.4	-1216.332	315.9	0.3271	69.266	0.49	30.4	17.22
2700	146.3	-1201.745	321.4	0.3339	65.431	0.47	30.5	17.24
2800	147.1	-1187.073	326.7	0.3407	61.736	0.46	30.6	17.22
2900	148.0	-1172.317	331.9	0.3475	58.192	0.44	30.7	17.14
3000	148.8	-1157.478	336.9	0.3544	54.806	0.42	30.8	17.02

<INT VdP (J)>

PRESSURE (GPa)

T(K)	1	5	10	20	30	40	50
298	28492	140801	277775	541974	795270	1039477	1275914
300	28493	140806	277782	541981	795272	1039472	1275898
400	28555	141059	278176	542413	795517	1039377	1275354
500	28617	141320	278592	542920	795917	1039545	1275204
600	28680	141584	279018	543468	796413	1039882	1275315
700	28743	141850	279452	544044	796975	1040339	1275613
800	28806	142118	279890	544639	797582	1040876	1276038
900	28870	142388	280332	545245	798212	1041459	1276537
1000	28935	142661	280777	545853	798847	1042055	1277063
1100	29000	142936	281222	546454	799469	1042635	1277571
1200	29067	143213	281666	547040	800062	1043170	1278022
1300	29136	143492	282107	547603	800608	1043634	1278379
1400	29205	143772	282542	548136	801093	1044006	1278610
1500	29276	144053	282968	548631	801503	1044263	1278686
1600	29348	144335	283386	549082	801828	1044390	1278586
1700	29421	144616	283791	549483	802058	1044373	1278290
1800	29496	144897	284182	549830	802184	1044201	1277785
1900	29573	145177	284559	550119	802202	1043866	1277064
2000	29651	145454	284918	550347	802108	1043366	1276121
2100	29730	145730	285260	550512	801900	1042700	1274958
2200	29811	146003	285583	550614	801580	1041870	1273581
2300	29893	146273	285886	550653	801150	1040883	1271997
2400	29976	146540	286171	550631	800615	1039746	1270219
2500	30061	146803	286435	550550	799980	1038469	1268263
2600	30148	147063	286680	550413	799253	1037066	1266145
2700	30236	147319	286907	550224	798443	1035549	1263884
2800	30325	147571	287116	549989	797559	1033934	1261502
2900	30416	147819	287309	549711	796612	1032236	1259018
3000	30508	148064	287486	549397	795612	1030471	1256455

ILMENITE(Mg)

$MgSiO_3$
ILMENITE(Mg) Formula wt. (g) = 100.39

Enthalpy (J/mol) = -1489500.000 This study
Entropy (J/mol.K) = 59.700 This study
Heat Capacity:
No. of *Cp* ranges = 1 Watanabe (1982)
1.2550000D+02 5.3030D-03 -6.2820D+06 0.0000D+00
7.6650D+08 0.0000D+00 -3.0400D+03
Volume (cc/mol) = 26.350 Ito and Matsui (1977)
Thermal expn:
2.27040D-05 6.82087D-09 -1.80870D-03 3.85100D-01 Ashida et al. (1988)
Isot.compressibility:
4.50400D-07 8.29870D-11 -3.87770D-15 1.20950D-17
P deriv. 4.00000D+00 Weidner and Ito (1985)
T deriv. 3.00000D-04 This study
Melting pt. and enthalpy of melting:
\- -

NOTE: The bulk modulus and elasticity for $MgSiO_3$ ilmenite have been reported by Weidner and Ito (1985) and heat capacity at 300-600 K has been measured by Watanabe (1982). Recently Ashida et al. (1988) have measured heat capacity over a somewhat wider range (160 - 550 K) and the thermal expansion at 295-875 K. McMillan and Ross (1987) calculated the entropy and heat capacity at 0-1800 K using vibrational models. All these estimates of *Cp* are in close agreement. We evaluated the enthalpy and entropy of ilmenite from phase equilibrium data for the reaction, spinel + stishovite = ilmenite (Ito and Navrotsky, 1985) and for the ilmenite-perovskite transition by Ito and Takahashi (1989), P(GPa) = 27.5-0.0025 T(K), a slightly lower curve than previously determined (Ito and Yamada,1982). Since the data for ilmenite have to be consistent with both transformations, very little adjustment can be done to the ilmenite data. The calculated enthalpy of the pyroxene-ilmenite transition at 298 K (59.9 kJ/mol) is in good agreement with the new calorimetric value of 59.03±4.26 kJ/mol (Ashida et al., 1988) which revises the earlier preliminary calorimetric value of 71.8 kJ/mol (Ito and Navrotsky, 1985).

$MgSiO_3$
ILMENITE(MG)

Thermochemical Data

T(K)	C_p(J)	H(kJ)	S(J)	$\alpha\ 10^{-4}$	K 10^4	γ	V(cc)	δ_T
298	75.1	-1489.50	59.7	0.2300	210.474	1.72	26.4	7.68
300	75.5	-1489.36	60.2	0.2300	210.405	1.71	26.4	7.68
400	92.7	-1480.87	84.5	0.2332	206.719	1.39	26.4	7.60
500	103.1	-1471.04	106.4	0.2404	203.072	1.27	26.5	7.45
600	109.7	-1460.37	125.8	0.2485	199.438	1.22	26.5	7.33
700	114.3	-1449.16	143.1	0.2568	195.795	1.20	26.6	7.27
800	117.6	-1437.56	158.6	0.2650	192.123	1.18	26.7	7.25
900	120.2	-1425.66	172.6	0.2731	188.407	1.18	26.7	7.27
1000	122.2	-1413.54	185.3	0.2810	184.637	1.18	26.8	7.33
1100	124.0	-1401.22	197.1	0.2888	180.802	1.18	26.9	7.41
1200	125.4	-1388.75	207.9	0.2965	176.897	1.18	27.0	7.51
1300	126.7	-1376.15	218.0	0.3041	172.920	1.18	27.1	7.63
1400	127.8	-1363.42	227.4	0.3116	168.870	1.17	27.1	7.76
1500	128.9	-1350.59	236.3	0.3190	164.751	1.17	27.2	7.90
1600	129.8	-1337.65	244.6	0.3264	160.567	1.17	27.3	8.04
1700	130.7	-1324.62	252.5	0.3337	156.325	1.17	27.4	8.18
1800	131.5	-1311.51	260.0	0.3410	152.033	1.16	27.5	8.32
1900	132.3	-1298.32	267.2	0.3482	147.703	1.15	27.6	8.45
2000	133.1	-1285.04	274.0	0.3554	143.344	1.15	27.7	8.58
2100	133.8	-1271.69	280.5	0.3625	138.969	1.14	27.8	8.69
2200	134.6	-1258.27	286.7	0.3697	134.591	1.13	27.9	8.79
2300	135.3	-1244.78	292.7	0.3768	130.222	1.11	28.0	8.89
2400	135.9	-1231.22	298.5	0.3839	125.876	1.10	28.1	8.96
2500	136.6	-1217.60	304.1	0.3909	121.563	1.09	28.2	9.03
2600	137.2	-1203.91	309.4	0.3980	117.297	1.07	28.3	9.08
2700	137.9	-1190.15	314.6	0.4050	113.089	1.05	28.4	9.12
2800	138.5	-1176.33	319.6	0.4121	108.948	1.04	28.6	9.14
2900	139.1	-1162.45	324.5	0.4191	104.884	1.02	28.7	9.15
3000	139.7	-1148.51	329.2	0.4261	100.905	1.00	28.8	9.15

			<INT VdP (J)>				
			PRESSURE(GPa)				
T(K)	1	5	10	20	30	40	50
298	26287	130243	257685	505203	744231	975982	1201378
300	26288	130248	257694	505218	744250	976003	1201400
400	26348	130524	258194	506050	745297	977180	1202642
500	26409	130806	258709	506913	746396	978437	1204000
600	26473	131100	259242	507813	747553	979778	1205472
700	26539	131403	259795	508748	748764	981193	1207043
800	26607	131717	260368	509717	750022	982671	1208695
900	26677	132041	260957	510716	751322	984202	1210413
1000	26750	132375	261564	511742	752656	985775	1212182
1100	26825	132718	262187	512792	754018	987382	1213990
1200	26902	133071	262824	513862	755405	989014	1215824
1300	26981	133432	263475	514951	756810	990664	1217675
1400	27062	133802	264140	516055	758230	992325	1219533
1500	27146	134180	264817	517174	759660	993991	1221390
1600	27231	134567	265506	518304	761097	995658	1223240
1700	27319	134961	266206	519444	762538	997321	1225077
1800	27409	135364	266916	520592	763980	998976	1226897
1900	27501	135774	267637	521747	765422	1000620	1228694
2000	27595	136191	268367	522908	766861	1002251	1230468
2100	27691	136616	269106	524074	768296	1003868	1232215
2200	27790	137048	269854	525244	769726	1005469	1233936
2300	27890	137487	270610	526417	771151	1007054	1235629
2400	27993	137933	271374	527592	772569	1008624	1237295
2500	28098	138386	272145	528771	773982	1010178	1238937
2600	28205	138846	272925	529952	775390	1011718	1240556
2700	28314	139312	273712	531136	776793	1013246	1242154
2800	28425	139786	274506	532323	778193	1014764	1243736
2900	28539	140266	275307	533513	779591	1016274	1245304
3000	28655	140752	276116	534708	780989	1017781	1246864

Mg-PEROVSKITE

$MgSiO_3$
Mg-PEROVSKITE Formula wt. (g) = 100.39

Enthalpy (J/mol) = -1450000.000 This study
Entropy (J/mol.K) = 59.500 This study
Heat Capacity:
No. of *Cp* ranges = 1 This study
1.5640000D+02 8.3080D-03 1.3030D+07 0.0000D+00
-2.0400D+09 0.0000D+00 -4.4330D+04
Volume (cc/mol) = 24.500 Yagi et al. (1978)
Thermal expn:
3.36337D-05 1.13200D-08 1.08700D-05 -1.35660D+00
This study
Isot.compressibility:
3.34127D-07 1.81300D-10 -5.91600D-14 2.98600D-17
This study
P deriv. 4.00000D+00
T deriv. 7.00000D-04 This study
Melting pt. and enthalpy of melting:
- -

NOTE: Data for $MgSiO_3$ perovskite include estimates of compression by X-ray diffraction (Yagi et al., 1982; Kudoh et al., 1987) and from elasticity (Yeganeh-Haeri et al.,1989) and of thermal expansion (Knittle et al., 1986). The reactions ilmenite = perovskite and spinel = perovskite + periclase have been studied by Ito and Yamada (1982) and recently by Ito and Takahashi (1989). Both these transitions appear to have negative P-T slopes. No C_p measurements but an estimate of *Cp* from vibrational models have been reported (Fei et al., 1990). Recently Ito, Akaogi and Navrotsky (Navrotsky, pers. comm.) attempted to measure the enthalpy of the perovskite by drop solution calorimetry. They obtained a value of H(ilmenite-perovskite) in the range 40 - 60 kJ/mol; the large uncertainty reflects the presence of significant amounts of other phases in the sample. Vibrational calculations confirm that the entropy of the ilmenite-perovskite transition is positive, implying a negative dP/dT, but they suggest that S may depend quite strongly on temperature. Thermal expansion is consistent with the data of Knittle et al. (1986).

$MgSiO_3$

Mg-PEROVSKITE

Thermochemical Data

T(K)	CP(J)	H(KJ)	S(J)	A E-4	K E4	GR	V(cc)	DT
298	79.8	-1450.00	59.5	0.2178	260.611	1.76	24.5	18.42
300	80.3	-1449.85	60.0	0.2199	260.417	1.77	24.5	18.22
400	98.5	-1440.77	86.0	0.2971	250.569	1.90	24.6	12.51
500	107.7	-1430.43	109.0	0.3389	241.710	1.94	24.6	10.31
600	114.3	-1419.32	129.3	0.3668	233.613	1.93	24.7	9.08
700	119.5	-1407.62	147.3	0.3880	226.096	1.92	24.8	8.30
800	124.0	-1395.44	163.5	0.4058	219.013	1.90	24.9	7.77
900	127.9	-1382.84	178.4	0.4216	212.249	1.88	25.0	7.42
1000	131.4	-1369.87	192.0	0.4361	205.708	1.86	25.1	7.20
1100	134.5	-1356.57	204.7	0.4497	199.316	1.84	25.2	7.07
1200	137.3	-1342.98	216.5	0.4628	193.015	1.82	25.4	7.02
1300	139.9	-1329.12	227.6	0.4756	186.763	1.80	25.5	7.02
1400	142.3	-1315.01	238.1	0.4880	180.528	1.78	25.6	7.07
1500	144.5	-1300.67	248.0	0.5002	174.294	1.76	25.7	7.16
1600	146.6	-1286.12	257.4	0.5122	168.049	1.73	25.9	7.26
1700	148.5	-1271.36	266.3	0.5241	161.795	1.71	26.0	7.38
1800	150.4	-1256.42	274.9	0.5360	155.537	1.68	26.1	7.50
1900	152.2	-1241.29	283.0	0.5477	149.290	1.66	26.3	7.63
2000	153.9	-1225.98	290.9	0.5594	143.068	1.62	26.4	7.75
2100	155.5	-1210.52	298.4	0.5710	136.894	1.59	26.6	7.86
2200	157.0	-1194.89	305.7	0.5826	130.787	1.56	26.7	7.96
2300	158.5	-1179.11	312.7	0.5942	124.771	1.52	26.9	8.04
2400	160.0	-1163.19	319.5	0.6057	118.868	1.48	27.0	8.11
2500	161.4	-1147.12	326.1	0.6172	113.098	1.44	27.2	8.16
2600	162.8	-1130.91	332.4	0.6287	107.480	1.40	27.4	8.19
2700	164.1	-1114.57	338.6	0.6402	102.031	1.35	27.6	8.21
2800	165.4	-1098.09	344.6	0.6516	96.764	1.31	27.7	8.20
2900	166.7	-1081.49	350.4	0.6630	91.691	1.27	27.9	8.18
3000	167.9	-1064.76	356.1	0.6745	86.820	1.22	28.1	8.14

<INT VdP (J)>

PRESSURE(GPa)

T(K)	1	5	10	20	30	40	50
298	24453	121360	240574	473233	699095	919007	1133637
300	24454	121364	240581	473241	699100	919007	1133629
400	24516	121641	241052	473906	699757	919505	1133850
500	24593	121989	241668	474874	700895	920681	1134967
600	24678	122380	242373	476032	702338	922294	1136669
700	24770	122803	243142	477327	704003	924226	1138799
800	24867	123251	243963	478729	705841	926406	1141263
900	24968	123721	244827	480220	707819	928787	1143996
1000	25074	124211	245728	481786	709914	931332	1146945
1100	25183	124719	246664	483416	712106	934012	1150071
1200	25296	125244	247630	485102	714381	936803	1153339
1300	25413	125785	248625	486837	716724	939684	1156719
1400	25534	126341	249646	488616	719126	942637	1160187
1500	25658	126912	250691	490431	721575	945647	1163722
1600	25785	127498	251759	492280	724063	948703	1167307
1700	25916	128097	252849	494157	726584	951792	1170925
1800	26051	128710	253959	496060	729131	954907	1174567
1900	26189	129336	255088	497986	731700	958040	1178223
2000	26331	129974	256236	499933	734288	961188	1181887
2100	26476	130626	257401	501899	736892	964347	1185554
2200	26625	131290	258584	503883	739511	967515	1189225
2300	26777	131966	259783	505884	742144	970693	1192898
2400	26933	132655	260998	507903	744793	973882	1196578
2500	27093	133355	262230	509940	747459	977086	1200267
2600	27256	134068	263478	511997	750144	980307	1203971
2700	27423	134793	264743	514074	752851	983552	1207697
2800	27594	135530	266025	516173	755585	986825	1211454
2900	27769	136280	267325	518296	758349	990133	1215249
3000	27947	137042	268642	520447	761147	993482	1219091

SANIDINE

$KAlSi_3O_8$
SANIDINE Formula wt. (g) = 278.34

Enthalpy (J/mol) = -3959560.000 Robie et al. (1978)
Entropy (J/mol.K) = 239.990 Chatterjee (1989)
Heat Capacity:
No. of *Cp* ranges = 1 Hemingway et al. (1981) (298-1600 K)
3.2527000D+02 8.3700D-03 -2.1373D+07 0.0000D+00
3.4796D+09 0.0000D+00 -4.3104D+03
Volume (cc/mol) = 109.050 Robie et al. (1978)
Thermal expn:
3.17208D-05 4.06954D-09 3.04450D-03 -2.34778D+00 This study
Isot.compressibility:
1.67862D-06 4.26840D-10 -7.64000D-14 2.86430D-17 This study
P deriv. 4.00000D+00
T deriv. 0.00000D+00
Melting pt. and enthalpy of melting:
\- \-

NOTE: Calorimetric data: H: -3959.56±3.37, S: 232.9 ±0.48, Robie et al. (1978). The equilibrium relations between sanidine and muscovite have been studied by many workers. Important reactions are as follows:

$$Ms = Cor + San + H_2O \quad (a)$$

$$Ms + Q = San + Al_2SiO_5 + H_2O \quad (b)$$

Experimental work has been done on (a) by Velde (1966) and Chatterjee and Johannes (1974) and on (b) by Evans (1965), Althaus et al. (1970), Day (1973) and Chatterjee and Johannes (1974). Krupka et al. (1979) attempted to calculate the enthalpy and free energy energy of muscovite from the experimental data of Chatterjee and Johannes from the above reactions. Their calculations were based on the fugacity values of steam from Chatterjee and Johannes (1974) and Fisher and Zen (1971) and the *Cp* muscovite determined by them and the unpublished data on sanidine listed in Robie et al.(1978). Here, the more general equation for water fugacity of Saxena and Fei (1987a,b) and the published *Cp* data on sanidine by Hemingway et al. (1981) are used. The enthalpy of sanidine is from Robie et al. (1978). The thermal expansion of muscovite is also optimized from a combination of the phase equilibrium and *Cp* data. The optimized S of sanidine is found to be 7.09 J/mol.K higher than that determined by Openshaw et al. (1976). This amount can be considered as the disorder S of sanidine. Sanidine shows an Al/Si disorder and since, at very low temperatures, its *Cp* is not well characterized, this treatment is reasonable. However, it should be kept in mind that the zero-point Al/Si disorder S of muscovite is also unknown.

$KAlSi_3O_8$

SANIDINE

Thermochemical Data

T(K)	C_p(J)	H(kJ)	S(J)	α 10^{-4}	K 10^4	γ	V(cc)	δ_T
298	204.2	-3959.56	240.0	0.1673	55.560	0.50	109.0	12.91
300	204.8	-3959.18	241.3	0.1700	55.538	0.50	109.1	12.70
400	238.6	-3936.96	305.0	0.2629	54.378	0.66	109.3	7.85
500	263.2	-3911.79	361.0	0.3045	53.290	0.68	109.6	6.51
600	279.8	-3884.58	410.6	0.3272	52.263	0.68	110.0	5.85
700	291.5	-3855.98	454.6	0.3413	51.287	0.67	110.3	5.44
800	300.0	-3826.39	494.1	0.3511	50.356	0.66	110.7	5.16
900	306.4	-3796.05	529.9	0.3587	49.462	0.66	111.1	4.95
1000	311.4	-3765.15	562.4	0.3649	48.598	0.65	111.5	4.79
1100	315.5	-3733.80	592.3	0.3702	47.760	0.64	111.9	4.68
1200	318.9	-3702.07	619.9	0.3751	46.942	0.64	112.3	4.60
1300	321.8	-3670.04	645.5	0.3796	46.140	0.63	112.8	4.54
1400	324.3	-3637.73	669.5	0.3839	45.350	0.63	113.2	4.50
1500	326.5	-3605.19	691.9	0.3881	44.570	0.62	113.6	4.49
1600	328.5	-3572.44	713.1	0.3922	43.796	0.62	114.1	4.49
1700	330.3	-3539.50	733.0	0.3962	43.026	0.62	114.5	4.51
1800	331.9	-3506.39	752.0	0.4001	42.258	0.61	115.0	4.54
1900	333.5	-3473.12	769.9	0.4040	41.489	0.61	115.4	4.59
2000	334.9	-3439.70	787.1	0.4080	40.719	0.60	115.9	4.64
2100	336.3	-3406.13	803.5	0.4118	39.947	0.60	116.4	4.70
2200	337.6	-3372.43	819.1	0.4157	39.171	0.59	116.9	4.77
2300	338.9	-3338.61	834.2	0.4196	38.392	0.59	117.4	4.85
2400	340.1	-3304.66	848.6	0.4235	37.609	0.58	117.9	4.93
2500	341.3	-3270.59	862.5	0.4274	36.822	0.58	118.4	5.01
2600	342.4	-3236.40	875.9	0.4313	36.031	0.57	118.9	5.10
2700	343.5	-3202.11	888.9	0.4351	35.237	0.57	119.4	5.19
2800	344.6	-3167.70	901.4	0.4390	34.440	0.56	119.9	5.28
2900	345.7	-3133.19	913.5	0.4429	33.642	0.56	120.4	5.37
3000	346.7	-3098.57	925.2	0.4468	32.842	0.55	121.0	5.45

<INT VdP (J)>

PRESSURE(GPa)

T(K)	1	5	10	20	30	40	50
298	108096	523753	1013223	1920033	2757666	3544497	4291490
300	108099	523761	1013230	1920020	2757621	3544415	4291367
400	108321	524528	1014171	1920410	2756828	3542100	4287328
500	108612	525636	1015787	1922125	2757986	3542344	4286439
600	108938	526916	1017749	1924532	2760180	3543962	4287255
700	109284	528301	1019922	1927366	2763021	3546446	4289154
800	109645	529756	1022237	1930493	2766308	3549530	4291807
900	110016	531264	1024658	1933835	2769922	3553055	4295015
1000	110396	532814	1027161	1937341	2773784	3556913	4298642
1100	110784	534398	1029729	1940974	2777836	3561024	4302584
1200	111178	536012	1032352	1944706	2782030	3565321	4306758
1300	111579	537652	1035018	1948512	2786328	3569751	4311091
1400	111986	539315	1037721	1952373	2790696	3574263	4315520
1500	112399	540998	1040453	1956270	2795101	3578814	4319988
1600	112817	542699	1043207	1960187	2799516	3583363	4324443
1700	113241	544417	1045979	1964107	2803915	3587874	4328834
1800	113670	546149	1048762	1968016	2808272	3592309	4333117
1900	114105	547893	1051552	1971901	2812565	3596637	4337248
2000	114544	549648	1054342	1975747	2816770	3600825	4341187
2100	114989	551413	1057130	1979542	2820867	3604845	4344897
2200	115440	553186	1059909	1983275	2824837	3608670	4348342
2300	115895	554964	1062676	1986933	2828660	3612273	4351491
2400	116355	556748	1065426	1990507	2832320	3615632	4354314
2500	116820	558536	1068155	1993987	2835802	3618725	4356783
2600	117290	560325	1070859	1997362	2839089	3621532	4358874
2700	117765	562115	1073535	2000624	2842169	3624035	4360566
2800	118245	563905	1076179	2003766	2845030	3626219	4361839
2900	118730	565693	1078787	2006780	2847660	3628069	4362676
3000	119219	567478	1081357	2009658	2850052	3629575	4363065

$KAlSi_3O_8$

SANIDINE

Thermochemical Data

DIAMOND

C
DIAMOND Formula wt. (g) = 12.01

Enthalpy (J/mol) = 1895.000 Robie et al. (1978)
Entropy (J/mol.K) = 2.360 Gustafson (1986)
Heat Capacity:
No. of *Cp* ranges = 1 Gustafson (1986)
2.8008000D+01 4.2760D-04 3.1118D+05 0.0000D+00
0.0000D+00 0.0000D+00 -7.7295D+03
Volume (cc/mol) = 3.417 Gustafson (1986)
Thermal expn:
2.43000D-06 1.00000D-08 0.00000D+00 0.00000D+00
This study
Isot.compressibility:
1.70000D-07 0.00000D+00 0.00000D+00 0.00000D+00
This study
P deriv. 4.00000D+00
T deriv. 0.00000D+00
Melting pt. and enthalpy of melting:
\- \-

Thermochemical Data

T(K)	C_p(J)	H(kJ)	S(J)	α 10^{-4}	K 10^4	γ	V(cc)	δ_T
298	5.7	1.895	2.4	0.0541	588.235	1.91	3.4	0.00
300	5.8	1.906	2.4	0.0543	588.235	1.88	3.4	0.00
400	10.8	2.757	4.8	0.0643	588.235	1.20	3.4	0.00
500	14.0	4.008	7.6	0.0743	588.235	1.07	3.4	0.00
600	16.2	5.527	10.4	0.0843	588.235	1.05	3.4	0.00
700	17.9	7.238	13.0	0.0943	588.235	1.07	3.4	0.00
800	19.2	9.094	15.5	0.1043	588.235	1.11	3.4	0.00
900	20.2	11.064	17.8	0.1143	588.235	1.16	3.4	0.00
1000	21.0	13.126	20.0	0.1243	588.235	1.21	3.4	0.00
1100	21.7	15.263	22.0	0.1343	588.235	1.27	3.4	0.00
1200	22.3	17.464	23.9	0.1443	588.235	1.34	3.4	0.00
1300	22.8	19.720	25.7	0.1543	588.235	1.41	3.4	0.00
1400	23.2	22.022	27.4	0.1643	588.235	1.48	3.5	0.00
1500	23.6	24.367	29.0	0.1743	588.235	1.56	3.5	0.00
1600	24.0	26.748	30.6	0.1843	588.235	1.64	3.5	0.00
1700	24.3	29.162	32.0	0.1943	588.235	1.73	3.5	0.00
1800	24.6	31.606	33.4	0.2043	588.235	1.81	3.5	0.00
1900	24.8	34.077	34.8	0.2143	588.235	1.91	3.5	0.00
2000	25.1	36.573	36.0	0.2243	588.235	2.00	3.5	0.00
2100	25.3	39.092	37.3	0.2343	588.235	2.11	3.5	0.00
2200	25.5	41.632	38.5	0.2443	588.235	2.21	3.5	0.00
2300	25.7	44.191	39.6	0.2543	588.235	2.33	3.5	0.00
2400	25.9	46.769	40.7	0.2643	588.235	2.45	3.5	0.00
2500	26.0	49.364	41.8	0.2743	588.235	2.58	3.5	0.00
2600	26.2	51.976	42.8	0.2843	588.235	2.72	3.5	0.00
2700	26.3	54.603	43.8	0.2943	588.235	2.87	3.6	0.00
2800	26.5	57.244	44.7	0.3043	588.235	3.04	3.6	0.00
2900	26.6	59.899	45.7	0.3143	588.235	3.21	3.6	0.00
3000	26.7	62.568	46.6	0.3243	588.235	3.41	3.6	0.00

C
DIAMOND

<INT VdP (J)>

PRESSURE(GPa)

T(K)	1	5	10	20	30	40	50
298	3409	16988	33839	67151	99981	132367	164340
300	3409	16988	33839	67152	99982	132368	164342
400	3411	16998	33859	67192	100041	132447	164440
500	3413	17010	33883	67238	100111	132538	164554
600	3416	17024	33910	67292	100190	132644	164684
700	3419	17039	33940	67352	100280	132762	164831
800	3422	17056	33974	67419	100379	132894	164995
900	3426	17075	34011	67492	100489	133039	165175
1000	3430	17095	34051	67573	100609	133198	165373
1100	3434	17117	34095	67660	100739	133370	165587
1200	3439	17141	34143	67755	100880	133556	165817
1300	3444	17166	34194	67856	101030	133756	166065
1400	3450	17194	34249	67964	101192	133969	166330
1500	3456	17223	34307	68079	101363	134196	166612
1600	3462	17254	34368	68201	101545	134437	166911
1700	3468	17287	34433	68331	101737	134692	167227
1800	3475	17321	34502	68467	101940	134960	167561
1900	3483	17357	34574	68610	102154	135243	167912
2000	3490	17395	34650	68761	102378	135540	168280
2100	3498	17435	34730	68919	102613	135851	168667
2200	3507	17477	34813	69084	102859	136177	169071
2300	3515	17521	34900	69256	103116	136517	169493
2400	3525	17566	34990	69436	103383	136871	169933
2500	3534	17614	35085	69624	103662	137240	170391
2600	3544	17663	35183	69818	103952	137624	170868
2700	3554	17714	35285	70021	104253	138023	171363
2800	3565	17767	35391	70230	104566	138437	171876
2900	3576	17822	35500	70448	104890	138865	172409
3000	3587	17879	35614	70673	105225	139309	172960

GRAPHITE

C
GRAPHITE Formula wt. (g) = 12.01

Enthalpy (J/mol) = 0.000 Robie et al. (1978)
Entropy (J/mol.K) = 5.742 Gustafson (1986)
Heat Capacity:
No. of *Cp* ranges = 1 Gustafson (1986)
2.7797000D+01 4.5830D-04 4.4755D+05 0.0000D+00
0.0000D+00 0.0000D+00 -7.3924D+03
Volume (cc/mol) = 5.298 Gustafson (1986)
Thermal expn. :
2.32000D-05 5.70000D-09 0.00000D+00 0.00000D+00
Isot.compressibility:
1.80000D-06 0.00000D+00 0.00000D+00 0.00000D+00
P deriv. 4.00000D+00
T deriv. 0.00000D+00
Melting pt. and enthalpy of melting:
- -

Thermochemical Data

T(K)	C_p(J)	H(kJ)	S(J)	α 10^{-4}	K 10^4	γ	V(cc)	δ_T
298	8.2	0.000	5.7	0.2490	55.556	0.90	5.3	0.00
300	8.3	0.015	5.8	0.2491	55.556	0.89	5.3	0.00
400	12.3	1.057	8.8	0.2548	55.556	0.62	5.3	0.00
500	15.0	2.432	11.8	0.2605	55.556	0.52	5.3	0.00
600	17.0	4.038	14.7	0.2662	55.556	0.47	5.3	0.00
700	18.5	5.815	17.5	0.2719	55.556	0.44	5.4	0.00
800	19.6	7.721	20.0	0.2776	55.556	0.43	5.4	0.00
900	20.5	9.732	22.4	0.2833	55.556	0.42	5.4	0.00
1000	21.3	11.826	24.6	0.2890	55.556	0.41	5.4	0.00
1100	22.0	13.990	26.7	0.2947	55.556	0.41	5.4	0.00
1200	22.5	16.213	28.6	0.3004	55.556	0.41	5.4	0.00
1300	23.0	18.487	30.4	0.3061	55.556	0.41	5.4	0.00
1400	23.4	20.805	32.1	0.3118	55.556	0.41	5.5	0.00
1500	23.8	23.162	33.8	0.3175	55.556	0.42	5.5	0.00
1600	24.1	25.555	35.3	0.3232	55.556	0.42	5.5	0.00
1700	24.4	27.978	36.8	0.3289	55.556	0.42	5.5	0.00
1800	24.7	30.430	38.2	0.3346	55.556	0.43	5.5	0.00
1900	24.9	32.908	39.5	0.3403	55.556	0.43	5.6	0.00
2000	25.1	35.410	40.8	0.3460	55.556	0.44	5.6	0.00
2100	25.3	37.933	42.0	0.3517	55.556	0.45	5.6	0.00
2200	25.5	40.477	43.2	0.3574	55.556	0.45	5.6	0.00
2300	25.7	43.041	44.3	0.3631	55.556	0.46	5.6	0.00
2400	25.9	45.621	45.4	0.3688	55.556	0.47	5.7	0.00
2500	26.1	48.219	46.5	0.3745	55.556	0.47	5.7	0.00
2600	26.2	50.833	47.5	0.3802	55.556	0.48	5.7	0.00
2700	26.4	53.461	48.5	0.3859	55.556	0.49	5.7	0.00
2800	26.5	56.104	49.5	0.3916	55.556	0.50	5.7	0.00
2900	26.6	58.760	50.4	0.3973	55.556	0.51	5.8	0.00
3000	26.8	61.430	51.3	0.4030	55.556	0.51	5.8	0.00

C
GRAPHITE

<INT VdP (J)>

	PRESSURE (GPa)						
T(K)	1	5	10	20	30	40	50
298	5252	25526	49661	95259	138388	179727	219659
300	5252	25527	49664	95264	138395	179735	219669
400	5266	25592	49789	95504	138744	180188	220223
500	5279	25658	49918	95751	139102	180653	220791
600	5293	25725	50049	96003	139469	181130	221374
700	5307	25795	50184	96262	139844	181618	221970
800	5322	25866	50322	96527	140229	182117	222581
900	5337	25938	50463	96798	140623	182629	223206
1000	5352	26013	50608	97075	141026	183152	223845
1100	5368	26089	50756	97359	141438	183687	224500
1200	5384	26166	50907	97649	141860	184235	225169
1300	5400	26246	51062	97945	142290	184794	225852
1400	5417	26327	51220	98249	142731	185366	226551
1500	5434	26410	51381	98558	143181	185950	227265
1600	5452	26495	51546	98874	143640	186547	227994
1700	5469	26581	51714	99197	144109	187156	228739
1800	5487	26670	51886	99527	144588	187778	229499
1900	5506	26760	52062	99863	145077	188413	230275
2000	5525	26852	52241	100207	145575	189060	231066
2100	5544	26946	52423	100557	146084	189721	231874
2200	5564	27041	52609	100914	146603	190395	232698
2300	5584	27139	52799	101278	147132	191082	233537
2400	5605	27238	52993	101649	147671	191783	234394
2500	5625	27340	53190	102028	148221	192497	235266
2600	5647	27443	53391	102414	148782	193224	236156
2700	5668	27549	53596	102807	149353	193966	237062
2800	5690	27656	53805	103207	149934	194722	237985
2900	5713	27765	54018	103615	150527	195491	238926
3000	5736	27877	54234	104031	151131	196275	239884

ANNITE

$KFe_3AlSi_3O_{10}(OH)_2$
ANNITE Formula wt. (g) = 511.89

Enthalpy (J/mol) = -5125500.000 Chatterjee (1989)
Entropy (J/mol.K) = 440.910 Chatterjee (1989)
Heat Capacity:
No. of *Cp* ranges = 1 Chatterjee (1989), refitted
6.2650000D+02 6.2630D-02 1.3276D+07 0.0000D+00
-7.9470D+08 0.0000D+00 -1.1670D+05
Volume (cc/mol) = 154.320 Robie and Bethke (1963)
Thermal expn:
1.00110D-04 4.82000D-09 -8.64784D-03 -4.92619D+00
This study
Isot.compressibility:
1.38972D-06 3.47480D-10 7.20870D-14 2.53600D-18
This study
P deriv. 4.00000D+00 This study
T deriv. 0.00000D+00
Melting pt. and enthalpy of melting:
- -

NOTE: Reversed distribution data of Fe and Mg between biotite and garnet are available from Ferry and Spear (1978) and Perchuk and Lavrent'eva (1983) These data are used to calculate the free energy and volume data of annite and the solution properties of the biotite solution. The relevant reaction is as follows:

Alm + Phl = Pyr + Ann (a)

The values for annite and the solution parameters are calculated assuming -ΔCp of reaction (a) as zero and from the following relation:-$\Delta G^o(P,T)$ + RT $\ln K_d$ + $RT\ln K_\gamma$= 0, the only unknown quantities in the term -$\Delta G^o(P,T)$ being the enthalpy, entropy and volume of annite at P and T and where K_d is from Ferry and Spear or Perchuk and Lavrent'eva. When the data of Ferry and Spear and Perchuk and Lavrent'eva are used together, convergence is not achieved. It is concluded that the two sets of data are inconsistent. When only the data of Ferry and Spear are used, the fit obtained is not satisfactory. Also,the solution parameters obtained suggest an extremely asymmetric solution. This could be due to the fact that the data are only in the Fe-rich end of the binary. When only the data of Perchuk and Lavrent'eva are used, a much better fit is obtained. Thus, their data are used in the calculations. An uncertainty of 700 J is found in the enthalpy of annite.

$KFe_3AlSi_3O_{10}(OH)_2$
ANNITE

Thermochemical Data

T(K)	C_p(J)	H(kJ)	S(J)	α 10^{-4}	K 10^4	γ	V(cc)	δ_T
298	373.1	-5125.50	440.9	0.1713	66.676	0.47	154.3	15.23
300	374.4	-5124.81	443.2	0.1799	66.644	0.50	154.3	14.50
400	430.4	-5084.41	559.1	0.4963	64.918	1.19	154.9	5.32
500	471.2	-5039.24	659.7	0.6552	63.219	1.43	155.8	4.07
600	502.8	-4990.48	748.5	0.7491	61.550	1.54	156.9	3.59
700	528.4	-4938.88	828.0	0.8108	59.911	1.58	158.1	3.34
800	549.9	-4884.93	900.0	0.8546	58.304	1.60	159.4	3.19
900	568.5	-4828.99	965.9	0.8876	56.731	1.61	160.8	3.09
1000	584.9	-4771.31	1026.6	0.9136	55.193	1.60	162.3	3.02
1100	599.7	-4712.06	1083.1	0.9348	53.690	1.60	163.8	2.96
1200	613.2	-4651.41	1135.8	0.9527	52.222	1.58	165.4	2.91
1300	625.6	-4589.46	1185.4	0.9681	50.791	1.57	167.0	2.87
1400	637.3	-4526.31	1232.2	0.9817	49.396	1.56	168.6	2.84
1500	648.3	-4462.03	1276.6	0.9939	48.038	1.54	170.3	2.81
1600	658.8	-4396.67	1318.7	1.0049	46.715	1.53	172.0	2.78
1700	668.8	-4330.29	1359.0	1.0151	45.429	1.51	173.7	2.75
1800	678.4	-4262.93	1397.5	1.0246	44.179	1.49	175.5	2.72
1900	687.6	-4194.63	1434.4	1.0335	42.963	1.48	177.3	2.70
2000	696.6	-4125.41	1469.9	1.0419	41.783	1.46	179.2	2.67
2100	705.4	-4055.31	1504.1	1.0500	40.637	1.44	181.0	2.65
2200	713.9	-3984.34	1537.1	1.0577	39.524	1.43	183.0	2.62
2300	722.3	-3912.53	1569.0	1.0650	38.445	1.41	184.9	2.60
2400	730.4	-3839.90	1600.0	1.0722	37.398	1.39	186.9	2.57
2500	738.5	-3766.45	1629.9	1.0791	36.382	1.38	188.9	2.55
2600	746.4	-3692.21	1659.0	1.0859	35.398	1.36	191.0	2.52
2700	754.2	-3617.18	1687.4	1.0925	34.443	1.35	193.1	2.50
2800	761.8	-3541.38	1714.9	1.0989	33.518	1.33	195.2	2.47
2900	769.4	-3464.82	1741.8	1.1052	32.621	1.31	197.4	2.45
3000	776.9	-3387.50	1768.0	1.1114	31.752	1.30	199.6	2.42

<INT VdP (J)>

	PRESSURE (GPa)						
T(K)	1	5	10	20	30	40	50
298	153190	745740	1448972	2762414	3984076	5136863	6234882
300	153195	745753	1448981	2762390	3983998	5136720	6234668
400	153716	747814	1452115	2766027	3986898	5138170	6234215
500	154585	751548	1458468	2775734	3998497	5150792	6247283
600	155647	756204	1466577	2788714	4014757	5169373	6267540
700	156836	761447	1475792	2803724	4033882	5191594	6292172
800	158114	767104	1485774	2820119	4054940	5216256	6319723
900	159463	773076	1496330	2837529	4077399	5242673	6349359
1000	160868	779301	1507341	2855725	4100930	5270422	6380570
1100	162323	785739	1518727	2874558	4125319	5299228	6413024
1200	163822	792364	1530437	2893929	4150423	5328908	6446498
1300	165361	799156	1542432	2913766	4176141	5359332	6480838
1400	166937	806102	1554686	2934019	4202400	5390409	6515934
1500	168549	813191	1567180	2954650	4229149	5422074	6551707
1600	170194	820416	1579898	2975633	4256349	5454279	6588102
1700	171873	827771	1592829	2996947	4283974	5486989	6625078
1800	173584	835254	1605967	3018578	4312002	5520179	6662606
1900	175327	842860	1619304	3040515	4340419	5553832	6700665
2000	177101	850587	1632836	3062750	4369215	5587936	6739242
2100	178907	858434	1646561	3085278	4398383	5622484	6778329
2200	180744	866400	1660475	3108095	4427919	5657470	6817922
2300	182612	874484	1674578	3131199	4457822	5692894	6858020
2400	184511	882686	1688869	3154590	4488091	5728756	6898624
2500	186442	891007	1703349	3178268	4518727	5765059	6939738
2600	188404	899446	1718016	3202235	4549734	5801807	6981369
2700	190399	908003	1732873	3226492	4581115	5839006	7023523
2800	192425	916681	1747921	3251043	4612874	5876662	7066208
2900	194484	925479	1763160	3275890	4645018	5914782	7109434
3000	196575	934398	1778594	3301037	4677550	5953374	7153211

PHLOGOPITE

$KMg_3AlSi_3O_{10}(OH)_2$
PHLOGOPITE Formula wt. (g) = 417.27

Enthalpy (J/mol) =	-6203900.000	Clemens et al. (1987), Chatterjee (1989)
Entropy (J/mol.K) =	334.600	Robie and Hemingway (1984a)
Heat Capacity:		
No. of *Cp* ranges = 1		Robie and Hemingway (1984a), refitted.

5.5442000D+02 2.7835D-02 -1.6458D+06 0.0000D+00
-1.0629D+08 0.0000D+00 -5.5447D+04

Volume (cc/mol) =	149.660	Robie et al. (1978)

Thermal expn:
3.13800D-05 1.15800D-08 7.03000D-03 -3.44600D+00
This study

Isot.compressibility:
1.51284D-06 4.94680D-10 -1.46670D-13 5.97850D-17
This study

P deriv. 4.00000D+00
T deriv. 0.00000D+00
Melting pt. and enthalpy of melting:
\- -

NOTE: Calorimetric data: H: -6203.9 ±6.73, S: 334.6 ±1.0; H, Clemens et al. (1987) (derived indirectly); S, Robie and Hemingway (1984a). Enthalpy of phlogopite was determined by Clemens et al. (1987) by solution calorimetry. Their value of -6215.33 KJ/mol was derived by considering a calorimetric cycle for the reaction:

Per + Br + San = Phl

and using the free energies of periclase, brucite and sanidine from Robie et al. (1978) and the heat capacity of phlogopite from Robie and Hemingway (1984a). The same procedure is repeated here. A different value for the enthalpy of phlogopite is obtained for the following reason. Saxena (1989) and Fei and Chatterjee (1988) using the experimental data on the reaction

Br = Per + H_2O

from Irving et al. (1977), which are up to a temperature of 1400 K, obtained an optimized heat capacity expression for brucite. The heat capacities of brucite obtained from this expression differ from the original data of King et al. (1975) (used by Clemens et al.) by 0.8 and 1.35 J/mol at 500 and 700 K. The entropy of phlogopite is from Robie and Hemingway (1984a). This includes a configurational entropy of 18.7 J/mol.K assuming total Al-Si disorder.

$KMg_3AlSi_3O_{10}(OH)_2$

PHLOGOPITE

Thermochemical Data

T(K)	C_p(J)	H(kJ)	S(J)	$\alpha\ 10^{-4}$	K 10^4	γ	V(cc)	δ_T
298	354.2	-6203.90	334.6	0.1965	60.647	0.50	149.7	13.06
300	355.7	-6203.24	336.8	0.2000	60.619	0.51	149.7	12.82
400	415.0	-6164.41	448.1	0.3205	59.134	0.69	150.1	7.49
500	450.0	-6121.03	544.8	0.3745	57.771	0.73	150.6	6.06
600	473.6	-6074.78	629.0	0.4047	56.505	0.74	151.2	5.35
700	491.0	-6026.51	703.4	0.4250	55.317	0.74	151.8	4.91
800	504.6	-5976.70	769.9	0.4405	54.191	0.74	152.5	4.61
900	515.7	-5925.67	830.0	0.4536	53.111	0.74	153.2	4.40
1000	525.1	-5873.62	884.8	0.4654	52.066	0.73	153.9	4.26
1100	533.2	-5820.70	935.2	0.4766	51.044	0.73	154.6	4.17
1200	540.4	-5767.01	982.0	0.4874	50.036	0.73	155.3	4.12
1300	546.9	-5712.64	1025.5	0.4980	49.034	0.73	156.1	4.10
1400	552.9	-5657.64	1066.2	0.5086	48.031	0.73	156.9	4.11
1500	558.4	-5602.07	1104.6	0.5191	47.023	0.73	157.7	4.15
1600	563.6	-5545.97	1140.8	0.5296	46.004	0.73	158.5	4.21
1700	568.5	-5489.35	1175.1	0.5401	44.971	0.73	159.4	4.28
1800	573.2	-5432.27	1207.7	0.5507	43.923	0.72	160.2	4.37
1900	577.7	-5374.72	1238.8	0.5613	42.857	0.72	161.1	4.47
2000	581.9	-5316.74	1268.6	0.5719	41.775	0.72	162.1	4.57
2100	586.1	-5258.34	1297.1	0.5826	40.675	0.72	163.0	4.67
2200	590.1	-5199.53	1324.4	0.5934	39.559	0.71	164.0	4.78
2300	594.0	-5140.32	1350.7	0.6042	38.430	0.71	164.9	4.89
2400	597.8	-5080.73	1376.1	0.6150	37.289	0.70	166.0	4.99
2500	601.6	-5020.76	1400.6	0.6259	36.140	0.70	167.0	5.09
2600	605.2	-4960.42	1424.2	0.6368	34.986	0.69	168.0	5.19
2700	608.8	-4899.72	1447.2	0.6478	33.830	0.68	169.1	5.28
2800	612.3	-4838.66	1469.4	0.6588	32.675	0.67	170.2	5.35
2900	615.8	-4777.25	1490.9	0.6698	31.526	0.66	171.4	5.42
3000	619.3	-4715.50	1511.8	0.6808	30.385	0.65	172.5	5.48

<INT VdP (J)>

PRESSURE (GPa)

T(K)	1	5	10	20	30	40	50
298	148458	721002	1397815	2656663	3823364	4921690	5966042
300	148463	721017	1397828	2656646	3823297	4921565	5965852
400	148834	722351	1399577	2657772	3822749	4918763	5960444
500	149328	724300	1402552	2661332	3825815	4920731	5960934
600	149885	726566	1406168	2666189	3830836	4925305	5964672
700	150482	729032	1410190	2671874	3837113	4931563	5970518
800	151108	731643	1414503	2678150	3844284	4939018	5977864
900	151758	734372	1419045	2684877	3852128	4947366	5986321
1000	152430	737199	1423776	2691959	3860490	4956391	5995610
1100	153122	740115	1428666	2699325	3869248	4965919	6005507
1200	153833	743112	1433693	2706915	3878301	4975806	6015821
1300	154563	746182	1438839	2714678	3887560	4985922	6026382
1400	155311	749322	1444087	2722566	3896942	4996150	6037036
1500	156079	752525	1449421	2730536	3906374	5006380	6047639
1600	156864	755789	1454828	2738547	3915784	5016511	6058059
1700	157668	759109	1460294	2746561	3925106	5026450	6068174
1800	158491	762482	1465805	2754542	3934278	5036107	6077870
1900	159332	765903	1471350	2762455	3943244	5045403	6087043
2000	160192	769369	1476917	2770268	3951950	5054264	6095598
2100	161071	772876	1482494	2777952	3960347	5062623	6103452
2200	161968	776422	1488070	2785478	3968392	5070420	6110531
2300	162884	780003	1493636	2792824	3976046	5077604	6116771
2400	163820	783615	1499181	2799965	3983276	5084133	6122120
2500	164774	787257	1504697	2806883	3990053	5089970	6126537
2600	165747	790924	1510177	2813561	3996355	5095088	6129992
2700	166740	794614	1515612	2819984	4002164	5099468	6132465
2800	167752	798325	1520997	2826143	4007467	5103099	6133947
2900	168783	802055	1526326	2832029	4012259	5105977	6134440
3000	169834	805800	1531594	2837637	4016537	5108108	6133954

HYDROUS CORDIERITE

$Mg_2Al_4Si_5O_{18}.H_2O$
HYDROUS CORDIERITE Formula wt. (g) = 602.98

Enthalpy (J/mol) =	-9430318.600	Bhattacharya and Sen (1985)
Entropy (J/mol.K) =	275.904	Bhattacharya and Sen (1985); Chatterjee (1989)

Heat Capacity:
No. of *Cp* ranges = 1 — Same as in Mg-cordierite

7.6548000D+02 -1.9181D-02 5.7814D+06 0.0000D+00
-5.4662D+08 0.0000D+00 -8.8644D+04

Volume (cc/mol) = 223.980 — Bhattacharya and Sen (1985)

Thermal expn:
1.52780D-04 0.00000D+00 0.00000D+00 0.00000D+00
Isot.compressibility:
1.00000D-06 0.00000D+00 0.00000D+00 0.00000D+00
P deriv. 0.00000D+00
T deriv. 0.00000D+00
Melting pt. and enthalpy of melting:
\- -

NOTE: For the reaction Cord + H_2O = Hydrous-cord, Bhattacharya and Sen (1985) give the following relations in the range 773-1073 K: $-\Delta G^o(1,T) = -271018.6 + 134.976T$ J/mol $-\Delta V^o(1,T) = -3.7656 \times 10^{-3} \times T - 2.1514$ J/bar.mol. Using these relations, the enthalpy, entropy, molar volume and thermal expansion of hydrous cordierite are calculated. The heat capacity is assumed to be the same as that of cordierite.

Thermochemical Data

T(K)	CP(J)	H(kJ)	S(J)
298	506.9	-9430.32	275.9
300	508.2	-9429.38	279.0
400	563.8	-9375.52	433.6
500	597.4	-9317.33	563.3
600	619.8	-9256.41	674.3
700	635.6	-9193.60	771.1
800	647.3	-9129.42	856.8
900	656.1	-9064.23	933.6
1000	662.9	-8998.27	1003.1
1100	668.2	-8931.71	1066.5
1200	672.3	-8864.67	1124.8
1300	675.5	-8797.28	1178.8
1400	678.1	-8729.59	1228.9
1500	680.0	-8661.68	1275.8
1600	681.5	-8593.60	1319.7
1700	682.6	-8525.39	1361.1
1800	683.4	-8457.09	1400.1
1900	683.9	-8388.72	1437.1
2000	684.2	-8320.32	1472.2
2100	684.2	-8251.90	1505.5
2200	684.1	-8183.48	1537.4
2300	683.9	-8115.08	1567.8
2400	683.5	-8046.71	1596.9
2500	683.0	-7978.38	1624.8
2600	682.3	-7910.12	1651.5
2700	681.6	-7841.92	1677.3
2800	680.8	-7773.80	1702.0

BRUCITE

$Mg(OH)_2$
BRUCITE Formula wt. (g) = 58.32

Enthalpy (J/mol) =	-925500.000	Holland and Powell (1990)
Entropy (J/mol.K) =	64.400	Holland and Powell (1990)
Heat Capacity:		
No. of *Cp* ranges = 1		King et al. (1967), refitted
1.0301000D+02 1.6533D-02 -2.7462D+06 0.0000D+00		
-1.3650D+02 0.0000D+00 -1.1410D+02		
Volume (cc/mol) =	24.630	Robie et al. (1978)
Thermal expn:		
8.75400D-05 1.80900D-08 -2.21800D-02 5.84300D-01		Saxena (1989)
Isot.compressibility:		
1.75100D-06 -5.48900D-12 2.04450D-13 1.105D-16		Saxena (1989)
P deriv. 4.71200D+00		Saxena (1989)
T deriv. 1.00000D-03		This study
Melting pt. and enthalpy of melting:		
- -		

NOTE: Calorimetric data: H: -924.5 ±1.4, S: 63.18±0.13, Robie et al. (1978). The equilibrium data on the following reaction were determined by Barnes and Ernst (1963), Irving et al. (1977) and Schramke et al. (1982):

$$Br = Per + H_2O \qquad (a)$$

Experimental data on this reaction are used to determine the free energy of brucite. The heat capacity of brucite IS available up to a temperature of 700 K (King et al.,1975). The equilibrium data of Irving et al. are up to a temperature of approximately 1400 K. Thus, it is necessary to extrapolate the heat capacity using the Fei-Saxena formulation. Enthalpy, entropy, the expansion and bulk modulus of brucite are optimized from the experimental data.

$Mg(OH)_2$
BRUCITE

Thermochemical Data

T(K)	C_p(J)	H(kJ)	S(J)	α 10^{-4}	K 10^4	γ	V(cc)	δ_T
298	76.7	-925.50	64.4	0.2511	56.482	0.46	24.6	3.28
300	77.1	-925.36	64.9	0.2553	56.474	0.46	24.6	3.25
400	92.2	-916.80	89.4	0.4298	55.910	0.65	24.7	2.75
500	100.1	-907.15	110.9	0.5456	55.152	0.76	24.8	2.85
600	105.1	-896.88	129.6	0.6305	54.195	0.84	25.0	3.09
700	108.8	-886.17	146.1	0.6971	53.044	0.89	25.2	3.37
800	111.8	-875.14	160.8	0.7520	51.705	0.93	25.3	3.67
900	114.4	-863.83	174.1	0.7990	50.195	0.96	25.5	3.96
1000	116.7	-852.27	186.3	0.8403	48.533	0.97	25.7	4.24
1100	118.8	-840.50	197.5	0.8776	46.742	0.98	26.0	4.50
1200	120.8	-828.51	208.0	0.9117	44.848	0.98	26.2	4.73
1300	122.8	-816.33	217.7	0.9434	42.879	0.98	26.4	4.94
1400	124.7	-803.96	226.9	0.9732	40.862	0.96	26.7	5.11
1500	126.5	-791.40	235.6	1.0015	38.824	0.95	27.0	5.25
1600	128.3	-778.66	243.8	1.0285	36.789	0.93	27.2	5.36
1700	130.1	-765.73	251.6	1.0545	34.778	0.90	27.5	5.43
1800	131.9	-752.64	259.1	1.0796	32.809	0.87	27.8	5.48
1900	133.6	-739.36	266.3	1.1040	30.897	0.85	28.1	5.51
2000	135.3	-725.92	273.2	1.1278	29.054	0.82	28.4	5.51
2100	137.1	-712.30	279.8	1.1510	27.289	0.78	28.8	5.49
2200	138.8	-698.51	286.2	1.1738	25.608	0.75	29.1	5.45
2300	140.5	-684.54	292.4	1.1961	24.013	0.72	29.4	5.40
2400	142.2	-670.41	298.4	1.2182	22.507	0.69	29.8	5.33
2500	143.9	-656.11	304.3	1.2399	21.090	0.66	30.2	5.25
2600	145.5	-641.64	310.0	1.2613	19.759	0.63	30.6	5.17
2700	147.2	-627.00	315.5	1.2825	18.513	0.60	30.9	5.07
2800	148.9	-612.20	320.9	1.3035	17.348	0.58	31.3	4.98
2900	150.6	-597.22	326.1	1.3242	16.262	0.55	31.8	4.87
3000	152.3	-582.08	331.3	1.3448	15.248	0.53	32.2	4.77

<INT VdP (J)>

PRESSURE(GPa)

T(K)	1	5	10	20	30	40	50
298	24418	118440	229515	436510	629150	811324	985313
300	24420	118445	229524	436524	629169	811347	985339
400	24503	118821	230211	437747	630876	813513	987953
500	24621	119357	231205	439581	633523	816976	992242
600	24763	120000	232399	441799	636751	821232	997548
700	24924	120723	233735	444275	640357	825990	1003489
800	25100	121507	235175	446929	644212	831070	1009827
900	25288	122340	236696	449710	648233	836355	1016404
1000	25489	123215	238281	452583	652368	841766	1023120
1100	25700	124127	239919	455528	656579	847256	1029909
1200	25920	125070	241603	458528	660846	852793	1036733
1300	26150	126042	243326	461576	665157	858363	1043571
1400	26389	127041	245087	464668	669509	863961	1050418
1500	26636	128066	246885	467805	673902	869590	1057280
1600	26891	129117	248719	470989	678344	875260	1064168
1700	27154	130192	250590	474224	682842	880983	1071101
1800	27426	131293	252500	477518	687408	886777	1078101
1900	27705	132419	254451	480877	692054	892660	1085190
2000	27993	133572	256447	484310	696794	898649	1092394
2100	28288	134753	258491	487824	701642	904765	1099738
2200	28592	135962	260586	491428	706610	911026	1107247
2300	28903	137201	262736	495130	711714	917451	1114943
2400	29223	138473	264945	498940	716964	924058	1122850
2500	29552	139777	267215	502864	722375	930863	1130990
2600	29889	141116	269552	506911	727957	937883	1139382
2700	30235	142492	271959	511088	733722	945133	1148046
2800	30590	143906	274439	515402	739680	952627	1156999
2900	30954	145360	276996	519860	745842	960377	1166259
3000	31328	146856	279633	524469	752217	968398	1175841

MUSCOVITE

$KAl_2(AlSi_3O_{10})(OH)_2$
MUSCOVITE Formula wt. (g) = 398.32

Enthalpy (J/mol) = -5976300.000 Chatterjee (1989)
Entropy (J/mol.K) = 306.400 Robie et al. (1978)
Heat Capacity:
No. of *Cp* ranges = 1 Krupka et al. (1979), refitted
4.6260000D+02 6.5980D-02 -3.3374D+07 0.0000D+00
4.7388D+09 0.0000D+00 1.2027D+04
Volume (cc/mol) = 140.820 Robie et al. (1978)
Thermal expn:
4.19392D-05 1.56675D-08 -3.47200D-03 -9.06530D-01
This study
Isot.compressibility:
1.15461D-06 1.19917D-10 1.07134D-13 7.17490D-18
This study
P deriv. 4.00000D+00
T deriv. 0.00000D+00
Melting pt. and enthalpy of melting:
- -

NOTE: Calorimetric data: H: -5976.74±3.24, S: 306.4 ±0.61, Robie et al. (1978) The equilibrium relations between sanidine and muscovite have been studied by many workers. Important reactions are as follows:

Ms = Cor + San + H2O (a)

Ms + Q = San + Al_2SiO_5 + H_2O (b)

Experimental work has been done on (a) by Velde (1966) and Chatterjee and Johannes (1974) and on (b) by Evans (1965), Althaus et al. (1970), Day (1973) and Chatterjee and Johannes (1974).

Krupka et al. (1979) attempted to calculate the enthalpy and free energy energy of muscovite from the experimental data of Chatterjee and Johannes in the above reactions. Their calculations were based on the fugacity values of steam from Chatterjee and Johannes (1974) and Fisher and Zen (1971) and the *Cp* muscovite determined by them and the unpublished data on sanidine listed in Robie et al. (1978). Here, the more general equation for water fugacity of Saxena and Fei (1987a,b) and the published *Cp* data on sanidine by Hemingway et al.(1981) are used. The enthalpy of sanidine is from Robie et al. (1978). The thermal expansion of muscovite is also optimized from a combination of the phase equilibrium and *Cp* data. The optimized S of sanidine is found to be 7.09 J/mol.K higher than that determined by Openshaw et al. (1976). This amount can be considered as the disorder S of sanidine. Sanidine shows an Al/Si disorder and since, at very low temperatures, its *Cp* is not well characterized, this treatment is reasonable. However, it should be kept in mind that the zero-point Al/Si disorder S of muscovite is also unknown.

$KAl_2(AlSi_3O_{10})(OH)_2$
MUSCOVITE

Thermochemical Data

T(K)	C_p(J)	H(kJ)	S(J)	α 10^{-4}	K 10^4	γ	V(cc)	δ_T
298	326.0	-5976.30	306.4	0.2477	83.328	0.90	140.8	6.25
300	327.2	-5975.70	308.4	0.2499	83.304	0.90	140.8	6.20
400	384.5	-5939.96	410.8	0.3386	81.955	1.03	141.2	5.06
500	424.1	-5899.40	501.2	0.3920	80.499	1.08	141.8	4.77
600	451.5	-5855.55	581.0	0.4303	78.947	1.10	142.4	4.70
700	471.7	-5809.35	652.2	0.4610	77.309	1.12	143.0	4.70
800	487.5	-5761.36	716.3	0.4872	75.598	1.13	143.7	4.73
900	500.6	-5711.93	774.5	0.5106	73.826	1.14	144.4	4.77
1000	512.0	-5661.29	827.8	0.5323	72.003	1.15	145.1	4.81
1100	522.1	-5609.58	877.1	0.5527	70.141	1.16	145.9	4.84
1200	531.4	-5556.90	922.9	0.5722	68.251	1.16	146.8	4.87
1300	540.0	-5503.33	965.8	0.5910	66.343	1.17	147.6	4.88
1400	548.3	-5448.91	1006.1	0.6093	64.426	1.17	148.5	4.89
1500	556.2	-5393.68	1044.2	0.6272	62.510	1.17	149.4	4.88
1600	563.8	-5337.68	1080.4	0.6448	60.601	1.17	150.4	4.87
1700	571.3	-5280.93	1114.8	0.6622	58.708	1.17	151.4	4.84
1800	578.6	-5223.44	1147.6	0.6793	56.837	1.16	152.4	4.81
1900	585.7	-5165.22	1179.1	0.6963	54.993	1.16	153.4	4.78
2000	592.8	-5106.29	1209.3	0.7131	53.181	1.15	154.5	4.73
2100	599.8	-5046.66	1238.4	0.7298	51.405	1.14	155.6	4.68
2200	606.8	-4986.33	1266.5	0.7464	49.668	1.14	156.8	4.63
2300	613.7	-4925.31	1293.6	0.7629	47.974	1.13	158.0	4.57
2400	620.5	-4863.60	1319.9	0.7794	46.324	1.12	159.2	4.51
2500	627.3	-4801.21	1345.4	0.7957	44.721	1.11	160.5	4.44
2600	634.1	-4738.13	1370.1	0.8121	43.164	1.10	161.8	4.37
2700	640.9	-4674.39	1394.1	0.8283	41.656	1.09	163.1	4.30
2800	647.6	-4609.96	1417.6	0.8445	40.196	1.08	164.5	4.23
2900	654.3	-4544.87	1440.4	0.8607	38.784	1.07	165.9	4.16
3000	661.0	-4479.10	1462.7	0.8768	37.421	1.06	167.3	4.08

	<INT VdP (J)> PRESSURE(GPa)						
T(K)	1	5	10	20	30	40	50
298	139991	684824	1336914	2566206	3719055	4812980	5859213
300	139997	684850	1336958	2566269	3719124	4813047	5859272
400	140403	686609	1339956	2570799	3724406	4818598	5864764
500	140904	688808	1343766	2576747	3731606	4826495	5872980
600	141469	691292	1348077	2583522	3739863	4835626	5882569
700	142083	693986	1352749	2590847	3748782	4845480	5892913
800	142739	696853	1357703	2598576	3758151	4855789	5903687
900	143431	699869	1362895	2606625	3767850	4866399	5914708
1000	144157	703021	1368298	2614940	3777805	4877218	5925871
1100	144916	706299	1373893	2623487	3787969	4888191	5937112
1200	145706	709697	1379667	2632244	3798315	4899286	5948398
1300	146526	713212	1385612	2641198	3808826	4910488	5959716
1400	147376	716839	1391724	2650339	3819494	4921790	5971062
1500	148256	720578	1397997	2659663	3830317	4933194	5982446
1600	149167	724428	1404430	2669169	3841296	4944709	5993884
1700	150106	728387	1411023	2678857	3852437	4956346	6005396
1800	151076	732456	1417774	2688730	3863748	4968123	6017007
1900	152076	736635	1424685	2698792	3875241	4980059	6028746
2000	153107	740925	1431756	2709048	3886928	4992173	6040645
2100	154168	745326	1438990	2719505	3898821	5004489	6052734
2200	155260	749839	1446388	2730170	3910936	5017031	6065050
2300	156383	754466	1453954	2741051	3923290	5029824	6077626
2400	157539	759208	1461689	2752157	3935898	5042895	6090500
2500	158727	764066	1469598	2763497	3948778	5056268	6103706
2600	159947	769043	1477684	2775080	3961949	5069973	6117282
2700	161201	774139	1485951	2786917	3975427	5084035	6131265
2800	162489	779357	1494402	2799019	3989232	5098483	6145690
2900	163811	784699	1503043	2811395	4003383	5113343	6160595
3000	165168	790166	1511877	2824059	4017899	5128643	6176014

ANTIGORITE

$Mg_{48}Si_{34}O_{85}(OH)_{62}$
ANTIGORITE Formula wt. (g) = 4536.06

Enthalpy (J/mol) = -71377000.000 Chatterjee (1989)
Entropy (J/mol.K) = 3672.800 Chatterjee (1989)
Heat Capacity:
No. of *Cp* ranges = 1 King et al. (1967)
7.7014000D+03 2.0910D-01 -3.8256D+07 0.0000D+00
1.1440D+10 0.0000D+00 -9.9281D+05
Volume (cc/mol) = 1749.130 Kunze (1961)
Thermal expn:
3.12430D-05 2.12040D-08 0.00000D+00 4.50930D-01
Isot.compressibility:
3.93200D-06 0.00000D+00 0.00000D+00 0.00000D+00
P deriv. 3.69510D+00
T deriv. 0.00000D+00
Melting pt. and enthalpy of melting:
\- -

NOTE: Calorimetric data: S: 3598.53±26.78, King et al. (1967). The stability of antigorite is defined by the following reaction:

$$Ant = 18Fo + 4Tc + 27H_2O$$

The equilibria for this reaction were experimentally determined by Evans et al. (1976). These data are used to optimize the free energy of antigorite. The heat capacity of antigorite was measured up to a temperature of 850 K by King et al. (1967). The optimized parameters for antigorite include enthalpy, entropy, thermal expansion and bulk modulus.

$Mg_{48}Si_{34}O_{85}(OH)_{62}$
ANTIGORITE

Thermochemical Data

T(K)	C_p(J)	H(kJ)	S(J)	α 10^{-4}	K 10^4	γ	V(cc)	δ_T
298	4435.1	-71377.000	3672.8	0.4264	25.432	0.43	1749.1	0.00
300	4453.4	-71368.778	3700.3	0.4261	25.432	0.43	1749.3	0.00
400	5242.7	-70881.008	5098.1	0.4254	25.432	0.36	1756.7	0.00
500	5758.8	-70329.255	6327.1	0.4365	25.432	0.34	1764.3	0.00
600	6118.9	-69734.386	7410.7	0.4522	25.432	0.34	1772.1	0.00
700	6384.7	-69108.590	8374.9	0.4701	25.432	0.34	1780.3	0.00
800	6590.2	-68459.435	9241.4	0.4891	25.432	0.34	1788.9	0.00
900	6754.9	-67791.895	10027.4	0.5088	25.432	0.35	1797.8	0.00
1000	6890.9	-67109.402	10746.4	0.5290	25.432	0.36	1807.2	0.00
1100	7005.8	-66414.417	11408.7	0.5494	25.432	0.37	1816.9	0.00
1200	7105.0	-65708.759	12022.6	0.5700	25.432	0.38	1827.1	0.00
1300	7192.1	-64993.814	12594.9	0.5908	25.432	0.40	1837.8	0.00
1400	7269.6	-64270.656	13130.7	0.6116	25.432	0.41	1848.9	0.00
1500	7339.6	-63540.139	13634.7	0.6325	25.432	0.42	1860.4	0.00
1600	7403.3	-62802.949	14110.5	0.6535	25.432	0.44	1872.4	0.00
1700	7462.0	-62059.648	14561.1	0.6745	25.432	0.46	1884.9	0.00
1800	7516.4	-61310.699	14989.1	0.6955	25.432	0.47	1897.8	0.00
1900	7567.2	-60556.491	15396.9	0.7166	25.432	0.49	1911.3	0.00
2000	7615.1	-59797.354	15786.3	0.7376	25.432	0.51	1925.2	0.00
2100	7660.3	-59033.565	16158.9	0.7587	25.432	0.53	1939.7	0.00
2200	7703.3	-58265.367	16516.3	0.7798	25.432	0.55	1954.7	0.00
2300	7744.4	-57492.968	16859.6	0.8010	25.432	0.57	1970.2	0.00
2400	7783.8	-56716.547	17190.1	0.8221	25.432	0.60	1986.2	0.00
2500	7821.6	-55936.266	17508.6	0.8433	25.432	0.62	2002.8	0.00
2600	7858.2	-55152.264	17816.1	0.8644	25.432	0.65	2020.0	0.00
2700	7893.6	-54364.665	18113.3	0.8856	25.432	0.68	2037.8	0.00
2800	7927.9	-53573.579	18401.0	0.9067	25.432	0.71	2056.1	0.00
2900	7961.4	-52779.107	18679.8	0.9279	25.432	0.74	2075.1	0.00
3000	7993.9	-51981.335	18950.2	0.9491	25.432	0.77	2094.6	0.00

<INT VdP (J)>

PRESSURE(GPa)

T(K)	1	5	10	20	30	40	50
298	1716686	8072106	15222747	27897481	39175820	49506243	59128111
300	1716821	8072742	15223947	27899681	39178909	49510147	59132774
400	1724119	8107061	15288668	28018289	39345468	49720626	59384161
500	1731556	8142029	15354611	28139138	39515173	49935082	59640298
600	1739263	8178267	15422950	28264377	39691044	50157328	59905739
700	1747299	8216055	15494213	28394974	39874439	50389083	60182537
800	1755698	8255547	15568688	28531459	40066101	50631285	60471813
900	1764479	8296839	15646558	28674166	40266501	50884530	60774277
1000	1773659	8340001	15727956	28823336	40475978	51149245	61090441
1100	1783247	8385089	15812986	28979163	40694802	51425771	61420713
1200	1793256	8432152	15901738	29141812	40923207	51714404	61765444
1300	1803694	8481232	15994295	29311434	41161403	52015412	62124954
1400	1814570	8532371	16090736	29488174	41409595	52329051	62499551
1500	1825892	8585611	16191138	29672173	41667980	52655570	62889532
1600	1837670	8640992	16295578	29863571	41936756	52995221	63295196
1700	1849912	8698555	16404133	30062511	42216124	53348256	63716845
1800	1862627	8758342	16516883	30269138	42506286	53714931	64154787
1900	1875824	8820397	16633908	30483601	42807451	54095513	64609337
2000	1889512	8884763	16755293	30706053	43119836	54490272	65080820
2100	1903703	8951487	16881124	30936653	43443663	54899489	65569571
2200	1918404	9020616	17011490	31175564	43779160	55323455	66075938
2300	1933627	9092198	17146484	31422956	44126567	55762471	66600279
2400	1949383	9166285	17286201	31679004	44486129	56216847	67142966
2500	1965683	9242930	17430740	31943889	44858102	56686908	67704386
2600	1982539	9322186	17580206	32217802	45242752	57172987	68284939
2700	1999962	9404111	17734704	32500938	45640353	57675434	68885039
2800	2017965	9488764	17894345	32793500	46051191	58194607	69505117
2900	2036561	9576204	18059245	33095698	46475562	58730882	70145620
3000	2055763	9666497	18229522	33407752	46913772	59284646	70807012

PARAGONITE

$NaAl_2(AlSi_3O_{10})(OH)_2$

PARAGONITE Formula wt. (g) = 382.21

Enthalpy (J/mol) =	-5951000.000	Chatterjee (1989)
Entropy (J/mol.K) =	277.100	Robie and Hemingway (1984a)
Heat Capacity:		
No. of *Cp* ranges = 1		Chatterjee (1989), refitted
5.0220900D+02 3.2215D-02 -1.9623D+07 0.0000D+00		
2.6177D+09 0.0000D+00 -2.1204D+04		
Volume (cc/mol) =	131.880	Flux and Chatterjee (1986)
Thermal expn:		
2.89858D-05 1.13394D-08 -3.97920D-04 -1.39857D+00		This study
Isot.compressibility:		
1.16204D-06 1.44442D-10 4.86890D-14 1.56880D-17		This study
P deriv. 4.07000D+00		This study
T deriv. 0.00000D+00		
Melting pt. and enthalpy of melting:		
- -		

NOTE: Calorimetric data: H: -5949.33±4.6, S: 277.1 ±0.9, Robie and Hemingway (1984a) (derived from ordered model).

Reactions involving jadeite, albite and paragonite are as follows:

$$Pg = Ab + Cor + H_2O \qquad (a)$$

$$Pg + Q = Ab + Al_2SiO_5 + H_2O \qquad (b)$$

$$Pg = Jd + Ky + Q \qquad (c)$$

Experimental work has been done on reaction (a) by Chatterjee (1970), on reaction (b) by Chatterjee (1972) and on reaction (c) by Holland (1979). Using the entropy of paragonite from Robie and Hemingway (1984a), the enthalpy and heat capacity of paragonite are optimized. The thermal expansion and bulk modulus are also optimized simultaneously. The resulting value of the bulk modulus was not different from that of muscovite. The values listed above perfectly reproduce the reactions (a), (b) (for sillimanite and andalusite) and (c).

$NaAl_2(AlSi_3O_{10})(OH)_2$

PARAGONITE

Thermochemical Data

T(K)	C_p(J)	H(kJ)	S(J)	α 10^{-4}	K 10^4	γ	V(cc)	δ_T
298	318.7	-5951.00	277.1	0.1530	82.655	0.52	131.9	9.60
300	320.1	-5950.41	279.1	0.1552	82.632	0.53	131.9	9.47
400	380.3	-5915.15	380.1	0.2379	81.393	0.68	132.1	6.53
500	418.4	-5875.08	469.4	0.2827	80.103	0.72	132.5	5.81
600	443.8	-5831.89	548.0	0.3124	78.764	0.75	132.9	5.54
700	462.1	-5786.55	617.9	0.3350	77.376	0.76	133.3	5.44
800	475.9	-5739.63	680.5	0.3537	75.942	0.77	133.8	5.42
900	487.0	-5691.46	737.2	0.3702	74.465	0.78	134.3	5.43
1000	496.2	-5642.29	789.0	0.3853	72.947	0.79	134.8	5.47
1100	504.1	-5592.26	836.7	0.3994	71.392	0.79	135.3	5.51
1200	511.1	-5541.49	880.9	0.4129	69.804	0.80	135.9	5.56
1300	517.4	-5490.07	922.0	0.4259	68.187	0.80	136.4	5.61
1400	523.1	-5438.04	960.6	0.4386	66.545	0.80	137.0	5.66
1500	528.4	-5385.46	996.9	0.4511	64.884	0.80	137.6	5.70
1600	533.5	-5332.36	1031.1	0.4633	63.209	0.80	138.3	5.74
1700	538.2	-5278.77	1063.6	0.4754	61.524	0.80	138.9	5.77
1800	542.8	-5224.72	1094.5	0.4874	59.834	0.80	139.6	5.80
1900	547.2	-5170.22	1124.0	0.4993	58.145	0.80	140.3	5.81
2000	551.5	-5115.28	1152.2	0.5112	56.459	0.80	141.0	5.83
2100	555.6	-5059.93	1179.2	0.5229	54.783	0.79	141.7	5.83
2200	559.6	-5004.17	1205.1	0.5346	53.120	0.79	142.5	5.83
2300	563.6	-4948.01	1230.1	0.5463	51.475	0.79	143.2	5.82
2400	567.5	-4891.45	1254.1	0.5579	49.850	0.78	144.0	5.80
2500	571.3	-4834.51	1277.4	0.5695	48.249	0.77	144.8	5.78
2600	575.1	-4777.19	1299.9	0.5811	46.675	0.77	145.7	5.75
2700	578.8	-4719.50	1321.6	0.5926	45.131	0.76	146.5	5.72
2800	582.5	-4661.44	1342.7	0.6042	43.619	0.75	147.4	5.68
2900	586.1	-4603.01	1363.3	0.6157	42.141	0.74	148.3	5.63
3000	589.7	-4544.22	1383.2	0.6272	40.698	0.74	149.2	5.58

<INT VdP (J)>

T(K)	PRESSURE(GPa) 1	5	10	20	30	40	50
298	131098	641235	1251713	2402590	3482157	4506888	5487303
300	131101	641248	1251733	2402610	3482166	4506880	5487273
400	131354	642289	1253388	2404732	3484143	4508340	5487972
500	131686	643703	1255746	2408131	3487898	4512034	5491324
600	132066	645334	1258499	2412214	3492571	4516845	5495969
700	132480	647117	1261516	2416722	3497783	4522277	5501300
800	132922	649015	1264728	2421520	3503332	4528071	5506997
900	133387	651010	1268095	2426528	3509103	4534072	5512872
1000	133874	653091	1271591	2431696	3515019	4540180	5518803
1100	134381	655248	1275201	2436987	3521028	4546327	5524704
1200	134908	657476	1278912	2442378	3527091	4552462	5530516
1300	135453	659773	1282716	2447848	3533180	4558549	5536192
1400	136016	662134	1286604	2453382	3539273	4564559	5541699
1500	136597	664558	1290573	2458970	3545352	4570469	5547013
1600	137196	667042	1294617	2464600	3551403	4576264	5552113
1700	137813	669586	1298732	2470265	3557417	4581932	5556988
1800	138448	672188	1302916	2475960	3563384	4587465	5561630
1900	139101	674848	1307167	2481678	3569301	4592856	5566036
2000	139771	677564	1311481	2487417	3575163	4598105	5570205
2100	140459	680336	1315857	2493173	3580968	4603212	5574141
2200	141166	683164	1320294	2498946	3586717	4608180	5577853
2300	141890	686047	1324790	2504734	3592411	4613014	5581348
2400	142633	688985	1329346	2510538	3598053	4617721	5584641
2500	143394	691977	1333960	2516357	3603647	4622311	5587745
2600	144174	695025	1338632	2522195	3609200	4626794	5590677
2700	144973	698127	1343363	2528052	3614717	4631183	5593457
2800	145790	701285	1348152	2533933	3620206	4635491	5596104
2900	146627	704497	1353000	2539841	3625676	4639734	5598640
3000	147483	707764	1357908	2545779	3631137	4643926	5601088

CUMMINGTONITE

$Mg_7Si_8O_{22}(OH)_2$

CUMMINGTONITE Formula wt. (g) = 780.84

Enthalpy (J/mol) = -12217100.000 Chatterjee (1989)
Entropy (J/mol.K) = 483.056 Chatterjee (1989)
Heat Capacity:
No. of *Cp* ranges = 1 Chatterjee (1989)
6.7146100D+02 4.9935D-02 -2.6698D+07 0.0000D+00
-2.4368D+09 0.0000D+00 4.0199D+04
Volume (cc/mol) = 269.682 Miyano and Klein (1983)

Thermal expn:
- - - -

Isot.compressibility:
- - - -

Melting pt. and enthalpy of melting:
- - .

NOTE: Calculated from the experimental data on Mg-Fe exchange between olivine and (Mg,Fe)-grunerite reaction from Miyano and Klein (1983):

Cumm + Fay = Grun + For

They give the following equilibrium relation:

$$LnK_d = -2328.2/T(K) + 1.27$$

Thermochemical Data

T(K)	CP(J)	H(kJ)	S(J)
298	428.9	-12217.10	483.1
300	433.5	-12216.30	485.7
400	587.0	-12164.01	635.1
500	650.5	-12101.74	773.8
600	683.0	-12034.91	895.6
700	702.3	-11965.57	1002.4
800	715.2	-11894.67	1097.1
900	724.8	-11822.65	1181.9
1000	732.5	-11749.78	1258.7
1100	739.0	-11676.19	1328.8
1200	744.9	-11601.99	1393.3
1300	750.4	-11527.22	1453.2
1400	755.6	-11451.92	1509.0
1500	760.6	-11376.11	1561.3
1600	765.5	-11299.81	1610.5
1700	770.3	-11223.02	1657.1
1800	775.0	-11145.76	1701.2
1900	779.7	-11068.02	1743.3
2000	784.5	-10989.81	1783.4
2100	789.1	-10911.13	1821.8
2200	793.8	-10831.98	1858.6
2300	798.5	-10752.36	1894.0
2400	803.2	-10672.27	1928.1
2500	808.0	-10591.71	1961.0
2600	812.7	-10510.68	1992.7
2700	817.4	-10429.18	2023.5
2800	822.1	-10347.20	2053.3
2900	826.9	-10264.76	2082.2
3000	831.6	-10181.83	2110.3

GRUNERITE

$Fe_7Si_8O_{22}(OH)_2$
GRUNERITE Formula wt. (g) = 1001.63

Enthalpy (J/mol) =	-963150.000	Chatterjee (1989)
Entropy (J/mol.K) =	714.600	Chatterjee (1989)
Heat Capacity:		
No. of *Cp* ranges = 1		Miyano and Klein (1983), refitted
1.1377000D+03 2.4812D-01 1.5574D+08 0.0000D+00		
-2.3990D+10 0.0000D+00 -4.0674D+05		
Volume (cc/mol) =	278.730	Klein and Waldbaum (1967)
Thermal expn:		
1.06700D-04 2.48000D-08 -3.06700D-02 1.65580D+00		This study
Isot.compressibility:		
1.44700D-06 5.04200D-10 -2.50500D-13 1.23500D-16		This study
P deriv. 4.00000D+00		This study
T deriv. 0.00000D+00		
Melting pt. and enthalpy of melting:		
- -		

NOTE: The grunerite breakdown reaction was studied by Mel'nik and Radchuk (1977) and Forbes (1977). Their results agree at the WM buffer. The reaction is as follows:

$$\text{Grun} = \text{Fay} + \text{Q} + H_2O$$

Experimental data on the above reaction are used to optimize the free energy of grunerite. Miyano and Klein (1983) determined the free energy of grunerite by the same method. However, since they used different free energy values of fayalite and quartz than the values used in this study, the procedure had to be repeated. Their heat capacity equation is accepted in this work while calculating the enthalpy and entropy of grunerite. The molar volume of grunerite is from Klein and Waldbaum (1967).

$Fe_7Si_8O_{22}(OH)_2$

GRUNERITE

Thermochemical Data

T(K)	C_p(J)	H(kJ)	S(J)	α 10^{-4}	K 10^4	γ	V(cc)	δ_T
298	694.3	-963.15	714.6	0.2985	63.358	0.76	278.7	8.23
300	698.3	-961.86	718.9	0.3030	63.329	0.77	278.7	8.09
400	818.6	-884.95	939.4	0.5029	61.862	1.09	279.9	4.47
500	879.3	-799.89	1128.9	0.6438	60.536	1.30	281.5	3.26
600	930.2	-709.38	1293.8	0.7506	59.311	1.45	283.5	2.66
700	978.2	-613.94	1440.8	0.8362	58.155	1.55	285.7	2.33
800	1024.3	-513.80	1574.5	0.9079	57.036	1.63	288.3	2.14
900	1068.4	-409.15	1697.7	0.9699	55.931	1.70	291.0	2.04
1000	1110.8	-300.17	1812.5	1.0249	54.819	1.75	293.9	2.00
1100	1151.6	-187.04	1920.3	1.0747	53.680	1.81	297.0	2.01
1200	1190.8	-69.91	2022.2	1.1205	52.501	1.85	300.3	2.05
1300	1228.6	51.07	2119.0	1.1633	51.270	1.90	303.7	2.11
1400	1265.3	175.77	2211.4	1.2036	49.980	1.94	307.3	2.20
1500	1300.8	304.08	2299.9	1.2419	48.627	1.98	311.1	2.30
1600	1335.5	435.91	2384.9	1.2786	47.208	2.01	315.1	2.40
1700	1369.3	571.15	2466.9	1.3139	45.726	2.04	319.2	2.52
1800	1402.3	709.73	2546.1	1.3481	44.185	2.06	323.5	2.63
1900	1434.7	851.59	2622.8	1.3814	42.594	2.08	327.9	2.74
2000	1466.5	996.65	2697.2	1.4138	40.960	2.09	332.5	2.85
2100	1497.8	1144.87	2769.5	1.4455	39.295	2.09	337.3	2.95
2200	1528.6	1296.20	2839.9	1.4766	37.610	2.09	342.3	3.04
2300	1559.0	1450.58	2908.5	1.5072	35.918	2.07	347.4	3.13
2400	1589.0	1607.98	2975.5	1.5373	34.229	2.05	352.7	3.20
2500	1618.7	1768.37	3041.0	1.5670	32.557	2.02	358.3	3.26
2600	1648.0	1931.71	3105.1	1.5963	30.910	1.99	364.0	3.30
2700	1677.1	2097.97	3167.8	1.6253	29.299	1.95	369.9	3.34
2800	1705.9	2267.13	3229.3	1.6540	27.732	1.90	376.0	3.36
2900	1734.5	2439.15	3289.7	1.6824	26.216	1.85	382.3	3.38
3000	1762.9	2614.02	3349.0	1.7106	24.755	1.79	388.9	3.38

			<INT VdP (J)> PRESSURE(GPa)				
T(K)	1	5	10	20	30	40	50
298	276586	1344755	2609791	4967270	7155802	9218320	11181031
300	276600	1344810	2609870	4967347	7155839	9218296	11180933
400	277685	1349289	2617117	4977311	7166384	9228183	11189363
500	279245	1356113	2628996	4996319	7190201	9255416	11219079
600	281159	1364681	2644302	5021986	7223760	9295351	11264353
700	283357	1374634	2662308	5052837	7264848	9345054	11321545
800	285796	1385741	2682533	5087881	7311958	9402502	11388128
900	288446	1397844	2704644	5126402	7363981	9466198	11462215
1000	291289	1410832	2728392	5167852	7420057	9534964	11542311
1100	294309	1424621	2753585	5211791	7479483	9607828	11627179
1200	297499	1439146	2780067	5257853	7541664	9683964	11715758
1300	300851	1454356	2807710	5305722	7606084	9762650	11807121
1400	304361	1470208	2836403	5355122	7672285	9843245	11900442
1500	308024	1486668	2866049	5405806	7739860	9925176	11994985
1600	311840	1503706	2896563	5457551	7808443	10007931	12090094
1700	315807	1521297	2927868	5510161	7877704	10091055	12185187
1800	319924	1539417	2959895	5563455	7947355	10174147	12279760
1900	324191	1558048	2992583	5617276	8017141	10256863	12373380
2000	328610	1577173	3025876	5671485	8086844	10338913	12465692
2100	333181	1596776	3059726	5725964	8156285	10420064	12556415
2200	337905	1616845	3094090	5780614	8225321	10500138	12645344
2300	342784	1637369	3128932	5835357	8293846	10579011	12732347
2400	347821	1658338	3164222	5890134	8361789	10656615	12817363
2500	353017	1679747	3199937	5944908	8429117	10732931	12900397
2600	358375	1701589	3236061	5999658	8495828	10807989	12981518
2700	363898	1723862	3272583	6054386	8561954	10881862	13060848
2800	369589	1746563	3309499	6109107	8627554	10954663	13138562
2900	375451	1769695	3346812	6163857	8692713	11026542	13214876
3000	381488	1793259	3384531	6218683	8757542	11097675	13290041

ANTHOPHYLLITE

$Mg_7Si_8O_{22}(OH)_2$
ANTHOPHYLLITE Formula wt. (g) = 780.84

Enthalpy (J/mol) = -12071032.000 Chatterjee (1989)
Entropy (J/mol.K) = 535.190 Chatterjee (1989)
Heat Capacity:
No. of *Cp* ranges = 1 Krupka et al. (1985b), refitted
1.1287000D+03 3.8860D-02 5.0430D+06 0.0000D+00
-7.6420D+08 0.0000D+00 -1.5100D+05
Volume (cc/mol) = 265.440 Chernosky et al. (1985)
Thermal expn:
6.44229D-05 9.94090D-09 4.71500D-03 -4.97167D+00
This study
Isot.compressibility:
1.86582D-06 4.38030D-10 2.30450D-14 3.27547D-17
This study
P deriv. 4.00000D+00
T deriv. 0.00000D+00
Melting pt. and enthalpy of melting:
\- -

NOTE: Calorimetric data: H: -12086.0 ±7.5, Weeks (1955); S: 537.0 ±2.7, Krupka et al. (1985a). The stability limits of anthophyllite and talc were recently studied in detail by Chernosky et al. (1985). The following reactions were studied:

Anth = 7En + Q + H_2O
Anth + Fo = 9En + H_2O
Tc + 4En = Anth
9Tc + 4Fo = 5Anth + $4H_2O$
7Tc = 3Anth + 4Q + $4H_2O$

Experimental data on these five reactions from many sources are used to optimize the thermodynamic properties of anthophyllite. Heat capacity data on anthophyllite in the temperature range 300-700 K from Krupka et al. (1985a,b) are also included in the optimization of thermal expansion and bulk modulus.

$Mg_7Si_8O_{22}(OH)_2$

ANTHOPHYLLITE

Thermochemical Data

T(K)	C_p(J)	H(kJ)	S(J)	α 10^{-4}	K 10^4	γ	V(cc)	δ_T
298	661.7	-12071.03	535.2	0.2727	50.017	0.55	265.4	8.45
300	664.8	-12069.80	539.3	0.2788	49.995	0.56	265.5	8.26
400	786.3	-11996.67	748.9	0.4911	48.856	0.83	266.5	4.70
500	860.2	-11914.08	932.8	0.5894	47.740	0.90	268.0	3.93
600	910.8	-11825.39	1094.4	0.6444	46.642	0.92	269.6	3.63
700	948.3	-11732.36	1237.7	0.6797	45.559	0.93	271.4	3.47
800	977.4	-11636.02	1366.3	0.7050	44.489	0.92	273.3	3.39
900	1001.1	-11537.05	1482.9	0.7247	43.429	0.92	275.3	3.35
1000	1020.8	-11435.93	1589.4	0.7411	42.379	0.91	277.3	3.33
1100	1037.8	-11332.98	1687.5	0.7554	41.337	0.90	279.4	3.33
1200	1052.6	-11228.45	1778.5	0.7683	40.302	0.90	281.5	3.33
1300	1065.7	-11122.52	1863.2	0.7803	39.275	0.89	283.7	3.34
1400	1077.5	-11015.35	1942.7	0.7917	38.254	0.88	286.0	3.36
1500	1088.3	-10907.05	2017.4	0.8027	37.240	0.88	288.2	3.38
1600	1098.3	-10797.71	2087.9	0.8133	36.234	0.87	290.6	3.40
1700	1107.5	-10687.42	2154.8	0.8238	35.236	0.86	293.0	3.42
1800	1116.2	-10576.23	2218.3	0.8340	34.247	0.85	295.4	3.45
1900	1124.3	-10464.20	2278.9	0.8441	33.268	0.84	297.9	3.47
2000	1132.1	-10351.37	2336.8	0.8542	32.299	0.84	300.4	3.49
2100	1139.5	-10237.79	2392.2	0.8642	31.342	0.83	303.0	3.51
2200	1146.5	-10123.49	2445.4	0.8741	30.397	0.82	305.7	3.53
2300	1153.3	-10008.49	2496.5	0.8840	29.466	0.81	308.4	3.55
2400	1159.9	-9892.83	2545.7	0.8938	28.550	0.80	311.1	3.56
2500	1166.2	-9776.53	2593.2	0.9037	27.649	0.79	313.9	3.57
2600	1172.4	-9659.60	2639.0	0.9135	26.765	0.78	316.8	3.58
2700	1178.3	-9542.06	2683.4	0.9233	25.899	0.77	319.7	3.59
2800	1184.2	-9423.93	2726.4	0.9331	25.050	0.76	322.7	3.59
2900	1189.9	-9305.23	2768.0	0.9429	24.220	0.75	325.8	3.59
3000	1195.5	-9185.96	2808.5	0.9526	23.410	0.74	328.9	3.59

				<INT VdP (J)>			
				PRESSURE (GPa)			
T(K)	1	5	10	20	30	40	50
298	262870	1269817	2449846	4625421	6626956	8502235	10279269
300	262883	1269861	2449902	4625457	6626941	8502152	10279109
400	263879	1273786	2455960	4633094	6634259	8508114	10283072
500	265263	1279554	2465538	4647312	6650961	8526083	10301526
600	266848	1286265	2476888	4664787	6672266	8549906	10327018
700	268559	1293544	2489285	4684136	6696175	8577007	10356420
800	270359	1301213	2502375	4704676	6721702	8606110	10388179
900	272230	1309175	2515966	4726033	6748297	8636500	10421423
1000	274159	1317372	2529942	4747977	6775625	8667738	10455615
1100	276142	1325770	2544231	4770361	6803463	8699534	10490396
1200	278173	1334345	2558781	4793079	6831657	8731683	10525515
1300	280250	1343081	2573558	4816057	6860092	8764031	10560783
1400	282372	1351966	2588535	4839236	6888679	8796464	10596056
1500	284538	1360993	2603691	4862570	6917348	8828887	10631220
1600	286747	1370153	2619008	4886023	6946044	8861228	10666185
1700	288999	1379443	2634475	4909563	6974721	8893426	10700878
1800	291294	1388858	2650079	4933167	7003341	8925433	10735239
1900	293632	1398395	2665810	4956811	7031873	8957209	10769223
2000	296013	1408050	2681662	4980480	7060293	8988724	10802794
2100	298437	1417822	2697626	5004158	7088581	9019956	10835928
2200	300906	1427709	2713697	5027835	7116722	9050888	10868608
2300	303418	1437708	2729870	5051501	7144707	9081510	10900824
2400	305975	1447820	2746141	5075151	7172531	9111820	10932577
2500	308577	1458042	2762508	5098780	7200190	9141817	10963873
2600	311225	1468375	2778968	5122388	7227686	9171509	10994723
2700	313919	1478817	2795519	5145973	7255026	9200907	11025146
2800	316659	1489368	2812161	5169539	7282217	9230027	11055165
2900	319447	1500029	2828895	5193091	7309270	9258889	11084810
3000	322282	1510799	2845720	5216633	7336199	9287515	11114114

TALC

$Mg_3Si_4O_{10}(OH)_2$
TALC Formula wt. (g) = 379.27

Enthalpy (J/mol) = -5900500.000 Chatterjee (1989)
Entropy (J/mol.K) = 260.180 Chatterjee (1989)
Heat Capacity:
No. of *Cp* ranges = 1 Krupka et al. (1985b), refitted
5.1658900D+02 3.3502D-02 -1.4491D+07 0.0000D+00
2.0616D+09 0.0000D+00 -3.5782D+04
Volume (cc/mol) = 136.100 Robie et al. (1978)
Thermal expn:
4.80310D-05 9.45330D-09 -7.67500D-03 -2.94600D-01
This study
Isot.compressibility:
1.55552D-06 -8.43720D-11 2.20309D-13 -1.37570D-17
This study
P deriv. 3.27290D+00 Chatterjee (1989)
T deriv. 0.00000D+00
Melting pt. and enthalpy of melting:
\- -

NOTE: Calorimetric data: H: -5915.9 ±4.33, Robie et al. (1978); S: 260.79±0.4, Robie and Stout (1963). The stability limit of talc was recently studied in detail by Chernosky et al. (1985). The following reactions studied involve talc exclusively:

$$Tc = 3En + Q + H_2O \qquad (a)$$

$$Tc + Fo = 5En + H_2O \qquad (b)$$

At first, experimental data on these two reactions from many sources are used to optimize the thermodynamic data on talc. It was found that the phase equilibrium data on the two reactions are not consistent with each other. There is no way to adjust thermodynamic data on talc to fit the experimental data on both reactions simultaneously. Thus, the higher temperature reaction (a) is chosen to optimize the properties of talc. The reason behind this choice is that this reaction is less influenced by kinetic factors. Thus values presented here do not reproduce the experimental data of Kitihara et al. (1966) on reaction (b). The heat capacity of talc is in the temperature range 300-700 K and is from Krupka et al. (1985a,b). These measured data are also included in the optimization of thermal expansion and bulk modulus.

$Mg_3Si_4O_{10}(OH)_2$

TALC

Thermochemical Data

T (K)	C_p (J)	H (kJ)	S (J)	α 10^{-4}	K 10^4	γ	V (cc)	δ_T
298	321.3	-5900.50	260.2	0.2179	64.533	0.60	136.1	1.28
300	322.7	-5899.90	262.2	0.2201	64.530	0.60	136.1	1.29
400	382.2	-5864.43	363.8	0.3078	64.262	0.71	136.5	1.78
500	420.3	-5824.18	453.5	0.3623	63.829	0.76	136.9	2.21
600	446.3	-5780.77	532.6	0.4009	63.242	0.80	137.5	2.60
700	465.4	-5735.14	602.9	0.4308	62.512	0.82	138.0	2.96
800	480.0	-5687.84	666.0	0.4554	61.653	0.84	138.6	3.27
900	491.9	-5639.22	723.2	0.4765	60.679	0.85	139.3	3.55
1000	501.9	-5589.52	775.6	0.4951	59.605	0.86	140.0	3.79
1100	510.5	-5538.89	823.8	0.5121	58.446	0.86	140.7	4.00
1200	518.1	-5487.45	868.6	0.5277	57.217	0.87	141.4	4.17
1300	525.0	-5435.29	910.3	0.5424	55.930	0.87	142.2	4.32
1400	531.3	-5382.48	949.5	0.5563	54.601	0.87	143.0	4.43
1500	537.2	-5329.05	986.3	0.5696	53.242	0.87	143.8	4.52
1600	542.7	-5275.06	1021.2	0.5824	51.863	0.87	144.6	4.59
1700	547.9	-5220.53	1054.2	0.5948	50.475	0.87	145.4	4.63
1800	552.9	-5165.48	1085.7	0.6069	49.086	0.86	146.3	4.65
1900	557.7	-5109.95	1115.7	0.6187	47.706	0.86	147.2	4.66
2000	562.3	-5053.95	1144.4	0.6303	46.340	0.85	148.1	4.65
2100	566.8	-4997.49	1172.0	0.6416	44.994	0.85	149.1	4.62
2200	571.2	-4940.59	1198.5	0.6528	43.674	0.84	150.1	4.58
2300	575.5	-4883.25	1224.0	0.6638	42.382	0.83	151.0	4.54
2400	579.7	-4825.49	1248.5	0.6747	41.121	0.82	152.1	4.48
2500	583.8	-4767.31	1272.3	0.6855	39.895	0.82	153.1	4.42
2600	587.9	-4708.72	1295.3	0.6961	38.705	0.81	154.2	4.35
2700	591.9	-4649.73	1317.5	0.7067	37.552	0.80	155.2	4.27
2800	595.9	-4590.34	1339.1	0.7172	36.436	0.80	156.4	4.20
2900	599.8	-4530.56	1360.1	0.7276	35.359	0.79	157.5	4.11
3000	603.6	-4470.39	1380.5	0.7380	34.320	0.78	158.6	4.03

<INT VdP (J)>

T (K)	PRESSURE (GPa) 1	5	10	20	30	40	50
298	135068	656649	1273333	2416850	3469741	4453293	5380883
300	135073	656675	1273381	2416936	3469859	4453440	5381055
400	135431	658347	1276489	2422462	3477401	4462723	5391874
500	135881	660420	1280291	2429069	3486247	4473429	5404162
600	136393	662746	1284501	2436219	3495631	4484582	5416746
700	136950	665251	1288975	2443646	3505178	4495708	5429063
800	137543	667893	1293634	2451202	3514686	4506558	5440822
900	138169	670644	1298428	2458805	3524043	4516997	5451870
1000	138822	673490	1303327	2466402	3533186	4526958	5462136
1100	139501	676418	1308313	2473967	3542085	4536412	5471600
1200	140203	679421	1313371	2481483	3550730	4545358	5480275
1300	140928	682493	1318495	2488946	3559124	4553815	5488193
1400	141675	685632	1323681	2496356	3567281	4561811	5495403
1500	142443	688836	1328926	2503720	3575221	4569385	5501965
1600	143232	692102	1334231	2511047	3582968	4576581	5507943
1700	144041	695430	1339597	2518348	3590551	4583446	5513410
1800	144871	698820	1345026	2525638	3598000	4590032	5518436
1900	145720	702273	1350522	2532932	3605347	4596391	5523097
2000	146590	705790	1356088	2540245	3612625	4602574	5527465
2100	147480	709370	1361729	2547593	3619868	4608634	5531613
2200	148390	713016	1367450	2554995	3627108	4614624	5535613
2300	149321	716728	1373254	2562467	3634379	4620593	5539534
2400	150272	720509	1379149	2570026	3641712	4626590	5543442
2500	151244	724360	1385139	2577690	3649141	4632664	5547402
2600	152237	728283	1391229	2585476	3656695	4638861	5551474
2700	153251	732280	1397426	2593400	3664405	4645224	5555719
2800	154287	736352	1403736	2601479	3672301	4651796	5560190
2900	155344	740502	1410164	2609730	3680409	4658619	5564942
3000	156424	744733	1416715	2618168	3688759	4665729	5570024

CHRYSOTILE

$Mg_3Si_2O_5(OH)_4$
CHRYSOTILE Formula wt. (g) = 277.12

Enthalpy (J/mol) = -4363000.000 Chatterjee (1989)
Entropy (J/mol.K) = 219.820 Chatterjee (1989)
Heat Capacity:
No. of *Cp* ranges = 1 King et al. (1967)
5.0768000D+02 2.8294D-02 1.0798D+07 0.0000D+00
-8.4309D+08 0.0000D+00 -9.7731D+04
Volume (cc/mol) = 108.500 Robie et al. (1978)
Thermal expn:
- - - -
Isot.compressibility:
1.71000D-06 0.00000D+00 0.00000D+00 0.00000D+00
P deriv. 4.56850D+00 Chatterjee (1989)
T deriv. 0.00000D+00
Melting pt. and enthalpy of melting:
- -

NOTE: Calorimetric data: H: -4364.3 ±3.4, S: 221.33±1.67, King et al. (1967). The following reactions involve chrysotile:

$$Chr + Br = 2Fo + 3H_2O$$
$$5Chr = 6Fo + Tc + 9H_2O$$

The experimental data on these two reactions are from Kitihara et al. (1966), Johannes (1968) and Chernosky (1982). Experimental data on the above reactions are used to optimize the free energy of chrysotile. The data of some other workers at low pressures on the second reaction were also included in the analysis.

$Mg_3Si_2O_5(OH)_4$
CHRYSOTILE

Thermochemical Data

T(K)	CP(J)	H(kJ)	S(J)
298	278.0	-4363.00	219.8
300	279.2	-4362.48	221.5
400	329.0	-4331.89	309.2
500	362.8	-4297.21	386.5
600	387.9	-4259.62	454.9
700	407.4	-4219.82	516.2
800	423.4	-4178.25	571.7
900	436.7	-4135.23	622.4
1000	448.2	-4090.97	669.0
1100	458.2	-4045.63	712.2
1200	467.2	-3999.35	752.5
1300	475.3	-3952.22	790.2
1400	482.7	-3904.32	825.7
1500	489.5	-3855.70	859.2
1600	495.9	-3806.43	891.0
1700	501.9	-3756.54	921.3
1800	507.5	-3706.07	950.1
1900	512.9	-3655.05	977.7
2000	518.0	-3603.51	1004.1
2100	522.9	-3551.46	1029.5
2200	527.7	-3498.93	1054.0
2300	532.2	-3445.93	1077.5
2400	536.7	-3392.49	1100.3
2500	541.0	-3338.60	1122.3
2600	545.2	-3284.29	1143.6
2700	549.3	-3229.56	1164.2
2800	553.3	-3174.43	1184.3
2900	557.3	-3118.90	1203.8
3000	561.2	-3062.98	1222.7

ILMENITE

$FeTiO_3$
ILMENITE Formula wt. (g) = 151.75

Enthalpy (J/mol) = -1235350.000		This study
Entropy (J/mol.K) = 106.350		This study
Heat Capacity:		
No. of *Cp* ranges = 1		Anovitz et al. (1985) refitted
1.5456000D+02 2.2920D-03 3.3850D+06 0.0000D+00		
-4.0660D+08 0.0000D+00 -2.3480D+04		
Volume (cc/mol) = 31.690		Robie et al. (1978)
Thermal expn:		
5.18806D-05 2.03268D-09 2.69020D-03 -3.01202D+00		This study
Isot.compressibility:		
5.16522D-07 1.36620D-10 2.84380D-14 5.69450D-18		This study
P deriv. 4.00000D+00		
T deriv. 0.00000D+00		
Melting pt. and enthalpy of melting:		
1640.00000 90667.00000		Robie et al. (1978)

NOTE: Calorimetric data: H: -1236.62±1.59, Robie et al. (1978); S: 108.9 ±0.5, Anovitz et al. (1985). The following reactions involving rutile and ilmenite were experimentally reversed by Bohlen et al. (1983a) and Bohlen and Liotta (1986), respectively

Alm + Ru = Ilm + Ky + Q (a)

garnet + Ru = Ilm + An + Q (b)

Reaction (b) involves the garnet composition Gr1Alm2. Enthalpy, entropy and *Cp* of rutile are calorimetric as given in Robie et al. (1978). Entropy and *Cp* of ilmenite are from Anovitz et al. (1985). The enthalpy and thermal expansion of ilmenite were calculated from reaction (a) by optimization of experimental equilibrium data. The enthalpy obtained is within the error limits of the enthalpy given in Robie et al. (1978). The data on ilmenite thus obtained, along with the solution data on the Gr-Alm solution, reproduce the experimental data on reaction (b) well and the experimental data on the reaction

Ilm = Fe + Ru + O_2 (c)

reasonably well.

$FeTiO_3$

ILMENITE

Thermochemical Data

T(K)	C_p(J)	H(kJ)	S(J)	α 10^{-4}	K 10^4	γ	V(cc)	δ_T
298	99.2	-1235.35	106.3	0.2763	178.592	1.60	31.7	10.03
300	99.5	-1235.17	107.0	0.2799	178.501	1.61	31.7	9.90
400	111.6	-1224.55	137.4	0.4059	173.586	2.08	31.8	6.93
500	119.0	-1213.00	163.2	0.4623	168.733	2.20	31.9	6.18
600	124.3	-1200.82	185.4	0.4922	163.945	2.22	32.1	5.89
700	128.3	-1188.17	204.8	0.5100	159.225	2.20	32.3	5.77
800	131.5	-1175.17	222.2	0.5216	154.575	2.17	32.4	5.72
900	134.2	-1161.89	237.8	0.5298	150.000	2.13	32.6	5.71
1000	136.4	-1148.36	252.1	0.5359	145.502	2.08	32.8	5.72
1100	138.2	-1134.63	265.2	0.5407	141.085	2.04	32.9	5.74
1200	139.9	-1120.72	277.3	0.5447	136.751	1.99	33.1	5.76
1300	141.3	-1106.66	288.5	0.5481	132.503	1.95	33.3	5.79
1400	142.6	-1092.47	299.1	0.5511	128.344	1.91	33.5	5.82
1500	143.7	-1078.15	308.9	0.5538	124.277	1.86	33.7	5.84
1600	144.8	-1063.72	318.2	0.5564	120.302	1.82	33.9	5.87
1700	145.7	-1049.20	327.0	0.5588	116.423	1.78	34.1	5.89
1800	146.6	-1034.58	335.4	0.5610	112.640	1.73	34.2	5.91
1900	147.4	-1019.88	343.4	0.5632	108.954	1.69	34.4	5.93
2000	148.2	-1005.09	350.9	0.5654	105.366	1.65	34.6	5.94
2100	148.9	-990.24	358.2	0.5675	101.877	1.61	34.8	5.95
2200	149.6	-975.31	365.1	0.5695	98.487	1.57	35.0	5.96
2300	150.2	-960.32	371.8	0.5716	95.195	1.53	35.2	5.96
2400	150.8	-945.27	378.2	0.5736	92.002	1.49	35.4	5.96
2500	151.4	-930.16	384.4	0.5756	88.906	1.46	35.6	5.96
2600	152.0	-914.99	390.3	0.5775	85.906	1.42	35.8	5.95
2700	152.5	-899.76	396.1	0.5795	83.002	1.38	36.0	5.94
2800	153.0	-884.49	401.6	0.5815	80.191	1.35	36.3	5.93
2900	153.5	-869.16	407.0	0.5834	77.473	1.31	36.5	5.91
3000	154.0	-853.79	412.2	0.5854	74.846	1.28	36.7	5.89

<INT VdP (J)>

T(K)	PRESSURE(GPa) 1	5	10	20	30	40	50
298	31602	156328	308758	603563	886990	1160834	1426418
300	31603	156335	308770	603580	887009	1160851	1426431
400	31712	156830	309647	604973	888673	1162610	1428151
500	31849	157457	310782	606857	891045	1165281	1430979
600	31999	158149	312041	608969	893737	1168357	1434291
700	32157	158878	313369	611200	896588	1171626	1437827
800	32320	159631	314737	613496	899520	1174986	1441459
900	32488	160401	316132	615828	902489	1178380	1445119
1000	32658	161182	317545	618180	905471	1181775	1448767
1100	32831	161972	318971	620540	908449	1185150	1452376
1200	33006	162771	320406	622901	911412	1188491	1455930
1300	33183	163576	321847	625258	914354	1191789	1459417
1400	33362	164386	323293	627608	917269	1195037	1462830
1500	33542	165202	324743	629948	920152	1198229	1466161
1600	33725	166023	326196	632275	923002	1201363	1469408
1700	33909	166847	327650	634588	925814	1204435	1472567
1800	34094	167676	329105	636886	928589	1207443	1475636
1900	34281	168509	330561	639167	931324	1210386	1478614
2000	34470	169345	332016	641430	934017	1213263	1481501
2100	34661	170185	333472	643675	936669	1216073	1484295
2200	34852	171028	334926	645901	939279	1218817	1486998
2300	35046	171875	336379	648108	941846	1221493	1489610
2400	35241	172724	337831	650295	944371	1224103	1492131
2500	35437	173577	339282	652461	946852	1226647	1494564
2600	35636	174432	340730	654608	949291	1229125	1496910
2700	35835	175291	342177	656733	951687	1231539	1499170
2800	36037	176152	343621	658838	954042	1233889	1501346
2900	36239	177016	345063	660922	956354	1236177	1503440
3000	36444	177882	346502	662986	958625	1238404	1505455

RUTILE

TiO_2
RUTILE Formula wt. (g) = 79.90

Enthalpy (J/mol) = -944750.000 Robie et al. (1978)
Entropy (J/mol.K) = 50.290 Robie et al. (1978)
Heat Capacity:
No. of *Cp* ranges = 1 Robie et al. (1978)
7.3590000D+01 1.0063D-02 6.0420D+05 -1.7085D-06
0.0000D+00 0.0000D+00 -8.3530D+03
Volume (cc/mol) = 18.820 Robie et al. (1978)
Thermal expn:
2.01800D-05 1.47930D-08 -2.72500D-03 5.17400D-02
This study
Isot.compressibility:
4.37630D-07 1.11980D-10 1.19310D-14 5.80950D-17
This study
P deriv. 6.91000D+00
T deriv. 7.50000D-04
Melting pt. and enthalpy of melting:
- -

NOTE: Calorimetric data: H: -944.75±1.26, S: 50.29±0.17, Robie et al. (1978). All thermochemical data are from calorimetric sources. The compressibility and the thermal expansion data respectively are consistent with those of Grimsditch and Ramdas (1976) and Fritz (1974) and as reported in Carmichael (1982).

TiO_2
RUTILE

Thermochemical Data

T(K)	C_p(J)	H(kJ)	S(J)	α 10^{-4}	K 10^4	γ	V(cc)	δ_T
298	55.2	-944.75	50.3	0.1603	211.141	1.16	18.8	17.72
300	55.3	-944.65	50.6	0.1611	211.030	1.16	18.8	17.66
400	60.2	-938.86	67.3	0.1961	204.897	1.27	18.9	15.61
500	63.9	-932.64	81.1	0.2233	198.466	1.33	18.9	14.88
600	66.8	-926.10	93.0	0.2466	191.695	1.37	18.9	14.70
700	69.1	-919.31	103.5	0.2675	184.574	1.39	19.0	14.77
800	71.0	-912.30	112.9	0.2869	177.118	1.41	19.0	14.98
900	72.7	-905.10	121.3	0.3053	169.369	1.41	19.1	15.24
1000	74.2	-897.76	129.1	0.3230	161.385	1.41	19.2	15.50
1100	75.5	-890.27	136.2	0.3402	153.241	1.40	19.2	15.73
1200	76.7	-882.66	142.8	0.3570	145.017	1.38	19.3	15.91
1300	77.7	-874.94	149.0	0.3735	136.799	1.36	19.4	16.03
1400	78.7	-867.12	154.8	0.3897	128.667	1.33	19.4	16.08
1500	79.5	-859.21	160.2	0.4058	120.698	1.30	19.5	16.06
1600	80.3	-851.22	165.4	0.4217	112.956	1.26	19.6	15.97
1700	81.1	-843.15	170.3	0.4374	105.497	1.22	19.7	15.83
1800	81.7	-835.01	175.0	0.4531	98.361	1.18	19.8	15.62
1900	82.3	-826.81	179.4	0.4687	91.580	1.14	19.9	15.37
2000	82.9	-818.55	183.6	0.4842	85.174	1.10	19.9	15.07
2100	83.3	-810.24	187.7	0.4996	79.150	1.06	20.0	14.74
2200	83.8	-801.88	191.6	0.5150	73.512	1.02	20.1	14.39
2300	84.2	-793.48	195.3	0.5303	68.253	0.97	20.3	14.01
2400	84.5	-785.04	198.9	0.5456	63.363	0.93	20.4	13.62
2500	84.8	-776.58	202.3	0.5608	58.828	0.90	20.5	13.23
2600	85.1	-768.08	205.7	0.5760	54.630	0.86	20.6	12.82
2700	85.3	-759.56	208.9	0.5912	50.750	0.82	20.7	12.42
2800	85.5	-751.02	212.0	0.6063	47.169	0.79	20.8	12.02
2900	85.6	-742.47	215.0	0.6215	43.868	0.76	21.0	11.63
3000	85.7	-733.91	217.9	0.6366	40.824	0.73	21.1	11.24

<INT VdP (J)>

PRESSURE (GPa)

T(K)	1	5	10	20	30	40	50
298	18775	93049	184210	361825	534245	702372	866834
300	18776	93051	184214	361829	534249	702373	866832
400	18808	93189	184439	362123	534517	702550	866870
500	18847	93352	184710	362501	534904	702882	867104
600	18889	93535	185016	362934	535362	703301	867438
700	18936	93735	185348	363403	535860	703761	867814
800	18987	93949	185702	363895	536376	704229	868189
900	19041	94176	186072	364400	536893	704684	868532
1000	19098	94414	186456	364911	537400	705109	868822
1100	19158	94662	186851	365423	537889	705491	869044
1200	19222	94920	187255	365931	538353	705824	869190
1300	19288	95186	187667	366433	538789	706106	869258
1400	19358	95460	188086	366927	539198	706336	869250
1500	19430	95743	188511	367413	539579	706518	869173
1600	19505	96033	188941	367891	539937	706659	869035
1700	19584	96330	189376	368365	540276	706767	868849
1800	19665	96634	189818	368835	540602	706850	868628
1900	19749	96945	190265	369305	540921	706921	868389
2000	19835	97263	190719	369780	541241	706990	868146
2100	19925	97589	191181	370261	541569	707070	867917
2200	20017	97922	191652	370756	541915	707173	867718
2300	20113	98262	192134	371266	542286	707311	867565
2400	20211	98611	192627	371798	542689	707496	867472
2500	20312	98968	193133	372356	543134	707738	867454
2600	20416	99333	193655	372944	543627	708047	867524
2700	20523	99708	194193	373566	544175	708435	867695
2800	20633	100093	194750	374227	544785	708909	867978
2900	20746	100489	195327	374931	545464	709478	868383
3000	20863	100895	195925	375682	546216	710149	868918

CALCITE

$CaCO_3$
CALCITE Formula wt. (g) = 100.09

Enthalpy (J/mol) =	-1206000.000	This study
Entropy (J/mol.K) =	91.700	Holland and Powell (1990)

Heat Capacity:
No. of *Cp* ranges = 1 — Holland and Powell (1990)

1.3997000D+02 3.5250D-03 1.5067D+06 1.7936D-06
0.0000D+00 0.0000D+00 -2.1830D+04

Volume (cc/mol) = 36.890

Thermal expn:
1.31250D-05 3.24920D-08 -4.07440D-03 6.51690D-01 — Holland and Powell (1990)

Isot.compressibility:
1.29160D-06 3.78450D-11 1.75400D-15 1.12950D-16 — Holland and Powell (1990)

P deriv. 4.00000D+00
T deriv. 0.00000D+00
Melting pt. and enthalpy of melting:
- -

NOTE: The experimental data on several reactions as used by Berman (1988) (i.e. Walter,1963a,b; Yin and Greenwood, 1983) are used for the estimation of enthalpy.

Thermochemical Data

T(K)	C_p(J)	H(kJ)	S(J)	$\alpha\ 10^{-4}$	$K\ 10^4$	γ	V(cc)	δ_T
298	84.9	-1206.00	91.7	0.1648	76.568	0.55	36.9	3.21
300	85.2	-1205.84	92.2	0.1653	76.560	0.55	36.9	3.21
400	96.5	-1196.72	118.4	0.2001	76.089	0.59	37.0	3.55
500	104.5	-1186.65	140.8	0.2383	75.467	0.64	37.0	3.94
600	110.5	-1175.88	160.4	0.2764	74.664	0.70	37.1	4.37
700	115.2	-1164.59	177.8	0.3138	73.654	0.76	37.2	4.84
800	119.0	-1152.87	193.5	0.3504	72.420	0.82	37.4	5.32
900	122.2	-1140.81	207.7	0.3865	70.951	0.87	37.5	5.79
1000	125.0	-1128.45	220.7	0.4219	69.245	0.91	37.7	6.24
1100	127.4	-1115.82	232.7	0.4570	67.309	0.96	37.8	6.65
1200	129.6	-1102.97	243.9	0.4917	65.159	0.99	38.0	7.02
1300	131.7	-1089.90	254.4	0.5262	62.817	1.03	38.2	7.34
1400	133.6	-1076.64	264.2	0.5604	60.315	1.05	38.4	7.61
1500	135.4	-1063.19	273.5	0.5944	57.686	1.07	38.6	7.82
1600	137.1	-1049.56	282.3	0.6282	54.967	1.09	38.9	7.97
1700	138.8	-1035.76	290.6	0.6619	52.194	1.09	39.1	8.07
1800	140.5	-1021.79	298.6	0.6955	49.404	1.10	39.4	8.11
1900	142.1	-1007.67	306.3	0.7290	46.630	1.09	39.7	8.11
2000	143.7	-993.38	313.6	0.7623	43.900	1.08	40.0	8.06
2100	145.2	-978.94	320.6	0.7957	41.240	1.07	40.3	7.98
2200	146.8	-964.34	327.4	0.8289	38.669	1.06	40.6	7.86
2300	148.4	-949.58	334.0	0.8621	36.203	1.04	40.9	7.72
2400	149.9	-934.66	340.3	0.8952	33.853	1.02	41.3	7.56
2500	151.5	-919.59	346.5	0.9283	31.625	0.99	41.7	7.37
2600	153.1	-904.36	352.5	0.9613	29.524	0.97	42.1	7.18
2700	154.7	-888.97	358.3	0.9943	27.550	0.94	42.5	6.98
2800	156.3	-873.43	363.9	1.0273	25.702	0.92	42.9	6.77
2900	157.9	-857.71	369.4	1.0602	23.976	0.89	43.4	6.55
3000	159.6	-841.84	374.8	1.0932	22.369	0.86	43.8	6.34

$CaCO_3$
CALCITE

<INT VdP (J)>

	PRESSURE (GPa)						
T(K)	1	5	10	20	30	40	50
298	36654	178995	348835	667910	966221	1248688	1518418
300	36655	179000	348844	667926	966241	1248712	1518446
400	36720	179297	349387	668888	967580	1250413	1520507
500	36799	179654	350046	670099	969346	1252770	1523498
600	36891	180071	350812	671514	971435	1255593	1527126
700	36997	180543	351672	673090	973756	1258733	1531168
800	37116	181068	352617	674792	976241	1262075	1535454
900	37248	181641	353635	676592	978832	1265529	1539856
1000	37393	182261	354720	678466	981487	1269026	1544278
1100	37551	182926	355864	680395	984172	1272518	1548649
1200	37720	183632	357063	682366	986864	1275969	1552922
1300	37903	184378	358311	684371	989550	1279361	1557072
1400	38098	185164	359605	686403	992223	1282684	1561090
1500	38305	185988	360945	688462	994884	1285945	1564982
1600	38525	186850	362330	690552	997542	1289155	1568768
1700	38757	187749	363760	692677	1000209	1292337	1572479
1800	39002	188686	365237	694847	1002903	1295520	1576154
1900	39259	189660	366765	697073	1005648	1298737	1579839
2000	39529	190674	368346	699369	1008468	1302025	1583584
2100	39812	191727	369987	701750	1011390	1305425	1587442
2200	40109	192822	371691	704232	1014443	1308976	1591466
2300	40418	193960	373465	706832	1017654	1312720	1595710
2400	40741	195144	375316	709567	1021054	1316698	1600227
2500	41078	196375	377251	712455	1024669	1320948	1605065
2600	41429	197656	379276	715513	1028527	1325507	1610272
2700	41794	198991	381399	718756	1032653	1330410	1615894
2800	42174	200381	383627	722202	1037073	1335691	1621970
2900	42569	201830	385968	725865	1041809	1341380	1628540
3000	42980	203342	388428	729761	1046883	1347506	1635639

MAGNESITE

$MgCO_3$
MAGNESITE Formula wt. (g) = 84.32

Enthalpy (J/mol) = -1106000.000 This study
Entropy (J/mol.K) = 69.000 This study
Heat Capacity:
No. of *Cp* ranges = 1 Holland and Powell (1990), refitted
1.2359000D+02 2.0986D-02 7.1220D+05 2.7041D-06
0.0000D+00 0.0000D+00 -1.8114D+04
Volume (cc/mol) = 28.030 Holland and Powell (1990)
Thermal expn:
2.55120D-05 4.10730D-08 -3.96020D-03 6.62290D-01
This study
Isot.compressibility:
8.62350D-07 8.42400D-11 1.16700D-14 6.21510D-17
This study
P deriv. 4.00000D+00
T deriv. 0.00000D+00
Melting pt. and enthalpy of melting:
\- -

NOTE: The data on magnesite have been evaluated by Trommsdorff and Connolly (1990). Their preferred value for enthalpy at 298.15 is -1111197.0. Such a value fits the reversals at low pressures and temperatures. We used additionally the phase equilibrium data at high P and T (Harker and Tuttle, 1955; Goldsmith and Heard, 1962; Irwing and Wyllie, 1975; Newton and Sharp, 1975; Haselton et al.,1978; Eggler et al., 1979) which lead to the adopted data.

$MgCO_3$
MAGNESITE

Thermochemical Data

T(K)	C_p(J)	H(kJ)	S(J)	α 10^{-4}	K 10^4	γ	V(cc)	δ_T
298	77.3	-1106.00	69.0	0.3193	112.341	1.32	28.0	3.79
300	77.7	-1105.86	69.5	0.3199	112.315	1.31	28.0	3.79
400	91.6	-1097.35	93.9	0.3618	110.878	1.25	28.1	3.78
500	101.4	-1087.67	115.4	0.4078	109.271	1.27	28.2	3.82
600	108.9	-1077.14	134.6	0.4540	107.467	1.32	28.4	3.91
700	115.2	-1065.93	151.9	0.4996	105.446	1.37	28.5	4.05
800	120.6	-1054.14	167.6	0.5445	103.196	1.42	28.6	4.21
900	125.4	-1041.83	182.1	0.5890	100.712	1.47	28.8	4.38
1000	129.9	-1029.06	195.5	0.6329	98.000	1.52	29.0	4.55
1100	134.1	-1015.87	208.1	0.6764	95.070	1.56	29.2	4.72
1200	138.1	-1002.26	219.9	0.7196	91.942	1.60	29.4	4.86
1300	141.9	-988.26	231.2	0.7625	88.642	1.64	29.6	5.00
1400	145.7	-973.88	241.8	0.8052	85.201	1.67	29.8	5.10
1500	149.4	-959.12	252.0	0.8478	81.651	1.69	30.1	5.19
1600	153.0	-944.00	261.7	0.8901	78.029	1.71	30.3	5.25
1700	156.7	-928.51	271.1	0.9324	74.370	1.73	30.6	5.29
1800	160.3	-912.66	280.2	0.9745	70.708	1.73	30.9	5.30
1900	163.9	-896.46	289.0	1.0165	67.077	1.73	31.2	5.29
2000	167.5	-879.89	297.5	1.0584	63.503	1.73	31.5	5.26
2100	171.1	-862.95	305.7	1.1003	60.013	1.72	31.9	5.21
2200	174.8	-845.66	313.8	1.1421	56.627	1.70	32.2	5.15
2300	178.4	-828.00	321.6	1.1838	53.361	1.68	32.6	5.07
2400	182.1	-809.98	329.3	1.2255	50.228	1.66	33.0	4.98
2500	185.8	-791.58	336.8	1.2672	47.237	1.63	33.4	4.88
2600	189.6	-772.81	344.1	1.3088	44.393	1.60	33.9	4.77
2700	193.4	-753.66	351.4	1.3503	41.698	1.57	34.3	4.65
2800	197.2	-734.14	358.5	1.3919	39.153	1.54	34.8	4.53
2900	201.0	-714.23	365.5	1.4334	36.757	1.50	35.3	4.41
3000	204.9	-693.93	372.3	1.4748	34.504	1.47	35.8	4.29

<INT VdP (J)>

PRESSURE(GPa)

T(K)	1	5	10	20	30	40	50
298	27907	137240	269354	521221	759911	988071	1207498
300	27908	137247	269368	521245	759942	988108	1207540
400	28002	137679	270160	522609	761744	990258	1209971
500	28108	138169	271057	524150	763777	992680	1212708
600	28227	138719	272060	525867	766032	995358	1215723
700	28360	139327	273166	527746	768488	998256	1218970
800	28505	139992	274370	529775	771118	1001339	1222398
900	28664	140714	275669	531941	773900	1004570	1225961
1000	28836	141491	277058	534232	776812	1007919	1229617
1100	29021	142322	278535	536638	779835	1011357	1233328
1200	29219	143207	280096	539149	782952	1014860	1237063
1300	29430	144146	281738	541756	786149	1018407	1240794
1400	29655	145137	283461	544453	789412	1021980	1244500
1500	29894	146181	285261	547233	792733	1025567	1248167
1600	30147	147278	287137	550091	796103	1029159	1251783
1700	30414	148428	289088	553023	799518	1032750	1255347
1800	30695	149630	291113	556028	802974	1036340	1258857
1900	30990	150886	293211	559105	806474	1039931	1262322
2000	31301	152195	295383	562252	810017	1043529	1265751
2100	31627	153558	297629	565472	813610	1047144	1269160
2200	31968	154975	299950	568767	817258	1050787	1272568
2300	32325	156448	302346	572141	820970	1054474	1275997
2400	32699	157977	304818	575598	824756	1058221	1279471
2500	33089	159563	307370	579143	828627	1062047	1283018
2600	33497	161208	310002	582784	832597	1065972	1286666
2700	33922	162912	312718	586528	836679	1070019	1290444
2800	34365	164677	315519	590382	840889	1074210	1294383
2900	34827	166505	318411	594355	845242	1078567	1298515
3000	35308	168397	321395	598457	849754	1083115	1302870

SIDERITE

$FeCO_3$
SIDERITE Formula wt. (g) = 115.858

Enthalpy (J/mol) = -761180.000 Holland and Powell (1990)
Entropy (J/mol.K) = 95.500 Holland and Powell (1990)
Heat Capacity:
No. of *Cp* ranges = 1 Holland and Powell (1990)
2.5739999D+02 -4.6200D-02 1.5230D+06 0.0000D+00
0.0000D+00 -3.0819D+03 0.0000D+00
Volume (cc/mol) = 29.380 Holland and Powell (1990)
Thermal expn:
2.45100D-05 0.00000D+00 0.00000D+00 0.00000D+00
Isot.compressibility:
9.99000D-07 0.00000D+00 0.00000D+00 0.00000D+00
P deriv. 4.00000D+00
T deriv. 0.00000D+00
Melting pt. and enthalpy of melting:
\- -

Thermochemical Data

T(K)	CP(J)	H(kJ)	S(J)
298	82.3	-761.18	95.5
300	82.5	-761.03	96.0
400	94.3	-752.15	121.5
500	102.6	-742.28	143.5
600	108.1	-731.73	162.7
700	111.7	-720.73	179.6
800	113.9	-709.44	194.7
900	115.0	-697.99	208.2
1000	115.3	-686.47	220.3
1100	114.9	-674.96	231.3
1200	114.1	-663.51	241.3
1300	112.8	-652.16	250.3
1400	111.1	-640.96	258.6
1500	109.2	-629.95	266.2
1600	107.0	-619.13	273.2
1700	104.6	-608.55	279.6
1800	102.1	-598.21	285.5
1900	99.3	-588.14	291.0
2000	96.5	-578.35	296.0
2100	93.5	-568.85	300.7
2200	90.4	-559.66	304.9
2300	87.2	-550.78	308.9
2400	83.9	-542.23	312.5
2500	80.5	-534.01	315.9
2600	77.1	-526.13	319.0
2700	73.6	-518.60	321.8
2800	70.0	-511.42	324.4
2900	66.4	-504.60	326.8
3000	62.7	-498.14	329.0

DOLOMITE

$CaMg(CO_3)_2$
DOLOMITE Formula wt. (g) = 184.41

Enthalpy (J/mol) = -2325720.000 Holland and Powell (1990)
Entropy (J/mol.K) = 155.200 Holland and Powell (1990)
Heat Capacity:
No. of *Cp* ranges = 1 Holland and Powell (1990), refitted
2.7380000D+02 1.3125D-02 2.8653D+06 8.0432D-06
0.0000D+00 0.0000D+00 -4.5578D+04
Volume (cc/mol) = 64.360 Holland and Powell (1990)
Thermal expn:
1.92720D-05 3.53110D-08 -3.04460D-03 7.06550D-01
This study
Isot.compressibility:
1.03680D-06 8.36600D-11 1.19300D-14 6.05900D-17
This study
P deriv. 4.00000D+00
T deriv. 0.00000D+00
Melting pt. and enthalpy of melting:
- -

Thermochemical Data

T(K)	C_p(J)	H(kJ)	S(J)	α 10^{-4}	K 10^4	γ	V(cc)	δ_T
298	157.8	-2325.72	155.2	0.2754	93.949	1.06	64.4	3.65
300	158.4	-2325.43	156.2	0.2757	93.931	1.06	64.4	3.65
400	184.3	-2308.21	205.5	0.3020	92.932	0.99	64.5	3.76
500	202.7	-2288.82	248.7	0.3366	91.812	1.00	64.8	3.85
600	216.6	-2267.83	287.0	0.3735	90.549	1.04	65.0	3.96
700	227.7	-2245.60	321.2	0.4108	89.127	1.08	65.2	4.11
800	237.0	-2222.35	352.2	0.4482	87.536	1.13	65.5	4.28
900	245.0	-2198.25	380.6	0.4854	85.769	1.18	65.8	4.46
1000	252.3	-2173.38	406.8	0.5224	83.824	1.22	66.2	4.64
1100	258.9	-2147.81	431.2	0.5593	81.706	1.27	66.5	4.82
1200	265.1	-2121.61	454.0	0.5960	79.424	1.31	66.9	4.98
1300	271.1	-2094.80	475.4	0.6325	76.992	1.34	67.3	5.14
1400	276.8	-2067.40	495.7	0.6689	74.429	1.38	67.8	5.27
1500	282.5	-2039.43	515.0	0.7052	71.755	1.40	68.2	5.38
1600	288.0	-2010.91	533.4	0.7414	68.995	1.43	68.7	5.46
1700	293.5	-1981.83	551.1	0.7775	66.174	1.45	69.2	5.53
1800	299.0	-1952.20	568.0	0.8136	63.315	1.46	69.8	5.57
1900	304.6	-1922.02	584.3	0.8496	60.445	1.47	70.4	5.59
2000	310.2	-1891.28	600.1	0.8855	57.585	1.47	71.0	5.58
2100	315.8	-1859.98	615.3	0.9214	54.758	1.47	71.6	5.56
2200	321.5	-1828.12	630.2	0.9572	51.982	1.46	72.3	5.52
2300	327.3	-1795.69	644.6	0.9930	49.273	1.45	73.0	5.46
2400	333.1	-1762.67	658.6	1.0287	46.644	1.44	73.8	5.39
2500	339.1	-1729.06	672.3	1.0644	44.107	1.42	74.5	5.30
2600	345.2	-1694.84	685.8	1.1001	41.669	1.40	75.3	5.21
2700	351.4	-1660.01	698.9	1.1358	39.335	1.38	76.2	5.10
2800	357.7	-1624.56	711.8	1.1715	37.111	1.35	77.1	4.99
2900	364.1	-1588.47	724.5	1.2071	34.996	1.32	78.0	4.88
3000	370.7	-1551.73	736.9	1.2427	32.992	1.29	79.0	4.76

$CaMg(CO_3)_2$
DOLOMITE

<INT VdP (J)>

PRESSURE (GPa)

T(K)	1	5	10	20	30	40	50
298	64023	313910	614213	1182987	1718688	2228523	2717217
300	64026	313925	614239	1183032	1718747	2228593	2717296
400	64207	314751	615734	1185562	1722050	2232496	2721677
500	64408	315666	617387	1188345	1725665	2236751	2726434
600	64632	316684	619218	1191410	1729626	2241388	2731594
700	64881	317807	621229	1194749	1733909	2246368	2737096
800	65153	319034	623415	1198341	1738474	2251629	2742860
900	65451	320364	625768	1202163	1743281	2257109	2748799
1000	65773	321795	628282	1206195	1748287	2262748	2754834
1100	66120	323325	630950	1210413	1753455	2268488	2760888
1200	66492	324954	633766	1214799	1758749	2274280	2766898
1300	66888	326680	636725	1219334	1764138	2280078	2772804
1400	67311	328501	639820	1224002	1769595	2285846	2778561
1500	67758	330417	643047	1228789	1775098	2291553	2784133
1600	68232	332427	646402	1233684	1780628	2297178	2789494
1700	68732	334531	649882	1238677	1786173	2302704	2794628
1800	69258	336728	653483	1243761	1791723	2308126	2799533
1900	69811	339017	657204	1248933	1797277	2313442	2804213
2000	70392	341400	661043	1254188	1802833	2318660	2808683
2100	71001	343876	665000	1259529	1808399	2323795	2812969
2200	71637	346446	669073	1264958	1813984	2328865	2817101
2300	72303	349110	673265	1270478	1819600	2333895	2821119
2400	72998	351869	677576	1276099	1825267	2338917	2825067
2500	73723	354725	682008	1281827	1831003	2343963	2828994
2600	74479	357678	686565	1287675	1836831	2349072	2832955
2700	75266	360731	691250	1293655	1842778	2354284	2837005
2800	76086	363885	696067	1299779	1848870	2359641	2841204
2900	76938	367142	701021	1306065	1855135	2365187	2845610
3000	77824	370505	706118	1312527	1861605	2370967	2850284

K-FELDSPAR

$KAlSi_3O_8$
K-FELDSPAR Formula wt. (g) = 278.34

Enthalpy (J/mol) = -3969620.000 Holland and Powell (1990)

Entropy (J/mol.K) = 214.000 Holland and Powell (1990)

Heat Capacity:
No. of *Cp* ranges = 1 Holland and Powell (1990)

4.4879999D+02 -1.0075D-02 -1.0073D+06 0.0000D+00
0.0000D+00 -3.9731D+03 0.0000D+00

Volume (cc/mol) = 108.800 Holland and Powell (1990)

Thermal expn:
1.89340D-05 0.00000D+00 0.00000D+00 0.00000D+00

Isot.compressibility:
1.83800D-06 0.00000D+00 0.00000D+00 0.00000D+00

P deriv. 4.00000D+00

T deriv. 0.00000D+00

Melting pt. and enthalpy of melting:
- -

Thermochemical Data

T (K)	C_p (J)	H (kJ)	S (J)
298	204.4	-3969.62	214.0
300	205.2	-3969.24	215.3
400	239.8	-3946.84	279.5
500	262.1	-3921.68	335.5
600	277.8	-3894.65	384.8
700	289.5	-3866.26	428.5
800	298.7	-3836.83	467.8
900	306.1	-3806.58	503.4
1000	312.1	-3775.66	536.0
1100	317.1	-3744.20	566.0
1200	321.3	-3712.27	593.7
1300	324.9	-3679.96	619.6
1400	328.0	-3647.31	643.8
1500	330.7	-3614.37	666.5
1600	333.0	-3581.19	687.9
1700	335.0	-3547.79	708.2
1800	336.7	-3514.20	727.4
1900	338.2	-3480.46	745.6
2000	339.6	-3446.56	763.0
2100	340.7	-3412.55	779.6
2200	341.7	-3378.43	795.5
2300	342.6	-3344.21	810.7
2400	343.3	-3309.91	825.3
2500	344.0	-3275.54	839.3
2600	344.5	-3241.12	852.8
2700	345.0	-3206.64	865.8
2800	345.4	-3172.12	878.4
2900	345.7	-3137.57	890.5
3000	345.9	-3102.99	902.2

LAWSONITE

$CaAl_2Si_2O_7(OH)_2.H_2O$
LAWSONITE Formula wt. (g) = 314.25

Enthalpy (J/mol) = -4868070.000 Holland and Powell (1990)
Entropy (J/mol.K) = 230.000 Holland and Powell (1990)
Heat Capacity:
No. of *Cp* ranges = 1 Holland and Powell (1990)
9.5009998D+02 -1.1321D-01 6.6456D+06 0.0000D+00
0.0000D+00 -1.2318D+04 0.0000D+00
Volume (cc/mol) = 101.320 Holland and Powell (1990)
Thermal expn:
2.46740D-05 0.00000D+00 0.00000D+00 0.00000D+00
Isot.compressibility:
8.88300D-07 0.00000D+00 0.00000D+00 0.00000D+00
P deriv. 4.00000D+00
T deriv. 0.00000D+00
Melting pt. and enthalpy of melting:
- -

Thermochemical Data

T(K)	C_p(J)	H(kJ)	S(J)
298	277.7	-4868.07	230.0
300	278.8	-4867.56	231.7
400	330.5	-4836.98	319.3
500	369.2	-4801.90	397.4
600	397.8	-4763.48	467.4
700	418.8	-4722.60	530.4
800	434.4	-4679.90	587.4
900	445.8	-4635.85	639.2
1000	454.0	-4590.84	686.7
1100	459.7	-4545.14	730.2
1200	463.3	-4498.98	770.4
1300	465.2	-4452.54	807.6
1400	465.8	-4405.98	842.1
1500	465.2	-4359.42	874.2
1600	463.6	-4312.97	904.2
1700	461.2	-4266.73	932.2
1800	458.0	-4220.76	958.5
1900	454.2	-4175.14	983.1
2000	449.9	-4129.93	1006.3
2100	445.1	-4085.18	1028.2
2200	439.8	-4040.93	1048.7
2300	434.1	-3997.23	1068.2
2400	428.1	-3954.12	1086.5
2500	421.8	-3911.62	1103.9
2600	415.2	-3869.77	1120.3
2700	408.3	-3828.60	1135.8
2800	401.2	-3788.12	1150.5
2900	393.8	-3748.37	1164.5
3000	386.3	-3709.36	1177.7

LIME

CAO

CAO Formula wt. (g) = 56.08

Enthalpy (J/mol) =	-634260.000	Holland and Powell (1990)
Entropy (J/mol.K) =	38.100	Holland and Powell (1990)
Heat Capacity		
No. of *Cp* ranges = 1		Holland and Powell (1990)
5.4250000D+01 1.2150D-03 3.0100D+05 0.0000D+00 -5.9500D+07 0.0000D+00 -3.6600D+03		
Volume (cc/mol) =	16.760	Holland and Powell (1990)
Thermal expn. 3.36200D-05 5.44000D-09 -1.20000D-05 -4.33700D-01		Saxena and Shen (1992)
Isot.bulk mod.(b0, b1*T, b2*T^2, b3*T^3): 8.60500D-07 1.25000D-10 3.22000D-14 3.28000D-18		Saxena and Shen (1992)
P deriv., T deriv.: 4.00000D+00 0.00000D+00		
Melting pt. and enthalpy of melting: - -		

Thermochemical Data

T(K)	C_p(J)	H(kJ)	S(J)	α 10^{-4}	K 10^4	γ	V(cc)	δ_T
298	43.5	-634.260	38.1	0.3032	111.023	1.31	16.8	5.31
300	43.6	-634.179	38.4	0.3039	110.989	1.31	16.8	5.30
400	46.5	-629.659	51.4	0.3306	109.187	1.33	16.8	5.03
500	48.3	-624.912	61.9	0.3458	107.358	1.33	16.9	4.96
600	49.4	-620.024	70.8	0.3566	105.507	1.32	16.9	4.95
700	50.3	-615.034	78.5	0.3653	103.637	1.32	17.0	4.96
800	51.0	-609.967	85.3	0.3728	101.751	1.32	17.1	4.99
900	51.6	-604.838	91.3	0.3797	99.853	1.32	17.1	5.02
1000	52.0	-599.657	96.8	0.3861	97.945	1.31	17.2	5.05
1100	52.5	-594.431	101.8	0.3923	96.031	1.31	17.3	5.08
1200	52.8	-589.166	106.4	0.3984	94.114	1.31	17.3	5.11
1300	53.2	-583.865	110.6	0.4043	92.198	1.30	17.4	5.14
1400	53.5	-578.534	114.6	0.4101	90.284	1.30	17.5	5.16
1500	53.7	-573.173	118.3	0.4158	88.377	1.30	17.5	5.18
1600	54.0	-567.784	121.7	0.4215	86.478	1.29	17.6	5.20
1700	54.3	-562.371	125.0	0.4271	84.590	1.29	17.7	5.21
1800	54.5	-556.934	128.1	0.4327	82.716	1.28	17.8	5.21
1900	54.7	-551.474	131.1	0.4383	80.858	1.28	17.8	5.22
2000	54.9	-545.993	133.9	0.4439	79.018	1.27	17.9	5.22
2100	55.1	-540.491	136.6	0.4494	77.198	1.27	18.0	5.22
2200	55.3	-534.969	139.1	0.4549	75.399	1.26	18.1	5.21
2300	55.5	-529.428	141.6	0.4604	73.624	1.26	18.2	5.20
2400	55.7	-523.868	144.0	0.4660	71.874	1.25	18.2	5.19
2500	55.9	-518.290	146.2	0.4715	70.151	1.24	18.3	5.17
2600	56.0	-512.695	148.4	0.4770	68.455	1.24	18.4	5.15
2700	56.2	-507.082	150.6	0.4824	66.787	1.23	18.5	5.13
2800	56.4	-501.452	152.6	0.4879	65.149	1.22	18.6	5.11
2900	56.5	-495.806	154.6	0.4934	63.541	1.22	18.7	5.08
3000	56.7	-490.143	156.5	0.4989	61.963	1.21	18.8	5.05

CAO

<INT VdP (J)>

	PRESSURE(GPa)						
T(K)	1	5	10	20	30	40	50
298	16685	82040	160987	311429	453944	590133	721081
300	16686	82044	160994	311441	453959	590150	721098
400	16738	82279	161411	312119	454805	591103	722114
500	16794	82530	161858	312848	455719	592137	723223
600	16851	82790	162322	313604	456666	593210	724373
700	16911	83058	162798	314378	457634	594302	725542
800	16972	83333	163284	315165	458614	595405	726719
900	17034	83613	163779	315962	459603	596513	727896
1000	17098	83897	164281	316768	460598	597624	729069
1100	17163	84187	164790	317580	461597	598734	730238
1200	17229	84481	165306	318399	462599	599842	731399
1300	17297	84780	165829	319224	463604	600948	732552
1400	17365	85084	166357	320055	464610	602051	733696
1500	17435	85392	166892	320890	465619	603151	734831
1600	17506	85704	167432	321731	466628	604247	735956
1700	17579	86020	167979	322576	467639	605339	737072
1800	17652	86341	168531	323426	468651	606427	738179
1900	17727	86666	169089	324281	469664	607512	739277
2000	17803	86996	169653	325141	470677	608593	740366
2100	17880	87330	170223	326005	471692	609671	741447
2200	17958	87668	170798	326873	472709	610745	742520
2300	18038	88011	171379	327746	473726	611817	743587
2400	18119	88358	171966	328624	474745	612887	744647
2500	18201	88709	172558	329507	475766	613955	745701
2600	18284	89065	173157	330395	476790	615022	746750
2700	18369	89426	173761	331287	477815	616088	747795
2800	18455	89790	174371	332185	478843	617153	748837
2900	18542	90160	174987	333088	479874	618219	749876
3000	18631	90534	175608	333996	480909	619286	750914

PSEUDO-WOLLASTONITE

$CaSiO_3$
PSEUDO-WOLLASTONITE Formula wt. (g) = 116.17

Enthalpy (J/mol) = -1629540.000 This study
Entropy (J/mol.K) = 84.460 This study
Heat Capacity:
No. of *Cp* ranges = 1 Holland and Powell (1990)
1.0710001D+02 1.7481D-02 -2.2965D+06 0.0000D+00
0.0000D+00 0.0000D+00 0.0000D+00
Volume (cc/mol) = 40.080 Holland and Powell (1990)
Thermal expn:
2.48500D-05 8.07100D-09 5.23860D-03 -2.36500D+00
This study
Isot.compressibility:
9.13500D-07 -8.94400D-11 1.27500D-13 1.35980D-17
This study
P deriv. 4.00000D+00
T deriv. 0.00000D+00
Melting pt. and enthalpy of melting:
- -

Thermochemical Data

T (K)	C_p (J)	H (kJ)	S (J)	α 10^{-4}	K 10^4	γ	V (cc)	δ_T
298	86.5	-1629.54	84.5	0.1822	111.293	0.94	40.1	-0.60
300	86.8	-1629.38	85.0	0.1846	111.295	0.95	40.1	-0.56
400	99.7	-1619.97	112.0	0.2639	111.235	1.20	40.2	0.80
500	106.7	-1609.62	135.0	0.2990	110.821	1.28	40.3	1.79
600	111.2	-1598.72	154.9	0.3185	110.051	1.31	40.4	2.70
700	114.6	-1587.42	172.3	0.3316	108.929	1.32	40.5	3.58
800	117.5	-1575.81	187.8	0.3416	107.468	1.32	40.7	4.43
900	120.0	-1563.93	201.8	0.3501	105.687	1.31	40.8	5.22
1000	122.3	-1551.81	214.6	0.3579	103.610	1.30	41.0	5.97
1100	124.4	-1539.48	226.3	0.3654	101.267	1.29	41.1	6.66
1200	126.5	-1526.93	237.2	0.3726	98.690	1.27	41.3	7.29
1300	128.5	-1514.18	247.4	0.3797	95.916	1.25	41.4	7.86
1400	130.4	-1501.24	257.0	0.3868	92.980	1.22	41.6	8.35
1500	132.3	-1488.10	266.1	0.3940	89.919	1.20	41.7	8.78
1600	134.2	-1474.78	274.7	0.4011	86.768	1.17	41.9	9.15
1700	136.0	-1461.27	282.9	0.4083	83.561	1.14	42.1	9.45
1800	137.9	-1447.58	290.7	0.4156	80.327	1.11	42.3	9.70
1900	139.7	-1433.70	298.2	0.4229	77.095	1.08	42.4	9.89
2000	141.5	-1419.64	305.4	0.4302	73.888	1.04	42.6	10.03
2100	143.3	-1405.40	312.4	0.4376	70.727	1.01	42.8	10.12
2200	145.1	-1390.98	319.1	0.4450	67.630	0.98	43.0	10.17
2300	146.9	-1376.39	325.6	0.4524	64.612	0.94	43.2	10.18
2400	148.7	-1361.61	331.9	0.4599	61.682	0.91	43.4	10.16
2500	150.4	-1346.65	338.0	0.4674	58.850	0.88	43.6	10.11
2600	152.2	-1331.52	343.9	0.4750	56.121	0.85	43.8	10.03
2700	154.0	-1316.21	349.7	0.4826	53.501	0.82	44.0	9.94
2800	155.8	-1300.73	355.3	0.4902	50.990	0.79	44.2	9.82
2900	157.5	-1285.06	360.8	0.4978	48.590	0.76	44.4	9.69
3000	159.3	-1269.22	366.2	0.5055	46.300	0.73	44.7	9.55

$CaSiO_3$
PSWO

			<INT VdP (J)> PRESSURE (GPa)				
T (K)	1	5	10	20	30	40	50
298	39902	196202	385019	744866	1085779	1411579	1724854
300	39903	196209	385033	744893	1085817	1411630	1724916
400	39995	196659	385913	746586	1088274	1414813	1728794
500	40108	197203	386956	748534	1091036	1418322	1733001
600	40231	197787	388057	750532	1093802	1421768	1737058
700	40361	198390	389176	752500	1096455	1424993	1740774
800	40494	199004	390294	754401	1098939	1427926	1744060
900	40632	199624	391401	756216	1101227	1430532	1746872
1000	40772	200249	392493	757935	1103304	1432790	1749187
1100	40915	200875	393567	759552	1105161	1434691	1750997
1200	41061	201504	394620	761063	1106794	1436232	1752299
1300	41210	202133	395652	762467	1108203	1437415	1753097
1400	41362	202764	396663	763763	1109390	1438245	1753403
1500	41516	203395	397650	764952	1110357	1438729	1753228
1600	41673	204027	398615	766035	1111110	1438877	1752590
1700	41834	204659	399558	767013	1111654	1438702	1751508
1800	41996	205292	400478	767890	1111996	1438216	1750002
1900	42162	205925	401375	768669	1112146	1437436	1748097
2000	42331	206559	402251	769352	1112111	1436376	1745817
2100	42503	207193	403106	769943	1111900	1435055	1743188
2200	42677	207828	403940	770447	1111526	1433489	1740236
2300	42855	208464	404754	770868	1110996	1431697	1736987
2400	43036	209100	405549	771211	1110324	1429698	1733470
2500	43220	209738	406325	771480	1109519	1427509	1729709
2600	43407	210376	407085	771681	1108593	1425150	1725733
2700	43597	211016	407828	771818	1107556	1422638	1721566
2800	43791	211657	408556	771898	1106421	1419991	1717233
2900	43987	212299	409270	771925	1105197	1417226	1712759
3000	44187	212944	409971	771904	1103896	1414361	1708167

PYROPHYLLITE

$Al_2Si_4O_{10}(OH)_2$
PYROPHYLLITE Formula wt. (g) = 360.32

Enthalpy (J/mol) = -5640960.000 Holland and Powell (1990)

Entropy (J/mol.K) = 239.400 Holland and Powell (1990)

Heat Capacity:
No. of *Cp* ranges = 1 Holland and Powell (1990), refitted

4.4308800D+02 3.9098D-02 -3.6205D+07 0.0000D+00
4.8694D+09 0.0000D+00 1.8121D+04

Volume (cc/mol) = 128.100 Holland and Powell (1990)

Thermal expn:
1.79869D-05 1.64094D-08 5.95190D-03 -1.77807D+00
This study

Isot.compressibility:
1.39402D-06 2.91563D-10 -2.65720D-14 3.77607D-17
This study

P deriv. 4.00000D+00
T deriv. 0.00000D+00
Melting pt. and enthalpy of melting:
- -

Thermochemical Data

T (K)	C_p(J)	H(kJ)	S(J)	α 10^{-4}	K 10^4	γ	V(cc)	δ_T
298	292.0	-5640.96	239.4	0.2284	67.586	0.68	128.1	8.46
300	293.3	-5640.42	241.2	0.2299	67.562	0.68	128.1	8.40
400	353.8	-5607.86	334.4	0.2832	66.277	0.69	128.4	6.75
500	393.0	-5570.38	417.9	0.3098	65.025	0.67	128.8	6.16
600	418.7	-5529.71	492.0	0.3281	63.794	0.65	129.2	5.84
700	436.7	-5486.89	558.0	0.3435	62.576	0.65	129.7	5.65
800	450.0	-5442.53	617.2	0.3578	61.365	0.65	130.1	5.52
900	460.4	-5396.99	670.8	0.3717	60.153	0.65	130.6	5.43
1000	469.0	-5350.51	719.8	0.3857	58.935	0.65	131.1	5.37
1100	476.3	-5303.24	764.8	0.3998	57.709	0.66	131.6	5.34
1200	482.8	-5255.28	806.6	0.4140	56.469	0.66	132.1	5.33
1300	488.6	-5206.70	845.4	0.4285	55.215	0.67	132.7	5.33
1400	494.1	-5157.56	881.9	0.4430	53.945	0.67	133.3	5.35
1500	499.2	-5107.90	916.1	0.4578	52.659	0.68	133.9	5.37
1600	504.0	-5057.74	948.5	0.4727	51.357	0.68	134.5	5.39
1700	508.7	-5007.10	979.2	0.4877	50.040	0.69	135.2	5.42
1800	513.2	-4956.01	1008.4	0.5028	48.710	0.69	135.8	5.45
1900	517.6	-4904.47	1036.3	0.5180	47.369	0.69	136.5	5.48
2000	521.9	-4852.49	1062.9	0.5334	46.021	0.69	137.2	5.51
2100	526.1	-4800.09	1088.5	0.5488	44.666	0.69	138.0	5.53
2200	530.3	-4747.26	1113.1	0.5643	43.310	0.69	138.8	5.55
2300	534.4	-4694.03	1136.7	0.5798	41.955	0.69	139.6	5.56
2400	538.5	-4640.38	1159.5	0.5954	40.606	0.69	140.4	5.57
2500	542.6	-4586.32	1181.6	0.6111	39.264	0.69	141.2	5.57
2600	546.6	-4531.86	1203.0	0.6268	37.934	0.69	142.1	5.56
2700	550.6	-4476.99	1223.7	0.6425	36.620	0.68	143.0	5.55
2800	554.6	-4421.73	1243.8	0.6583	35.323	0.68	143.9	5.53
2900	558.6	-4366.07	1263.3	0.6742	34.048	0.67	144.9	5.51
3000	562.6	-4310.01	1282.3	0.6900	32.797	0.67	145.9	5.47

$Al_2Si_4O_{10}(OH)_2$
PYROPHYLL

	<INT VdP (J)>						
	PRESSURE (GPa)						
T(K)	1	5	10	20	30	40	50
298	127174	619294	1203656	2295723	3311995	4271296	5185248
300	127179	619313	1203684	2295750	3312008	4271288	5185213
400	127493	620557	1205579	2297953	3313765	4272164	5184931
500	127856	622032	1207916	2300994	3316731	4274602	5186551
600	128246	623640	1210502	2304490	3320341	4277864	5189164
700	128659	625345	1213266	2308297	3324377	4281655	5192406
800	129092	627136	1216177	2312346	3328729	4285827	5196084
900	129543	629005	1219220	2316598	3333332	4290285	5200078
1000	130013	630950	1222384	2321024	3338137	4294960	5204295
1100	130502	632969	1225664	2325604	3343104	4299793	5208658
1200	131011	635061	1229053	2330317	3348200	4304733	5213098
1300	131539	637225	1232547	2335148	3353393	4309734	5217553
1400	132086	639461	1236140	2340078	3358652	4314751	5221967
1500	132654	641767	1239829	2345094	3363950	4319746	5226288
1600	133242	644143	1243607	2350180	3369263	4324683	5230468
1700	133851	646588	1247471	2355322	3374566	4329527	5234466
1800	134481	649101	1251417	2360508	3379839	4334251	5238245
1900	135132	651682	1255441	2365727	3385062	4338829	5241774
2000	135804	654329	1259538	2370968	3390220	4343239	5245026
2100	136499	657042	1263706	2376221	3395299	4347464	5247981
2200	137215	659820	1267940	2381479	3400287	4351491	5250625
2300	137954	662663	1272239	2386736	3405176	4355310	5252949
2400	138716	665570	1276600	2391986	3409960	4358916	5254947
2500	139500	668540	1281020	2397226	3414636	4362308	5256623
2600	140308	671573	1285497	2402453	3419203	4365487	5257983
2700	141139	674669	1290031	2407666	3423663	4368462	5259038
2800	141994	677827	1294621	2412865	3428021	4371240	5259804
2900	142873	681047	1299266	2418054	3432283	4373836	5260301
3000	143776	684330	1303965	2423234	3436459	4376265	5260554

TREMOLITE

$Ca_2Mg_5Si_8O_{22}(OH)_2$
TREMOLITE Formula wt. (g) = 812.39

Enthalpy (J/mol) = -12313600.000 This study
Entropy (J/mol.K) = 550.000 Holland and Powell (1990)

Heat Capacity:
No. of *Cp* ranges = 1 Holland and Powell (1990),refitted
1.0407500D+03 6.9000D-02 -3.2361D+07 0.0000D+00
5.0535D+09 0.0000D+00 -6.9176D+04
Volume (cc/mol) = 272.700 Holland and Powell (1990)

Thermal expn:
7.46263D-05 5.23070D-09 -1.31680D-02 -5.68020D-01
This study

Isot.compressibility:
1.82570D-06 4.87490D-11 2.34910D-13 -2.81080D-17
This study

P deriv. 4.00000D+00
T deriv. 0.00000D+00
Melting pt. and enthalpy of melting:
- -

Thermochemical Data

T(K)	C_p(J)	H(kJ)	S(J)	α 10^{-4}	K 10^4	γ	V(cc)	δ_T
298	655.9	-12313.60	550.0	0.2563	53.753	0.58	272.7	3.80
300	658.5	-12312.38	554.1	0.2599	53.743	0.58	272.7	3.77
400	772.1	-12240.48	760.1	0.4025	53.164	0.77	273.6	2.95
500	847.9	-12159.23	941.1	0.4863	52.485	0.84	274.9	2.83
600	900.4	-12071.67	1100.6	0.5424	51.721	0.89	276.3	2.86
700	938.9	-11979.62	1242.4	0.5832	50.883	0.91	277.8	2.93
800	968.8	-11884.18	1369.8	0.6146	49.984	0.93	279.5	3.01
900	993.0	-11786.05	1485.4	0.6400	49.035	0.94	281.3	3.09
1000	1013.3	-11685.71	1591.1	0.6612	48.048	0.94	283.1	3.16
1100	1030.8	-11583.49	1688.5	0.6794	47.033	0.95	285.0	3.21
1200	1046.4	-11479.61	1778.9	0.6954	46.000	0.95	287.0	3.25
1300	1060.4	-11374.27	1863.2	0.7096	44.958	0.95	289.0	3.28
1400	1073.3	-11267.58	1942.3	0.7225	43.913	0.94	291.1	3.29
1500	1085.2	-11159.64	2016.7	0.7344	42.872	0.94	293.2	3.29
1600	1096.5	-11050.55	2087.1	0.7454	41.842	0.93	295.4	3.28
1700	1107.2	-10940.36	2153.9	0.7558	40.827	0.93	297.6	3.26
1800	1117.4	-10829.13	2217.5	0.7655	39.831	0.92	299.9	3.23
1900	1127.2	-10716.89	2278.2	0.7748	38.857	0.92	302.2	3.19
2000	1136.7	-10603.70	2336.2	0.7836	37.908	0.91	304.6	3.15
2100	1145.9	-10489.56	2391.9	0.7921	36.986	0.90	307.0	3.10
2200	1154.9	-10374.52	2445.4	0.8003	36.093	0.90	309.4	3.04
2300	1163.7	-10258.59	2497.0	0.8082	35.230	0.89	311.9	2.98
2400	1172.3	-10141.79	2546.7	0.8159	34.397	0.88	314.5	2.91
2500	1180.7	-10024.14	2594.7	0.8234	33.596	0.88	317.1	2.84
2600	1189.0	-9905.65	2641.2	0.8308	32.825	0.87	319.7	2.77
2700	1197.2	-9786.34	2686.2	0.8379	32.087	0.87	322.4	2.69
2800	1205.3	-9666.20	2729.9	0.8450	31.379	0.86	325.1	2.61
2900	1213.4	-9545.27	2772.3	0.8519	30.702	0.86	327.9	2.53
3000	1221.3	-9423.54	2813.6	0.8587	30.055	0.85	330.7	2.45

$Ca_2Mg_5Si_8O_{22}(OH)_2$
TREMOLITE

	<INT VdP (J)>						
	PRESSURE(GPa)						
T(K)	1	5	10	20	30	40	50
298	270238	1308153	2528557	4786099	6868786	8823560	10678273
300	270251	1308207	2528649	4786243	6868964	8823761	10678490
400	271141	1312110	2535487	4797382	6883204	8840389	10697028
500	272325	1317357	2544782	4812817	6903273	8864185	10723938
600	273698	1323431	2555540	4830682	6926510	8891752	10755130
700	275204	1330073	2567265	4850058	6951618	8921443	10788631
800	276812	1337137	2579680	4870441	6977894	8952378	10823398
900	278504	1344533	2592619	4891536	7004935	8984062	10858854
1000	280267	1352202	2605977	4913162	7032507	9016215	10894683
1100	282093	1360107	2619684	4935211	7060476	9048689	10930726
1200	283975	1368221	2633698	4957621	7088772	9081415	10966922
1300	285910	1376525	2647989	4980356	7117367	9114377	11003273
1400	287893	1385008	2662540	5003404	7146261	9147596	11039821
1500	289923	1393661	2677342	5026763	7175474	9181117	11076640
1600	291997	1402478	2692390	5050446	7205040	9214998	11113818
1700	294116	1411457	2707685	5074470	7235001	9249312	11151457
1800	296277	1420596	2723230	5098857	7265405	9284137	11189665
1900	298479	1429895	2739031	5123634	7296307	9319553	11228555
2000	300724	1439355	2755095	5148831	7327761	9355647	11268242
2100	303010	1448978	2771433	5174479	7359826	9392504	11308839
2200	305337	1458767	2788055	5200611	7392560	9430208	11350459
2300	307706	1468725	2804971	5227261	7426022	9468846	11393215
2400	310117	1478856	2822195	5254462	7460270	9508502	11437215
2500	312569	1489164	2839739	5282249	7495364	9549257	11482567
2600	315064	1499654	2857616	5310657	7531361	9591194	11529374
2700	317601	1510330	2875840	5339722	7568319	9634393	11577738
2800	320181	1521197	2894425	5369479	7606295	9678931	11627760
2900	322805	1532262	2913386	5399962	7645346	9724887	11679535
3000	325473	1543529	2932738	5431208	7685526	9772335	11733161

GLAUCOPHANE

$Na_2Mg_3Al_2Si_8O_{22}(OH)_2$
GLAUCOPHANE Formula wt. (g) = 863.72

Enthalpy (J/mol) = -11954100.000 Holland and Powell (1990)

Entropy (J/mol.K) = 535.000 Holland and Powell (1990)

Heat Capacity:
No. of *Cp* ranges = 1 Gillet et al. (1989), refitted

9.5061000D+02 6.9754D-02 -6.0531D+07 0.0000D+00
8.3164D+09 0.0000D+00 1.1304D+04

Volume (cc/mol) = 262.500 Gillet et al. (1989)

Thermal expn:
4.41600D-05 9.72700D-09 -1.54300D-02 2.22600D+00
This study

Isot.compressibility:
1.45630D-06 1.19790D-10 8.47000D-14 4.20520D-18
This study

P deriv. 4.00000D+00
T deriv. 0.00000D+00
Melting pt. and enthalpy of melting:
\- -

Thermochemical Data

T(K)	C_p(J)	H(kJ)	S(J)	α 10^{-4}	K 10^4	γ	V(cc)	δ_T
298	642.2	-11954.10	535.0	0.2035	66.682	0.56	262.5	5.62
300	644.7	-11952.91	539.0	0.2038	66.668	0.56	262.5	5.62
400	758.4	-11882.38	741.1	0.2339	65.875	0.54	263.1	5.34
500	832.5	-11802.57	918.9	0.2707	65.024	0.56	263.7	4.99
600	881.7	-11716.71	1075.3	0.3046	64.120	0.59	264.5	4.76
700	916.3	-11626.72	1213.9	0.3347	63.167	0.62	265.4	4.62
800	942.2	-11533.74	1338.1	0.3613	62.170	0.65	266.3	4.53
900	962.6	-11438.46	1450.3	0.3852	61.133	0.67	267.3	4.48
1000	979.5	-11341.33	1552.6	0.4068	60.060	0.69	268.3	4.46
1100	993.8	-11242.65	1646.6	0.4267	58.957	0.70	269.5	4.44
1200	1006.5	-11142.62	1733.6	0.4452	57.827	0.72	270.6	4.43
1300	1018.0	-11041.39	1814.7	0.4625	56.676	0.73	271.9	4.43
1400	1028.5	-10939.06	1890.5	0.4789	55.508	0.74	273.1	4.42
1500	1038.3	-10835.71	1961.8	0.4945	54.326	0.75	274.5	4.42
1600	1047.7	-10731.41	2029.1	0.5095	53.134	0.76	275.9	4.41
1700	1056.6	-10626.19	2092.9	0.5239	51.938	0.76	277.3	4.40
1800	1065.2	-10520.10	2153.5	0.5378	50.739	0.77	278.8	4.39
1900	1073.5	-10413.16	2211.3	0.5514	49.541	0.77	280.3	4.38
2000	1081.7	-10305.40	2266.6	0.5646	48.348	0.77	281.9	4.36
2100	1089.6	-10196.83	2319.6	0.5774	47.162	0.78	283.5	4.34
2200	1097.5	-10087.48	2370.5	0.5901	45.986	0.78	285.1	4.31
2300	1105.2	-9977.34	2419.4	0.6024	44.822	0.78	286.8	4.29
2400	1112.8	-9866.44	2466.6	0.6146	43.672	0.78	288.6	4.26
2500	1120.4	-9754.78	2512.2	0.6266	42.538	0.77	290.4	4.22
2600	1127.8	-9642.37	2556.3	0.6384	41.421	0.77	292.2	4.19
2700	1135.3	-9529.21	2599.0	0.6501	40.323	0.77	294.1	4.15
2800	1142.6	-9415.32	2640.4	0.6617	39.245	0.77	296.1	4.11
2900	1149.9	-9300.69	2680.6	0.6731	38.189	0.77	298.0	4.07
3000	1157.2	-9185.33	2719.7	0.6844	37.154	0.76	300.1	4.03

$Na_2Mg_3Al_2Si_8O_{22}(OH)_2$
GLAUCOPHANE

<INT VdP (J)>

	PRESSURE (GPa)						
T(K)	1	5	10	20	30	40	50
298	260579	1268516	2464730	4698935	6777026	8737962	10605740
300	260588	1268555	2464795	4699027	6777126	8738060	10605829
400	261132	1270837	2468557	4704392	6783047	8743918	10611221
500	261767	1273525	2473040	4710960	6790536	8751641	10618724
600	262495	1276625	2478259	4718761	6799637	8761283	10628408
700	263306	1280096	2484129	4727630	6810114	8772543	10639907
800	264193	1283891	2490562	4737400	6821727	8785114	10652852
900	265147	1287974	2497485	4747931	6834276	8798740	10666936
1000	266162	1292314	2504839	4759114	6847605	8813224	10681923
1100	267234	1296890	2512578	4770861	6861591	8828412	10697629
1200	268360	1301681	2520668	4783106	6876142	8844185	10713918
1300	269536	1306674	2529078	4795795	6891183	8860456	10730685
1400	270760	1311858	2537787	4808888	6906659	8877155	10747851
1500	272030	1317222	2546776	4822352	6922527	8894232	10765360
1600	273346	1322760	2556031	4836160	6938754	8911648	10783172
1700	274705	1328465	2565540	4850294	6955315	8929379	10801259
1800	276108	1334333	2575293	4864738	6972193	8947405	10819606
1900	277553	1340359	2585284	4879482	6989376	8965717	10838205
2000	279041	1346542	2595507	4894516	7006857	8984310	10857057
2100	280570	1352878	2605957	4909836	7024633	9003185	10876166
2200	282140	1359365	2616630	4925438	7042703	9022347	10895546
2300	283752	1366003	2627525	4941321	7061070	9041805	10915210
2400	285405	1372790	2638640	4957486	7079740	9061570	10935177
2500	287100	1379726	2649975	4973935	7098720	9081655	10955469
2600	288836	1386811	2661529	4990670	7118018	9102077	10976110
2700	290613	1394044	2673303	5007697	7137645	9122854	10997125
2800	292432	1401426	2685298	5025020	7157612	9144004	11018543
2900	294294	1408958	2697515	5042645	7177933	9165549	11040392
3000	296198	1416641	2709957	5060579	7198621	9187509	11062702

ZOISITE

$Ca_2Al_3Si_3O_{12}(OH)$
ZOISITE Formula wt. (g) = 454.37

Enthalpy (J/mol) =	-6899000.000	This study
Entropy (J/mol.K) =	294.000	This study
Heat Capacity:		
No. of *Cp* ranges = 1		Holland and Powell (1990)

6.9750000D+02 1.3263D-02 -3.7941D+06 0.0000D+00
0.0000D+00 -5.3179D+03 0.0000D+00

Volume (cc/mol) =	135.580	Holland and Powell (1990)

Thermal expn:
2.54500D-05 0.00000D+00 0.00000D+00 0.00000D+00
Isot.compressibility:
5.88000D-07 0.00000D+00 0.00000D+00 0.00000D+00
P deriv. 4.00000D+00
T deriv. 0.00000D+00
Melting pt. and enthalpy of melting:
\- \-

Thermochemical Data

T(K)	C_p(J)	H(kJ)	S(J)
298	350.8	-6899.00	294.0
300	352.3	-6898.35	296.2
400	413.2	-6859.80	406.7
500	451.1	-6816.45	503.2
600	477.8	-6769.94	588.0
700	498.0	-6721.10	663.2
800	514.2	-6670.47	730.8
900	527.5	-6618.36	792.1
1000	538.8	-6565.03	848.3
1100	548.6	-6510.65	900.1
1200	557.3	-6455.35	948.3
1300	565.0	-6399.23	993.2
1400	572.0	-6342.37	1035.3
1500	578.4	-6284.85	1075.0
1600	584.3	-6226.71	1112.5
1700	589.8	-6168.00	1148.1
1800	594.9	-6108.77	1182.0
1900	599.6	-6049.04	1214.3
2000	604.2	-5988.85	1245.1
2100	608.4	-5928.22	1274.7
2200	612.5	-5867.17	1303.1
2300	616.4	-5805.72	1330.4
2400	620.1	-5743.89	1356.7
2500	623.7	-5681.70	1382.1
2600	627.1	-5619.16	1406.6
2700	630.4	-5556.28	1430.4
2800	633.7	-5493.07	1453.4
2900	636.8	-5429.55	1475.7
3000	639.8	-5365.72	1497.3

CLINOCHLORITE

$Mg_5Al_2Si_3O_{10}(OH)_8$
CLINOCHLORITE Formula wt. (g) = 555.81

Enthalpy (J/mol) = -8928250.000 This study
Entropy (J/mol.K) = 421.000 Holland and Powell (1990)

Heat Capacity:
No. of *Cp* ranges = 1 Holland and Powell (1990)

1.1618000D+03 1.0133D-02 -7.6573D+06 0.0000D+00
0.0000D+00 -9.6909D+03 0.0000D+00

Volume (cc/mol) = 210.900 Holland and Powell (1990)

Thermal expn:
2.79800D-05 0.00000D+00 0.00000D+00 0.00000D+00
Holland and Powell (1990)

Isot.compressibility:
1.81000D-06 0.00000D+00 0.00000D+00 0.00000D+00
Holland and Powell (1990)

P deriv. 4.00000D+00
T deriv. 0.00000D+00
Melting pt. and enthalpy of melting:
- -

Thermochemical Data

T (K)	C_p (J)	H (kJ)	S (J)
298	517.4	-8928.25	421.0
300	520.3	-8927.29	424.2
400	633.5	-8869.07	590.9
500	702.8	-8802.02	740.3
600	751.0	-8729.20	872.9
700	787.0	-8652.22	991.5
800	815.3	-8572.05	1098.5
900	838.4	-8489.33	1195.9
1000	857.8	-8404.49	1285.3
1100	874.4	-8317.86	1367.8
1200	888.9	-8229.68	1444.5
1300	901.7	-8140.14	1516.2
1400	913.1	-8049.39	1583.4
1500	923.4	-7957.56	1646.8
1600	932.7	-7864.74	1706.7
1700	941.3	-7771.03	1763.5
1800	949.3	-7676.50	1817.5
1900	956.6	-7581.20	1869.1
2000	963.5	-7485.19	1918.3
2100	969.9	-7388.52	1965.5
2200	975.9	-7291.23	2010.7
2300	981.6	-7193.36	2054.2
2400	987.0	-7094.92	2096.1
2500	992.1	-6995.97	2136.5
2600	997.0	-6896.51	2175.5
2700	1001.6	-6796.58	2213.2
2800	1006.1	-6696.20	2249.8
2900	1010.3	-6595.38	2285.1
3000	1014.4	-6494.14	2319.4

ARAGONITE

$CaCO_3$

ARAGONITE Formula wt. (g) = 100.09

Enthalpy (J/mol) = -1208160.000 Holland and Powell (1990)

Entropy (J/mol.K) = 88.000 Holland and Powell (1990)

Heat Capacity:

No. of *Cp* ranges = 1 Holland and Powell (1990)

8.4200005D+01 4.2844D-02 -1.3975D+06 0.0000D+00

0.0000D+00 0.0000D+00 0.0000D+00

Volume (cc/mol) = 34.150 Holland and Powell (1990)

Thermal expn:

2.43100D-05 0.00000D+00 0.00000D+00 0.00000D+00

Holland and Powell (1990)

Isot.compressibility:

1.55300D-06 0.00000D+00 0.00000D+00 0.00000D+00

Holland and Powell (1990)

P deriv. 4.00000D+00

T deriv. 0.00000D+00

Melting pt. and enthalpy of melting:

\- -

Thermochemical Data

T(K)	C_p(J)	H(kJ)	S(J)
298	81.3	-1208.16	88.0
300	81.5	-1208.01	88.5
400	92.6	-1199.25	113.6
500	100.0	-1189.61	135.1
600	106.0	-1179.29	153.9
700	111.3	-1168.42	170.6
800	116.3	-1157.04	185.8
900	121.0	-1145.17	199.8
1000	125.6	-1132.84	212.8
1100	130.2	-1120.04	225.0
1200	134.6	-1106.80	236.5
1300	139.1	-1093.12	247.5
1400	143.5	-1078.99	257.9
1500	147.8	-1064.42	268.0
1600	152.2	-1049.42	277.7
1700	156.6	-1033.98	287.0
1800	160.9	-1018.11	296.1
1900	165.2	-1001.81	304.9
2000	169.5	-985.07	313.5
2100	173.9	-967.90	321.9
2200	178.2	-950.30	330.1
2300	182.5	-932.27	338.1
2400	186.8	-913.80	345.9
2500	191.1	-894.91	353.6
2600	195.4	-875.59	361.2
2700	199.7	-855.83	368.7
2800	204.0	-835.65	376.0
2900	208.3	-815.03	383.2
3000	212.6	-793.99	390.4

MARGARITE

$CaAl_4Si_2O_{10}(OH)_2$
MARGARITE Formula wt. (g) = 398.20

Enthalpy (J/mol) = -6244640.000 Holland and Powell (1990)
Entropy (J/mol.K) = 265.000 Holland and Powell (1990)
Heat Capacity:
No. of *Cp* ranges = 1 Holland and Powell (1990)
7.4440002D+02 -1.6800D-02 -2.0744D+06 0.0000D+00
0.0000D+00 -6.7832D+03 0.0000D+00
Volume (cc/mol) = 129.640 Holland and Powell (1990)
Thermal expn:
2.78000D-05 0.00000D+00 0.00000D+00 0.00000D+00
Holland and Powell (1990)
Isot.compressibility:
1.02000D-06 0.00000D+00 0.00000D+00 0.00000D+00
Holland and Powell (1990)
P deriv. 4.00000D+00
T deriv. 0.00000D+00
Melting pt. and enthalpy of melting:
- -

NOTE:

Thermochemical Data

T(K)	C_p(J)	H(kJ)	S(J)
298	323.2	-6244.64	265.0
300	324.7	-6244.04	267.0
400	385.6	-6208.27	369.5
500	424.3	-6167.65	460.0
600	451.6	-6123.78	539.9
700	472.0	-6077.55	611.1
800	487.9	-6029.52	675.2
900	500.6	-5980.07	733.4
1000	511.0	-5929.48	786.7
1100	519.7	-5877.93	835.9
1200	527.0	-5825.58	881.4
1300	533.2	-5772.57	923.8
1400	538.5	-5718.97	963.6
1500	543.1	-5664.88	1000.9
1600	547.1	-5610.37	1036.1
1700	550.6	-5555.48	1069.3
1800	553.6	-5500.26	1100.9
1900	556.3	-5444.76	1130.9
2000	558.6	-5389.01	1159.5
2100	560.6	-5333.05	1186.8
2200	562.4	-5276.90	1212.9
2300	563.9	-5220.58	1238.0
2400	565.3	-5164.12	1262.0
2500	566.4	-5107.53	1285.1
2600	567.4	-5050.84	1307.3
2700	568.2	-4994.06	1328.7
2800	568.9	-4937.20	1349.4
2900	569.5	-4880.28	1369.4
3000	569.9	-4823.31	1388.7

MERWINITE

$Ca_3MgSi_2O_8$
MERWINITE Formula wt. (g) = 328.72

Enthalpy (J/mol) = -4537320.000 Holland and Powell (1990)
Entropy (J/mol.K) = 252.200 Holland and Powell (1990)
Heat Capacity:
No. of *Cp* ranges = 1 Holland and Powell (1990)
4.4160001D+02 2.2280D-03 -2.0029D+06 0.0000D+00
0.0000D+00 -2.8882D+03 0.0000D+00
Volume (cc/mol) = 98.470 Holland and Powell (1990)
Thermal expn:
3.55000D-05 0.00000D+00 0.00000D+00 0.00000D+00
Holland and Powell (1990)
Isot.compressibility:
8.02000D-07 0.00000D+00 0.00000D+00 0.00000D+00
Holland and Powell (1990)
P deriv. 4.00000D+00
T deriv. 0.00000D+00
Melting pt. and enthalpy of melting:
- -

Thermochemical Data

T(K)	C_p(J)	H(kJ)	S(J)
298	252.5	-4537.32	252.2
300	253.3	-4536.85	253.8
400	285.6	-4509.76	331.5
500	305.5	-4480.14	397.5
600	319.5	-4448.85	454.5
700	329.9	-4416.36	504.6
800	338.1	-4382.94	549.2
900	344.9	-4348.78	589.4
1000	350.5	-4314.01	626.0
1100	355.3	-4278.71	659.7
1200	359.5	-4242.97	690.8
1300	363.2	-4206.83	719.7
1400	366.5	-4170.34	746.7
1500	369.5	-4133.54	772.1
1600	372.2	-4096.45	796.1
1700	374.6	-4059.11	818.7
1800	376.9	-4021.53	840.2
1900	379.0	-3983.73	860.6
2000	381.0	-3945.73	880.1
2100	382.8	-3907.54	898.7
2200	384.5	-3869.17	916.6
2300	386.1	-3830.64	933.7
2400	387.6	-3791.95	950.2
2500	389.1	-3753.11	966.0
2600	390.5	-3714.14	981.3
2700	391.8	-3675.03	996.1
2800	393.0	-3635.79	1010.4
2900	394.2	-3596.43	1024.2
3000	395.3	-3556.95	1037.5

ULVOSPINEL

Fe_2TiO_4
ULVOSPINEL Formula wt. (g) = 223.59

Enthalpy (J/mol) = -1507330.000 Holland and Powell (1990)
Entropy (J/mol.K) = 168.900 Holland and Powell (1990)
Heat Capacity:
No. of *Cp* ranges = 1 Holland and Powell (1990)
-1.0260000D+02 1.4252D-01 -9.1445D+06 0.0000D+00
0.0000D+00 5.2707D+03 0.0000D+00
Volume (cc/mol) = 46.820 Holland and Powell (1990)
Thermal expn:
4.10000D-05 0.00000D+00 0.00000D+00 0.00000D+00
Holland and Powell (1990)
Isot.compressibility:
5.55000D-07 0.00000D+00 0.00000D+00 0.00000D+00
Holland and Powell (1990)
P deriv. 4.00000D+00
T deriv. 0.00000D+00
Melting pt. and enthalpy of melting:
- -

Thermochemical Data

T (K)	C_p (J)	H (kJ)	S (J)
298	142.3	-1507.33	168.9
300	142.9	-1507.07	169.8
400	160.8	-1491.71	213.8
500	167.8	-1475.25	250.5
600	172.7	-1458.22	281.6
700	177.7	-1440.70	308.6
800	183.5	-1422.65	332.7
900	190.1	-1403.98	354.6
1000	197.4	-1384.61	375.0
1100	205.5	-1364.47	394.2
1200	214.2	-1343.48	412.5
1300	223.4	-1321.60	430.0
1400	233.1	-1298.78	446.9
1500	243.2	-1274.97	463.3
1600	253.6	-1250.13	479.4
1700	264.4	-1224.23	495.1
1800	275.3	-1197.25	510.5
1900	286.6	-1169.15	525.7
2000	298.0	-1139.93	540.7
2100	309.6	-1109.54	555.5
2200	321.4	-1077.99	570.2
2300	333.4	-1045.25	584.7
2400	345.4	-1011.31	599.1
2500	357.7	-976.16	613.5
2600	370.0	-939.78	627.8
2700	382.4	-902.16	642.0
2800	394.9	-863.30	656.1
2900	407.5	-823.18	670.2
3000	420.2	-781.80	684.2

CLINOZOISITE

$Ca_2Al_3Si_3O_{12}(OH)$
CLINOZOISITE Formula wt. (g) = 454.37

Enthalpy (J/mol) = -6900990.000 Holland and Powell (1990)
Entropy (J/mol.K) = 291.000 Holland and Powell (1990)
Heat Capacity:
No. of *Cp* ranges = 1 Holland and Powell (1990)
6.9750000D+02 1.3263D-02 -3.7941D+06 0.0000D+00
0.0000D+00 -5.3179D+03 0.0000D+00
Volume (cc/mol) = 136.730 Holland and Powell (1990)
Thermal expn:
2.56000D-05 0.00000D+00 0.00000D+00 0.00000D+00
Holland and Powell (1990)
Isot.compressibility:
5.85000D-07 0.00000D+00 0.00000D+00 0.00000D+00
Holland and Powell (1990)
P deriv. 4.00000D+00
T deriv. 0.00000D+00 This study
Melting pt. and enthalpy of melting:
- -

Thermochemical Data

T(K)	C_p(J)	H(kJ)	S(J)
298	350.8	-6900.99	291.0
300	352.3	-6900.34	293.2
400	413.2	-6861.79	403.7
500	451.1	-6818.44	500.2
600	477.8	-6771.93	585.0
700	498.0	-6723.09	660.2
800	514.2	-6672.46	727.8
900	527.5	-6620.35	789.1
1000	538.8	-6567.02	845.3
1100	548.6	-6512.64	897.1
1200	557.3	-6457.34	945.3
1300	565.0	-6401.22	990.2
1400	572.0	-6344.36	1032.3
1500	578.4	-6286.84	1072.0
1600	584.3	-6228.70	1109.5
1700	589.8	-6169.99	1145.1
1800	594.9	-6110.76	1179.0
1900	599.6	-6051.03	1211.3
2000	604.2	-5990.84	1242.1
2100	608.4	-5930.21	1271.7
2200	612.5	-5869.16	1300.1
2300	616.4	-5807.71	1327.4
2400	620.1	-5745.88	1353.7
2500	623.7	-5683.69	1379.1
2600	627.1	-5621.15	1403.6
2700	630.4	-5558.27	1427.4
2800	633.7	-5495.06	1450.4
2900	636.8	-5431.54	1472.7
3000	639.8	-5367.71	1494.3

Ca-TSCHERMAK

$CaAl_2SiO_6$
CA-TSCHERMAK Formula wt. (g) = 218.13

Enthalpy (J/mol) =	-3296800.000	Chatterjee (1989)
Entropy (J/mol.K) =	144.500	Chatterjee (1989)
Heat Capacity:		
No. of *Cp* ranges = 1		Saxena and Chatterjee (1986), refitted

2.8008000D+02 2.6850D-03 -5.3440D+06 0.0000D+00
9.6110D+08 0.0000D+00 -2.7110D+04

Volume (cc/mol) =	63.560	Robie et al. (1978)
Thermal expn:		

3.61989D-05 4.75570D-09 8.73824D-03 -4.02769D+00
This study

Isot.compressibility:
7.99509D-07 2.61810D-10 -4.49603D-14 3.68373D-17
This study

P deriv.	4.20000D+00	As in clinoenstatite
T deriv.	4.00000D-04	This study

Melting pt. and enthalpy of melting:
\- -

NOTE: Calorimetric data: H: -3296.62±1.63, Charlu et al. (1978); S: 135.3, Haselton et al. (1984). Ca-Tschermak is related to grossular and anorthite by the following equilibrium reactions:

Gr + Cor = CaTs (a)

An + Ge + Cor = CaTs (b)

Reaction (a) was studied by Gasparik (1984b), Hays (1966) and Wood (1978), and reaction (b) was studied by Hays (1966). Free energy of Ca-tschermak was calculated by Saxena and Chatterjee (1986) using Gasparik's (1984b) data on reaction (a) and (b) which yielded $-\Delta C'p=2220824(\pm130)/T^2$ with a corresponding $-\Delta H_f(298)$ of -3300902 J/mol, where, $-\Delta C'p$ is the difference between the experimental and the calculated heat capacities This $-\Delta H_f(298)$ is comparable to -3297994 J/mol [calculated from $-\Delta H_f(970)$] determined by Charlu et al. (1978). The enthalpy and entropy needed adjustment in this study while the heat capacity of Saxena and Chatterjee (1986) is retained. The data on *Cp* listed above represent the addition of $-\Delta C'p$ to the *Cp* expression of Haselton et al. (1984). Thermal expansion data are consistent with those of Haselton et al. (1984).

$CaAl_2SiO_6$

CA-TSCHERMAK

Thermochemical Data

T(K)	C_p(J)	H(kJ)	S(J)	α 10^{-4}	K 10^4	γ	V(cc)	δ_T
298	166.1	-3296.80	144.5	0.2162	114.345	0.95	63.6	12.95
300	166.7	-3296.49	145.5	0.2200	114.286	0.96	63.6	12.72
400	195.0	-3278.31	197.7	0.3477	111.186	1.29	63.7	7.79
500	213.5	-3257.82	243.3	0.3994	108.251	1.33	64.0	6.63
600	226.1	-3235.80	283.4	0.4243	105.445	1.31	64.3	6.15
700	235.1	-3212.71	319.0	0.4379	102.735	1.28	64.5	5.94
800	241.9	-3188.85	350.8	0.4463	100.096	1.25	64.8	5.84
900	247.1	-3164.39	379.6	0.4522	97.506	1.22	65.1	5.83
1000	251.3	-3139.46	405.9	0.4567	94.949	1.19	65.4	5.87
1100	254.7	-3114.16	430.0	0.4605	92.410	1.16	65.7	5.95
1200	257.6	-3088.54	452.3	0.4639	89.880	1.14	66.0	6.07
1300	260.0	-3062.66	473.0	0.4672	87.351	1.11	66.3	6.20
1400	262.1	-3036.56	492.4	0.4704	84.817	1.09	66.6	6.36
1500	263.9	-3010.25	510.5	0.4737	82.278	1.06	66.9	6.52
1600	265.6	-2983.78	527.6	0.4770	79.733	1.04	67.3	6.70
1700	267.0	-2957.14	543.7	0.4803	77.182	1.02	67.6	6.88
1800	268.4	-2930.37	559.0	0.4837	74.631	0.99	67.9	7.07
1900	269.6	-2903.47	573.6	0.4872	72.082	0.97	68.2	7.25
2000	270.7	-2876.46	587.4	0.4907	69.542	0.94	68.6	7.43
2100	271.7	-2849.34	600.7	0.4943	67.016	0.92	68.9	7.60
2200	272.7	-2822.12	613.3	0.4980	64.511	0.90	69.3	7.76
2300	273.5	-2794.81	625.5	0.5017	62.034	0.87	69.6	7.91
2400	274.4	-2767.42	637.1	0.5055	59.590	0.85	70.0	8.05
2500	275.2	-2739.94	648.3	0.5094	57.188	0.82	70.3	8.17
2600	275.9	-2712.39	659.1	0.5133	54.833	0.80	70.7	8.28
2700	276.6	-2684.76	669.6	0.5172	52.529	0.77	71.0	8.38
2800	277.3	-2657.07	679.6	0.5212	50.283	0.75	71.4	8.46
2900	277.9	-2629.31	689.4	0.5252	48.099	0.72	71.8	8.52
3000	278.5	-2601.48	698.8	0.5293	45.979	0.70	72.2	8.57

	<INT VdP (J)> PRESSURE(GPa)						
T(K)	1	5	10	20	30	40	50
298	63286	311326	611276	1183826	1727295	2247575	2748641
300	63288	311336	611290	1183837	1727295	2247558	2748604
400	63467	312086	612501	1185413	1728766	2248633	2749086
500	63699	313098	614236	1188054	1731878	2251957	2752460
600	63955	314229	616207	1191186	1735768	2256378	2757279
700	64224	315421	618300	1194578	1740083	2261417	2762929
800	64501	316648	620463	1198119	1744647	2266828	2769095
900	64783	317899	622668	1201747	1749362	2272469	2775586
1000	65070	319165	624898	1205423	1754162	2278245	2782273
1100	65360	320442	627142	1209122	1759001	2284088	2789063
1200	65653	321728	629395	1212823	1763845	2289945	2795882
1300	65949	323021	631649	1216512	1768666	2295774	2802673
1400	66249	324318	633900	1220177	1773441	2301540	2809389
1500	66551	325619	636145	1223806	1778152	2307216	2815988
1600	66855	326922	638380	1227391	1782784	2312778	2822441
1700	67163	328227	640603	1230925	1787325	2318209	2828721
1800	67473	329533	642812	1234403	1791764	2323493	2834809
1900	67787	330840	645004	1237820	1796095	2328620	2840692
2000	68103	332145	647179	1241173	1800313	2333583	2846358
2100	68422	333450	649333	1244459	1804414	2338379	2851805
2200	68744	334754	651468	1247677	1808397	2343007	2857031
2300	69068	336056	653582	1250828	1812265	2347469	2862040
2400	69396	337356	655674	1253913	1816020	2351770	2866838
2500	69726	338653	657746	1256933	1819665	2355916	2871434
2600	70060	339948	659797	1259892	1823208	2359917	2875841
2700	70396	341241	661828	1262792	1826655	2363784	2880073
2800	70735	342531	663840	1265639	1830014	2367527	2884145
2900	71077	343819	665835	1268437	1833294	2371161	2888077
3000	71422	345104	667814	1271192	1836504	2374700	2891885

SPINEL

MGAL$_2$O$_4$
SPINEL Formula wt. (g) = 142.27

Enthalpy (J/mol) = -2305456.440		Shi et al. (1992)
Entropy (J/mol.K) = 82.896		Shi et al. (1992)
Heat Capacity		
No. of *Cp* ranges = 1		Robie et al. (1978)
2.0785000D+02 4.4900D-03 6.7109D+06 0.0000D+00		
-9.5984D+08 0.0000D+00 -3.9681D+04		
Volume (cc/mol) = 39.710		Robie et al., (1978)
Thermal expn.		
6.96922D-05 -1.08069D-09 -3.07991D-02 5.03946D+00		Saxena and Shen (1992)
Isot.bulk mod.		
4.80091D-07 4.89213D-11 5.98120D-14 -3.75550D-18		Saxena and Shen (1992)
P deriv., T deriv.:		
4.89000D+00 9.00000D-04		Wang and Simmons (1972)
Melting pt. and enthalpy of melting:		
2408.00000	-	Robie et al. (1978)

NOTE: Calorimetric data: H: -2299.38+0.7, S: 80.63+0.42; H, Charlu et al. (1975), S, Robie et al. (1978). Experiments involving spinel, orthopyroxene and pyrope were done by Haselton (1979), Lane and Ganguly (1980), Danckwerth and Newton (1978) and Gasparik and Newton (1984) on the reaction: En + Sp = Pyr + Fo. Free energy of spinel is calculated from the above reaction. Heat capacity of spinel is from Robie et al. (1978). Enthalpy and entropy of spinel are adjusted in order to reproduce the above reaction. The adjusted entropy value (81.5 J/mol.K) lies close to the value in Robie et al. (1978) (80.63 J/mol.K). However, the enthalpy value [ΔHf(970)-oxides = -30.659 KJ/mol] is much lower than the calorimetric value [ΔHf(970)-oxides=-22.51+/-0.75 KJ/mol] of Charlu et al. (1975). Wood and Holloway (1984) pointed out that Charlu et al. prepared their sample at 1698 K. Hence, the sample they used had a built-in Mg-Al disorder as the experiment was done at 970 K. The energy due to disorder is reflected in their result. Wood et al. (1986) studied the disordering behavior of spinel and concluded that above 1173 K disordering is negligible. Thus, the difference between the value of Charlu et al. (1975) and this study (appox 8 KJ/mol) could be due to the difference between the disorder state of spinel at 1173 and 970 K. Thus, as also concluded by Wood and Holloway (1984), the enthalpy of spinel should be much less than the value of Charlu et al. (1975). The value obtained here is similar to that obtained by Wood and Holloway (1984). The compressibility data (Saxena and Shen,1992) are consistent with those of Wang and Simmons (1972).

$MGAL_2O_4$
SPINEL

Thermochemical Data

T(K)	C_p(J)	H(kJ)	S(J)	$\alpha\ 10^{-4}$	K 10^4	γ	V(cc)	δ_T
298	115.4	-2305.456	82.9	0.2276	200.042	1.58	39.7	7.35
300	115.9	-2305.242	83.6	0.2270	199.980	1.57	39.7	7.38
400	137.4	-2292.456	120.2	0.2376	196.468	1.37	39.8	7.85
500	149.9	-2278.048	152.3	0.2771	192.665	1.45	39.9	7.36
600	158.6	-2262.601	180.5	0.3171	188.621	1.55	40.0	6.94
700	165.2	-2246.396	205.5	0.3522	184.381	1.64	40.2	6.66
800	170.5	-2229.605	227.9	0.3820	179.990	1.71	40.3	6.47
900	174.8	-2212.337	248.2	0.4072	175.491	1.76	40.5	6.35
1000	178.4	-2194.673	266.8	0.4285	170.920	1.80	40.6	6.27
1100	181.5	-2176.672	284.0	0.4467	166.312	1.82	40.8	6.21
1200	184.3	-2158.378	299.9	0.4623	161.698	1.83	41.0	6.16
1300	186.7	-2139.827	314.7	0.4758	157.104	1.84	41.2	6.12
1400	188.9	-2121.047	328.6	0.4875	152.554	1.83	41.4	6.08
1500	190.8	-2102.061	341.7	0.4978	148.066	1.83	41.6	6.04
1600	192.6	-2082.887	354.1	0.5068	143.657	1.81	41.8	5.99
1700	194.3	-2063.541	365.8	0.5148	139.341	1.80	42.0	5.95
1800	195.8	-2044.037	377.0	0.5219	135.128	1.78	42.2	5.90
1900	197.2	-2024.386	387.6	0.5282	131.027	1.75	42.5	5.84
2000	198.5	-2004.597	397.8	0.5339	127.043	1.73	42.7	5.78
2100	199.8	-1984.679	407.5	0.5390	123.180	1.70	42.9	5.72
2200	201.0	-1964.639	416.8	0.5436	119.443	1.67	43.1	5.66
2300	202.1	-1944.484	425.8	0.5477	115.832	1.64	43.4	5.59
2400	203.2	-1924.218	434.4	0.5514	112.347	1.61	43.6	5.52
2500	204.2	-1903.848	442.7	0.5548	108.987	1.58	43.9	5.45
2600	205.2	-1883.377	450.7	0.5578	105.752	1.55	44.1	5.38
2700	206.1	-1862.809	458.5	0.5606	102.639	1.52	44.4	5.31
2800	207.1	-1842.148	466.0	0.5631	99.646	1.49	44.6	5.23
2900	207.9	-1821.398	473.3	0.5654	96.769	1.46	44.9	5.15
3000	208.8	-1800.560	480.4	0.5674	94.006	1.43	45.1	5.08

<INT VdP (J)>

PRESSURE(GPa)

T(K)	1	5	10	20	30	40	50
298	39611	196182	388027	760593	1120624	1470125	1810563
300	39613	196189	388041	760616	1120653	1470159	1810600
400	39701	196595	388777	761852	1122239	1471993	1812610
500	39801	197056	389618	763291	1124136	1474264	1815203
600	39918	197595	390607	765012	1126454	1477105	1818526
700	40049	198205	391733	766994	1129157	1480462	1822503
800	40194	198878	392978	769199	1132188	1484254	1827030
900	40350	199604	394323	771592	1135495	1488413	1832017
1000	40517	200376	395752	774144	1139034	1492881	1837392
1100	40691	201186	397255	776832	1142773	1497614	1843100
1200	40873	202030	398820	779637	1146685	1502577	1849096
1300	41062	202905	400441	782547	1150750	1507745	1855349
1400	41256	203805	402111	785548	1154953	1513096	1861832
1500	41456	204729	403824	788634	1159280	1518614	1868527
1600	41661	205675	405577	791795	1163723	1524289	1875418
1700	41870	206640	407367	795028	1168274	1530109	1882494
1800	42083	207623	409190	798327	1172926	1536068	1889744
1900	42300	208623	411045	801690	1177676	1542160	1897162
2000	42521	209638	412929	805112	1182520	1548379	1904741
2100	42744	210667	414842	808593	1187454	1554722	1912478
2200	42971	211711	416781	812130	1192476	1561186	1920367
2300	43201	212768	418746	815721	1197584	1567769	1928406
2400	43434	213837	420736	819366	1202777	1574467	1936592
2500	43669	214918	422751	823063	1208054	1581279	1944922
2600	43907	216011	424789	826813	1213412	1588204	1953394
2700	44148	217116	426851	830613	1218851	1595240	1962006
2800	44391	218231	428935	834464	1224370	1602386	1970757
2900	44636	219357	431042	838364	1229968	1609640	1979645
3000	44883	220494	433172	842314	1235645	1617001	1988668

Appendix II: Summary of Reactions in the Mg-Fe-Si-O System

Reactions	References
1. Fe (*bcc*) = Fe (*fcc*)	Bundy (1965)
2. Fe (*bcc*) = Fe (*hcp*)	Bundy (1965)
3. Fe (*fcc*) = Fe (*hcp*)	Bundy (1965)
4. $4\{FeO\} + O_2 = 4\{FeO_{1.5}\}$	Kubaschewski (1982)
5. $5\{FeO\} + O_2 = 2\{FeO_{1.5}\} + Fe_3O_4$	Kubaschewski (1982)
6. $7\{FeO\} + O_2 = 6\{FeO_{1.5}\} + Fe$	Kubaschewski (1982)
7. Fe_3O_4 (*mgt*) = Fe_3O_4 (*high-pressure*)	Mao et al. (1974)
	Huang and Bassett (1986)
8. Fe_2O_3 (*hmt*) = Fe_2O_3 (*high-pressure hmt*)	Mao and Bell (1977)
9. SiO_2 (*qz*) = SiO_2 (*coe*)	Bohlen and Boettcher (1982)
10. SiO_2 (*coe*) = SiO_2 (*st*)	Yagi and Akimoto (1976)
11. Mg_2SiO_4 (α) = Mg_2SiO_4 (β)	Katsura and Ito (1989)
	Fukizawa (1982)
	Suito (1977)
12. Mg_2SiO_4 (β) = Mg_2SiO_4 (γ)	Katsura and Ito (1989)
	Sawamoto (1986b)
	Ohtani (1979)
	Suito (1977)
13. Mg_2SiO_4 (γ) = $MgSiO_3$ (*pv*) + MgO (*per*)	Ito and Takahashi (1989)
	Ito and Yamada (1982)
14. Mg_2SiO_4 (α) = Mg_2SiO_4 (*melt*)	Davis and England (1964)
	Ohtani and Kumazawa(1981)
15. $2MgSiO_3$ (*px*) = Mg_2SiO_4 (β) + SiO_2 (*st*)	Akaogi and Akimoto (1977)
	Sawamoto (1987)
	Ito and Navrotsky (1985)
16. $MgSiO_3$ (*px*) = $MgSiO_3$ (*gt*)	Sawamoto (1987)
17. $2MgSiO_3$ (*gt*) = Mg_2SiO_4 (β) + SiO_2 (*st*)	Sawamoto (1987)
18. $2MgSiO_3$ (*il*) = Mg_2SiO_4 (γ) + SiO_2 (*st*)	Ito and Navrotsky (1985)

19. $MgSiO_3$ (*il*) = $MgSiO_3$ (*pv*)	Ito and Takahashi (1989)
	Ito and Yamada (1982)
	Sawamoto (1987)
20. $MgSiO_3$ (*px*) = $MgSiO_3$ (*melt*)	Boyd et al. (1964)
	Kato and Kumazawa (1983)
21. Fe_2SiO_4 (α) = Fe_2SiO_4 (γ)	Yagi et al. (1987)
	Akimoto et al. (1965, 1967)
22. $2FeSiO_3$ (*px*) = Fe_2SiO_4 (γ) + SiO_2 (*st*)	Akimoto and Syono (1970)
23. Fe_2SiO_4 (γ) = FeO (mw) + SiO2 (st)	Yagi et al. (1979)
	Ohtani (1979)
24. $(Mg,Fe)_2SiO_4$ (α) = $(Mg,Fe)_2SiO_4$ (γ)	Akimoto and Fujisawa(1968)
	Kawada (1977)
	Akimoto (1987)
25. $(Mg,Fe)_2SiO_4$ (α) = $(Mg,Fe)_2SiO_4$ (β)	Katsura and Ito (1989)
	Akimoto (1987)
26. $(Mg,Fe)_2SiO_4$ (β) = $(Mg,Fe)_2SiO_4$ (γ)	Katsura and Ito (1989)
	Akimoto (1987)
27. $(Mg,Fe)_2SiO_4$ (γ) = $(Mg,Fe)SiO_3$ (*pv*) + (Mg,Fe)O	Ito and Takahashi (1989)
28. $(Mg,Fe)_2SiO_4$ (γ) = 2(Mg,Fe)O (*mw*) + SiO_2 (*st*)	Ito and Takahashi (1989)
29. $(Mg,Fe)SiO_3$ (*pv*) = (Mg,Fe)O (*mw*) + SiO_2 (*st*)	Ito and Takahashi (1989)
30. Mg-α + Fe-*mw* = Mg-*mw* + Fe-α	This study
	Nafziger and Muan (1967)
31. Mg-β + Fe-*mw* = Mg-*mw* + Fe-β	This study
32. Mg-γ + Fe-*mw* = Mg-*mw* + Fe-γ	This study
	Ito (1984)
	Yagi et al. (1979)
33. Mg-*pv* + Fe-*mw* = Mg-*mw* + Fe-*pv*	This study
	Ito (1984)
	Yagi et al. (1979)
34. Mg-α + Fe-*px* = Mg-*px* + Fe-α	Matsui and Nishizawa (1974)
35. Fe_2SiO_4 (α) + SiO_2 (*qz*) = $2FeSiO_3$ (*px*)	Lindsley et al. (1964)
36. $(Mg,Fe)_2SiO_4$ (α) + SiO_2 (*qz*) = $2(Mg,Fe)SiO_3$ (*px*)	Bohlen and Boettcher (1981)

Species Index (Text)

Species Index (Thermodynamic Data)

MIX
Papier aus verantwortungsvollen Quellen
Paper from responsible sources
FSC® C105338

If you have any concerns about our products,
you can contact us on
ProductSafety@springernature.com

In case Publisher is established outside the EU,
the EU authorized representative is:
Springer Nature Customer Service Center GmbH
Europaplatz 3, 69115 Heidelberg, Germany

Printed by Libri Plureos GmbH
in Hamburg, Germany